Virtual Work
in
Structural Analysis

'It is rare that anything more than the simplest algebraic expressions, scarcely worth dignifying with the name of mathematics, serve any useful purpose'

F. W. Lanchester, 1937
Aeronautical Engineer Extraordinary

Virtual Work in Structural Analysis

Glyn A. O. Davies

Department of Aeronautics
Imperial College of Science and Technology

A Wiley–Interscience Publication

175 YEARS OF PUBLISHING
1807 1982

JOHN WILEY & SONS
Chichester · New York · Brisbane · Toronto · Singapore

Library of Congress Cataloging in Publication Data:

Davies, Glyn A. O.
Virtual work in structural analysis.

'A Wiley–Interscience publication.'
Includes bibliographical references and index.
1. Structures, Theory of. 2. Virtual work.
I. Title.
TA646.D38 624.1'71 81-15926

ISBN 0 471 10112 5 (cloth)
ISBN 0 471 10113 3 (paper) AACR2

British Library Cataloguing in Publication Data:

Davies, Glyn A. O.
Virtual work in structural analysis.
1. Virtual work
2. Structures, Theory of
I. Title
624.1'71 TA645

ISBN 0 471 10112 5 (cloth)
ISBN 0 471 10113 3 (paper)

Typeset by Speedlith Ltd., Manchester
and Printed by Page Bros. (Norwich) Ltd.

To
Helen

ACKNOWLEDGEMENTS

In 1954 Professors Argyris and Kelsey published their classic *Energy Theorems and Structural Analysis*. It is this book which inspired the present text, and for this inspiration I will be ever grateful to my former colleagues. I am also indebted to all my colleagues in the Department of Aeronautics at Imperial College for many discussions and arguments on the merits of work–energy methods, in particular to Antony Chan and Dennis Hitchings for patience and forbearance. I wish to especially thank Miriam Pook for faultless typing, long hours, and decoding illegible drafts. Finally, because this is a teaching text, I am grateful to countless graduates, both in London and Sydney, who thought I made a subtle subject easy and exciting; also to those who thought an easy subject had been made difficult. But not dull: that would be unforgivable.

GLYN A. O. DAVIES
London, June 1981

CONTENTS

PREFACE AND HISTORY OF VIRTUAL WORK

In analysing structures of any kind, the student should realize – and the expert should occasionally remind himself – that there are only three fundamental conditions to be satisfied. Firstly, the structure and all parts of it should be in *equilibrium* – where the word 'equilibrium' implies that dynamics may also be included if necessary as reversed inertia forces. Secondly, the structure must not fall apart, or more precisely the deformations should be continuous throughout the stucture – the so-called *compatibility* requirement. Thirdly, the material properties should be obeyed, usually in the form of an empirical law describing the strain induced by stress, temperature, creep, or other effects.

The third requirement is based on experimental investigation and subsequent mathematical modelling, and its precise form may facilitate or complicate the analysis. The first two requirements, however, are purely analytical: they use the laws of mechanics, kinematics, or physics and are unrelated to the material properties. It is these two laws of equilibrium and compatibility of deformation which the principle of virtual work can indirectly replace. But why replace? Are not the equations of equilibrium and compatibility both eminently respectable and derivable? There are three reasons why work or energy arguments may be preferable. Firstly, they may be easier, both conceptually and algebraically, for particular forms of structure. Many students find energy arguments both natural and systematic, and preferable to direct methods. Indeed, there is a real danger that the attraction and routine nature of energy methods may cloud the physical understanding of a problem. Secondly, they are extremely powerful for obtaining approximate solutions. Since virtually all but the simplest structure involves some necessary approximation, this aspect alone would ensure their permanent popularity. Finally, in common with all disciplines, structural analysis has been transformed by the digital computer, and the use of *finite elements* is now commonplace in all branches of structural engineering. Since computers much prefer scalar quantities to vectors, work and energy become a natural formulation.

The *principle of virtual work*, upon which all else follows, is introduced in Chapter 1 and involves 'virtual' work which is simply the product of force times displacement and is not related to the real work done by forces moving over displacements. The consequences of this improbable axiom are then found to be the equations of equilibrium or compatibility, but only when one introduces the concepts of *virtual displacements* and *virtual forces* – these also are unrelated to the real ones. The traditional extremum or minimum energy theorems are avoided, and variational methods involving small incremental displacements and forces are not used; instead the virtual systems are visualized as hypothetical alternatives which turn out to have specific constraints upon them. The structural engineer is so familiar with the idea of small strains and displacements that it seems absurd to ask him to contemplate even smaller ones for purely mathematical reasons. Alternative virtual systems can be used with great effect to reduce algebraic effort. Also avoided are the multiplicity of special energy theorems which litter the literature and which can involve many sorts of energy and often the stress–strain law.

The principle of virtual work is shown to be simple and yet quite general, capable of delivering all that the various extremum theorems can do. The technique is illustrated initially on simple frameworks and then extended immediately to arbitrary three-dimensional structures. This extension is greatly facilitated by introducing *ab initio* matrix notation for forces, displacements, stress, and strain. (The reader unaccustomed to matrix notation will find an introduction and illustrations in mechanics in Appendix A.) The two *principles of virtual displacements and virtual forces* are extracted from the virtual-work theorem, and their duality – and differences – emphasized. Many examples are given in Chapters 2 and 3 of the applications of these two principles to discrete and continuous structures, including gross deformations and buckling problems. The power of the work–energy approach when using approximate methods is demonstrated in Chapter 4 on a variety of structures from simple rods to shells. In Chapters 5 and 6 the limitations imposed by geometrically complex structures are recognized, and the use of *finite-element* idealization is developed: here the concise matrix notation really comes into its own, both as an aid to demonstrating the principles and methods and as a convenient shorthand in which to address computers. Finally as an appendix, the more traditional minimum-energy theorems are described so that the student of virtual work will be able to relate to the alternative approach: it is shown that all these theorems are particular cases of the principle of virtual work.

It is interesting to speculate why it is that extremum energy methods seem to have such a measure of popularity in the literature, certainly in the application of approximate methods. It is probably due to the fact that the most famous extremum principle – that of *minimum total potential energy* – can be argued as a fundamental law of physics in the same fashion as the two laws of thermodynamics. It is then tempting to seek a grand minimum-energy theorem unifying all the laws in applied and structural mechanics – certainly Leonhard Euler took this view. However, this particular holy grail is illusory, as we shall show; moreover the continual quest for minimum energy does result in having to invent some pretty curious energies.

Both the principle of virtual work and extremum energy theorems have a long and honourable history, mostly as a substitute for the equations of motion or equilibrium. Archimedes is generally credited with first using work arguments in his study of levers, although he, and later Jordanus in the thirteenth century and Leonardo da Vinci in the fifteenth, probably used work primarily as a measure of lever efficiency, or why one does not get something for nothing. Nevertheless, they satisfied moment equilibrium correctly. However, in 1594 Galileo unambiguously used work arguments with *actual* displacements to solve the forces in lever and pulley systems. In 1717 Johann Bernoulli, in a letter to Varignon, first used the concept of *virtual*, but kinematically possible, *displacements*; and in 1740 Maupertuis defined work for general finite displacements, and first stated the minimal property of potential energy for stable equilibrium. Maupertuis' successor at the Berlin Academy was the legendary Swiss genius Leonhard Euler who had become enthused with minimal principles as a natural law. He used minimum potential energy in many problems of statics (for example, the catenary) and obtained in this way the displaced curve of a strut – successfully including the correct expression for finding strain energy provided by Daniel Bernoulli. Later Lagrange used D'Alembert's principle to convert dynamics to statics and then used the principle of virtual displacements, together with generalized forces, to derive his famous equations governing the laws of dynamics in terms of kinetic and potential energy. It was left to Hamilton to formalize Lagrange's equations as a minimal principle. Apart from Euler's work, most of the energy arguments so far mentioned were as substitutes for the equations governing the statics or dynamics of rigid body systems. (The interested reader may pursue this history in Ernst Mach's *The Science of Mechanics*.[1]) Brief mention will now be made of the development of work–energy methods applied to deformable bodies. Again the interested reader may consult that admirable treatise by Stephen Timoshenko, *The History of the Strength of Materials*.[2]

Lamé was the first in 1852 to prove a work equation, named after his colleague Clapeyron, for deformable bodies, and he used expressions in terms of final forces, displacements, stresses, and strains. However, he proved his equation for *actual* quantities and as we shall see this is valid only for linearly elastic systems, for which forces and displacements are proportional. Lamé's equation was applied by Clerk-Maxwell[3] in 1864 (and independently by Mohr in 1875) to the solution of redundant frameworks using the dummy unit-load technique. Later in 1875 the extremum version of this technique was published by Castigliano, who actually attributed the idea – of 'least work' – to General L. F. Menabrea in a dubious proof dated 1857. For some reason Castigliano's formulation became the most widely known in English-speaking countries, possibly because – although it was a compatibility condition – it appeared to be similar to the familiar principle of minimum potential energy. However, the resemblance is misleading, is only valid for linear elasticity, and when generalized for arbitrary elasticity it leads to the first of the hypothetical ('complementary') potential energies referred to previously. This generalization was in fact first made by Engesser[4] in 1889.

The principle of minimum potential energy is equally applicable when displacements are large, as was realized by Euler, but the duality is not preserved in the case of complementary energy and attempts to do so have been obscure (Westergaard[8]), or difficult (Libove[9]). The difficulty arises in expressing the current *actual* values in terms of their nonlinear history and this normally has to be done numerically by iteration – except in special cases such as initial buckling problems or the 'shallow arch' type of structure (see Chapter 2).

Since this text takes a simple work statement and extracts from it various sets of equations, mention must be made of the theorem of Hellinger[10] and Reissner[11] which is shown in Appendix B to be equivalent to *two* of our work statements, and which therefore always delivers two independent sources of information even though the theorem itself appears to be a single extremum statement. Both Reissner[12] and Washizu[13] showed how an extremum energy theorem can be extended to finite large displacements, but Washizu also concedes that the complementary energy version is not useful since stresses cannot be uncoupled from displacements.

The use of energy methods in structural mechanics received a boost this century with the development of approximate methods of analysis, and in turn this development became a positive explosion with the application of *discrete* (finite-element) modelling married to the digital computer. The ramifications of finite-element methods are still being felt and they have freed analysts at long last from the constraint imposed by purely geometrical complexity. Many of today's structures would not have been designed had not their strength or stiffness been predictable by finite-element analysis. One only has to think of the complexity of offshore oil rigs or modern reinforced-concrete shell roofs, a nuclear rector, or a supersonic aircraft. The environment of these quoted examples is pretty hostile, too.

As already mentioned, traditional extremum energy theorems are all shown to be versions of virtual work in the Appendix B, including those theorems attributed to Castigliano. It is a curious fact that the seemingly ubiquitous Castigliano is almost unquoted in Europe or Russia.

1

VIRTUAL WORK

1.1 Introduction

Before we have even started, let us digress and look at the traditional energy approach used in examining the equilibrium of structures. The idea is a simple one. A body in equilibrium has several forces acting upon it, $R_1, R_2, R_3, \ldots$, and we then deliberately increase these forces by *small* amounts $\delta R_1, \delta R_2, \delta R_3, \ldots$. Figure 1.1 shows solitary forces since these are easier to handle at this stage. The displacements of these loaded points, which had previously been $u_1, u_2, u_3, \ldots$ in the same direction as the forces, are consequently increased by small amounts $\delta u_1, \delta u_2, \delta u_3, \ldots$. A small amount of work δW is therefore done during this disturbance, and in the absence of motion or other dissipating mechanisms is presumably stored as 'internal' elastic strain energy δU_i. Thus,

$$\delta U_i = \delta W \tag{1.1}$$

The traditional energy approach is to write this equation as

$$\delta(U_i - W) = 0 \tag{1.2}$$

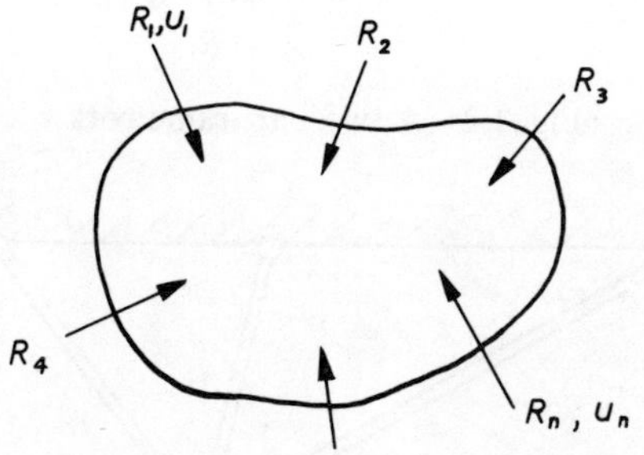

Fig. 1.1 Loads and displacements of an arbitrary structure

It may be shown that if the quantity $(U_i - W)$ in (1.2) has a zero first-order increment when small changes in U_i and W occur, then that quantity has an extremum. From this stems the well-known *principle of minimum potential energy*. In our case U_i is the potential elastic energy, or *strain energy*, and $-W$ is the potential of the applied forces. The principle of minimum potential energy can be shown to be equivalent to the equations of equilibrium, and these are not usually sufficient to solve structural problems except for simple structures like Fig. 1.2 – a two-bar truss. However, few structures are solvable only from the equations of equilibrium like the above. Such *statically determinate* structures are usually avoided since they are prone to fall apart after a local failure; that is, they may become mechanisms. Most structures are *statically indeterminate* or redundant, since they have an excess of members above that required simply to equilibrate the applied loads. In aeronautics parlance they are 'fail-safe'. The above example could easily be redesigned to be a fail-safe structure like Fig. 1.3. It is clear that we must now go further than simply equating the resultant of the bar forces to the applied loads R_1 and R_2. Two such equations are not enough to solve the three-bar forces. The further requirement is that the three members strain in such a fashion that their ends move together with the loaded joint. Strains derived from such a displacement are said to be kinematically *compatible.* It is tempting therefore to look for another energy theorem which can be used as alternative to satisfying *compatibility*. Unfortunately, it turns out to be necessary to invent a quite unnatural hypothetical 'complementary energy', and this was first done by Engesser[4] in 1889. If it is desired to include strains due to reasons other than stress (like temperature) then further hypothetical energies are needed, and the apparent simplicity of minimum potential energy has been lost. If it is necessary

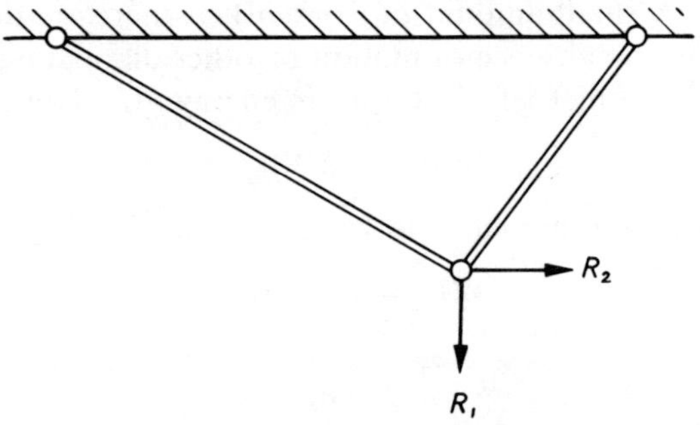

Fig. 1.2 A two-bar framework

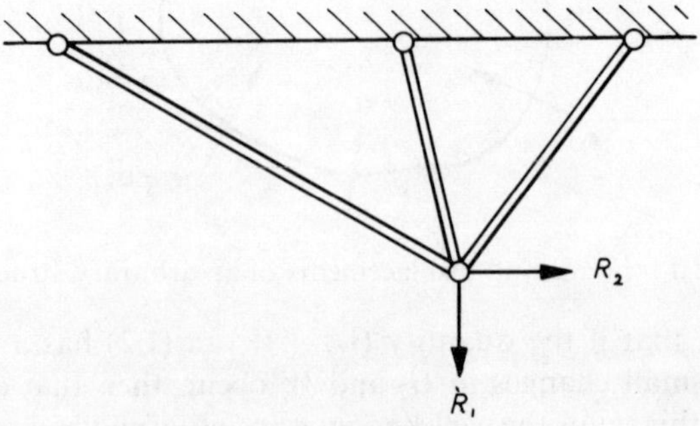

Fig. 1.3 A three-bar framework

to invent hypothetical energies then we will plunge in at the deep end and define a single hypothetical or *virtual* work, but before doing so we note how another virtual concept has crept into the evaluation of δW in (1.1).

If we require the *actual* work done when a force increasing from R to $R + \delta R$ moves over a small displacement δu, we may assume the force–displacement relationship is approximately linear and take the average value in evaluating

$$W = \tfrac{1}{2}\left[R + (R + \delta R)\right]\delta u$$

and ignoring second-order quantities simplify this to

$$W = R\,\delta u \tag{1.3}$$

The small increment in force δR, which we had postulated was actually responsible for the displacement δu, has actually disappeared in (1.3). Thus δu can be *totally unrealated to* R in any way; it is merely a small displacement which we choose to allow existing forces to do work over. It is said to be a *virtual displacement*. Armed with this concept of a fictitious displacement, we now consider the equilibrium of a rigid body ($\delta U_i = 0$) and prove $\delta W = 0$ as a simple work equation.

If a body is in equilibrium, the net resultant of all the forces acting upon it – including supporting reactions – must be zero, and if we 'allow' such a body to undergo a small virtual displacement δu of some description, obviously the product of zero and δu is zero, i.e. $\delta W = 0$. This idea was first exploited by Archimedes in – probably – trying to fathom how a large weight could be lifted by a small force; he realized that you never get something for nothing and he turned to 'work' as a likely explanation. Imagine a small rotation of the lever in Fig. 1.4 through an angle $\delta\theta$ about the fulcrum, then equating the work done by R to that done in lifting the weight W,

$$R(L\,\delta\theta) = W(l\,\delta\theta) \tag{1.4}$$

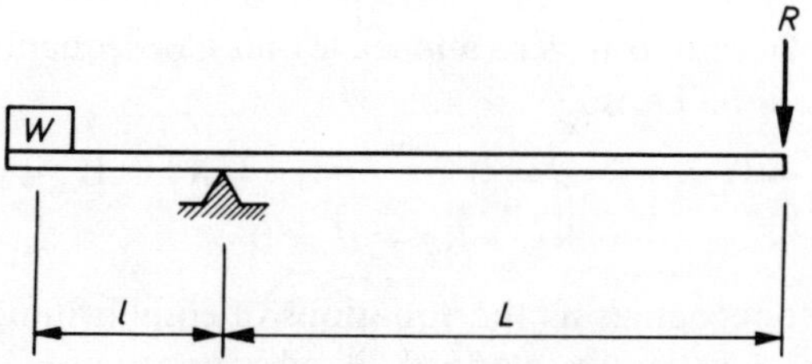

Fig. 1.4 Archimedes' lever

hence $R = Wl/L$, and this explained the mystery of the lever. We recognize (1.4) of course as an equation of moment equilibrium about the fulcrum. Now equation (1.4) can be imagined as $(RL - Wl) = 0$ in line with our idea that the total work done by a set of forces in equilibrium is zero. This trivial example can be extended to more complex systems; for example, suppose that the present lever had also applied forces W_1, W_2, and H; a moment M; and two supports as shown in Fig. 1.5. We now allow a small virtual displacement pattern consisting of a vertical component δv, a horizontal δu, and a rotation $\delta\theta$ about the left-hand support, shown exaggerated in Fig. 1.6. This displacement pattern is clearly fictitious. Equating the total virtual work to zero,

$$M\delta\theta - W_1(\delta v + l\,\delta\theta) + R_1\,\delta v + R_3\,\delta u + W_2(l\,\delta\theta - \delta v) + R_4\,\delta u + R_2(\delta v - 2l\delta\theta) + H\,\delta u = 0$$

Collecting terms,

$$(M - W_1 l + W_2 l - R_2 2l)\,\delta\theta + (-W_1 + R_1 - W_2 + R_2)\,\delta v + (R_3 + R_4 + H)\,\delta u = 0 \tag{1.5}$$

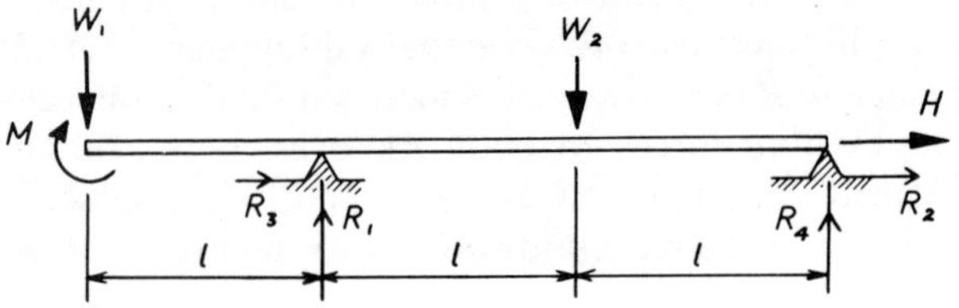

Fig. 1.5 A supported beam

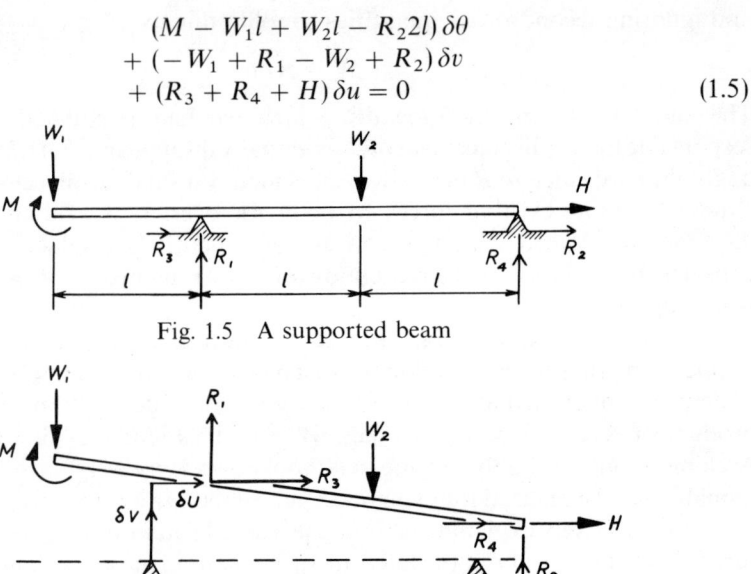

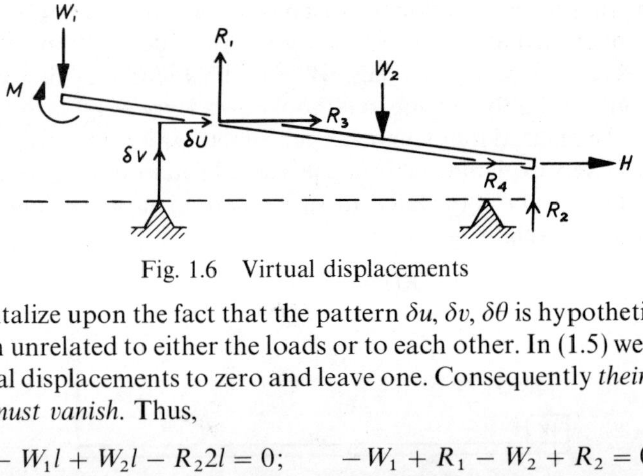

Fig. 1.6 Virtual displacements

We now capitalize upon the fact that the pattern δu, δv, $\delta\theta$ is hypothetical, that is they are each unrelated to either the loads or to each other. In (1.5) we are free to set two virtual displacements to zero and leave one. Consequently *their individual coefficients must vanish.* Thus,

$$M - W_1 l + W_2 l - R_2 2l = 0; \qquad -W_1 + R_1 - W_2 + R_2 = 0;$$
$$R_3 + R_4 + H = 0$$

These equations we recognize as the equations of equilibrium; firstly moments about the support, secondly vertical equilibrium, and lastly horizontal equilibrium. In fact each equation *corresponds* to its particular virtual displacement in both sense and direction.

We cannot emphasize too strongly the seemingly simple idea, which will be crucial to future progress, that the virtual displacements are hypothetical and unrelated to each other. As we will be considering, for the most part, structures whose displacements are small anyway, it is illogical to retain the 'δ' prefix, and from now on we adopt 'bars' to distinguish virtual displacements $\bar{u}$, $\bar{v}$, $\bar{\theta}$, etc.

When we look at buckling problems or very flexible structures like shallow arches, we will have to recognize that the real displacements u result in significant changes in the structural configuration: then we will be forced to adopt small virtual displacements δu. In this class of problems the difference in magnitude between δu and u is both necessary and illuminating.

Shortly we will use *virtual forces* also, and show how compatibility equations can also be derived indirectly, but first let us refresh our ideas on the equilibrium, compatibility, and stress–strain laws governing simple pinjointed frameworks, which are just about the simplest form of structure which can nevertheless exhibit the characteristics of all structures, albeit at a simple level.

1.2 Direct solution of pinjointed frameworks

Consider the simple two-bar frame, previously referred to, and shown again in Fig. 1.7, where the joint loaded by force components W and H responds by deflecting through displacement components u and v. The bar forces are denoted by N and the bar elongations by Δ. To determine the stresses and deformations of this simple framework involves the same analytical steps as are needed for all structures. We therefore follow these steps pedantically, but thoroughly.

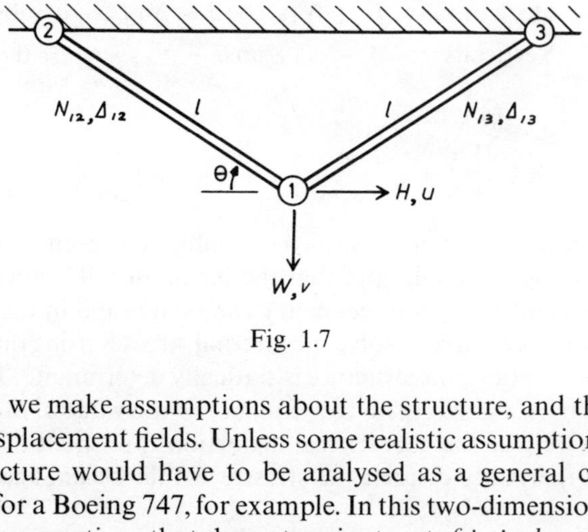

Fig. 1.7

Step 1 Firstly, we make assumptions about the structure, and the nature of its stress and/or displacement fields. Unless some realistic assumptions are possible, then every structure would have to be analysed as a general continuum – an expensive start for a Boeing 747, for example. In this two-dimensional framework we make the assumption that bars terminate at *frictionless* pinned joints, consequently the force system at the end of a bar has a resultant force but no resultant couple. Define this force by two components P_1 and P_2, for example:

Step 2 Secondly, we use equations of equilibrium for both internal and applied forces. For the isolated bar in Fig. 1.8 we have:

$$\begin{aligned} &\text{Horizontally, } P_1 - P_4 = 0 \\ &\text{Vertically, } P_3 + P_2 = 0 \\ &\text{Moments about the left end, } P_2 l = 0 \end{aligned}$$

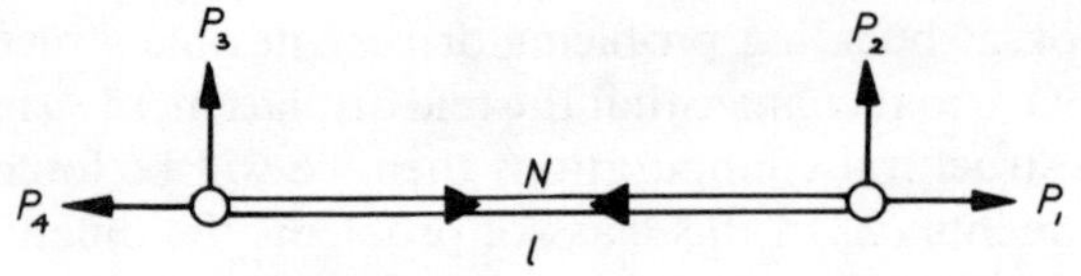

Fig. 1.8 Bar forces

Therefore $P_2 = P_3 = 0$; $P_1 = P_4 = N$. So the stresses in a pin-ended bar are described by just a single axial force, N.

Real pinjointed frameworks are almost unknown in practice, but it can be shown that for 'pinjointed' frameworks which are not mechanisms, the effect of having rigid joints is to introduce 'shear forces' P_2 and P_3 which are of order $N(k/l)^2$, where k is the radius of gyration of the section, so that $k/l \ll 1$ for slender bars.

The next assumption is that the stress in a slender bar of constant cross-sectional area (A) is uniform, in which case the stress

$$\sigma = \frac{N}{A} \tag{1.6}$$

Equilibrium arguments are not exhausted, for we still have the meeting of forces at the joint.

$$\begin{aligned} &\text{Horizontally: } H + N_{13}\cos\theta - N_{12}\cos\theta = 0 \\ &\text{Vertically: } \quad W - N_{13}\sin\theta - N_{12}\sin\theta = 0 \end{aligned} \tag{1.7}$$

whence

$$N_{12} = \frac{W}{2\sin\theta} + \frac{H}{2\cos\theta}; \qquad N_{13} = -\frac{H}{2\cos\theta} + \frac{W}{2\sin\theta} \tag{1.8}$$

It has been assumed here that the strains are unlikely to even approach 1 per cent in most engineering materials, and that the inclination θ is not close to zero, in which case the small changes in geometry can be ignored in the equations (1.8). We have therefore been able to solve the internal stresses using only the equations of equilibrium or statics – the structure is statically determinate. The stress–strain law has not been used and is of no consequence to the value of the stresses in such a structure. As long as the geometrical distortions are small the material could flow, creep, or thermally distort; the stresses would be unaffected.

Step 3 If we need to know the deflections, then, thirdly, we must use the *equations of compatibility* which relate strain to the displacements which are assumed to be continuous everywhere. These compatibility arguments are purely geometrical and again do not involve the stress–strain law. The cause of the strains is immaterial and they may be due to stress or other effects. We may have to invoke further assumptions about the strain field, but having assumed stress variations this is unlikely. In our simple bars of uniform cross-section both stress and strain are constant, so the latter is simply Δ/l,

$$\varepsilon = \Delta/l \tag{1.9}$$

Finally, like joint equilibrium, we have yet to insist that the displacements of the joint are compatible with the bar strains/elongations. As the displacements are small, a bar is not stretched by a displacement component at right angles to it; therefore we simply have

$$\begin{aligned} \Delta_{12} &= \quad u\cos\theta + v\sin\theta \\ \Delta_{13} &= -u\cos\theta + v\sin\theta \end{aligned} \tag{1.10}$$

Step 4 The last link to be forged between displacements (1.10) and the forces (1.8) is the *stress–strain law*, using (1.9) and (1.6) to convert displacements and forces to strain and stress. We take in this example the simple linear Hooke's law

$$\varepsilon = \sigma/E \tag{1.11}$$

On using (1.11) in equations (1.8) to (1.10), we find

$$u\cos\theta + v\sin\theta = \Delta_{12} = l\varepsilon_{12} = \frac{l}{EA}\left(\frac{W}{2\sin\theta} + \frac{H}{2\cos\theta}\right)$$

and

$$-u\cos\theta + v\sin\theta = \Delta_{13} = \frac{l}{EA}\left(\frac{W}{2\sin\theta} - \frac{H}{2\cos\theta}\right)$$

giving

$$u = Hl/2AE\cos^2\theta; \qquad v = Wl/2AE\sin^2\theta \tag{1.12}$$

Our analytical progress can be summarized as in Fig. 1.9, and this summary has all the ingredients necessary for any analysis.

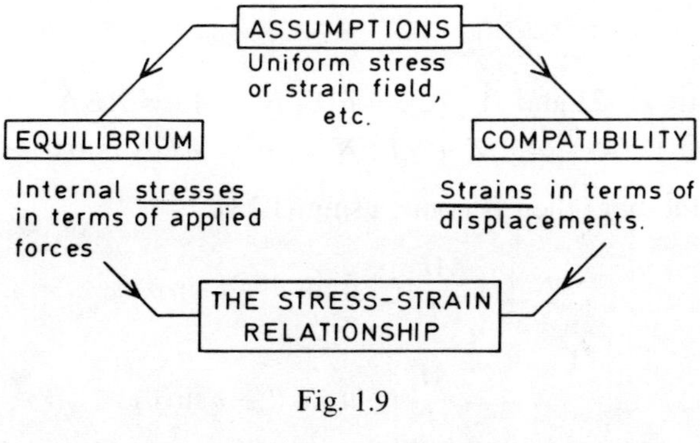

Fig. 1.9

In our example we moved successively from the assumptions to the stresses using equilibrium and thence, via the stress–strain law, to the equations of compatibility and the displacements. This methodical progress is only possible if the structure is statically determinate. In statically indeterminate structures we shall find that *all* the equations of equilibrium and compatibility have to be used before anything can be solved – the idea can be quickly demonstrated by modifying our simple frame to include another vertical member (1–4), having the same area A and modulus E as the others, and a coefficient of expansion α per degree (Fig. 1.10).

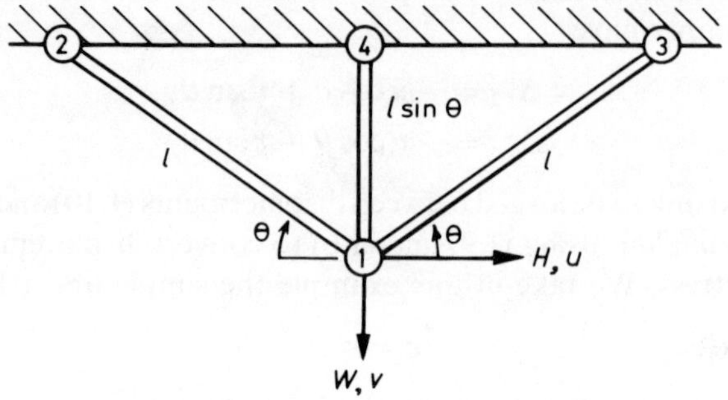

Fig. 1.10 Statically indeterminate framework

We now proceed schematically through Fig. 1.9 in the reverse manner to the statically determinate solution. The reason for this is that we are about to obtain a *displacement solution* in which the first variables to be solved will be the displacements u and v. The strains and stresses then follow; that is, compatibility → strain–stress → equilibrium. The compatibility equations (1.10) now contain one more equation:

$$\begin{aligned} \Delta_{12} &= \quad u\cos\theta + v\sin\theta \\ \Delta_{13} &= -u\cos\theta + v\sin\theta \\ \Delta_{14} &= v \end{aligned} \tag{1.13}$$

We now invoke the stress–strain law, and for interest we take the middle bar to be heated through T degrees.

Thus

for bars (1–2) and (1–3), $\varepsilon = \sigma/E$; $N = A\sigma = AE\,\Delta/l$
for bar (1–4), $\varepsilon = \sigma/E + \alpha T$; $N = A\sigma = AE\,\Delta/l\sin\theta - AE\alpha T$

The forces in the bars then become, using (1.13),

$$\begin{aligned} N_{12} &= \frac{AE}{l}(\quad u\cos\theta + v\sin\theta) \\ N_{13} &= \frac{AE}{l}(-u\cos\theta + v\sin\theta) \\ N_{14} &= \frac{AEv}{l\sin\theta} - AE\alpha T \end{aligned} \tag{1.14}$$

The internal stresses are now known in terms of the two unknown displacements u and v. But there are two remaining equations of equilibrium:

$$\begin{aligned} N_{12}\cos\theta - N_{13}\cos\theta - H &= 0 \\ N_{12}\sin\theta + N_{13}\sin\theta + N_{14} - W &= 0 \end{aligned} \tag{1.15}$$

On substituting (1.14) into (1.15) we obtain a solution for the displacements,

$$u = \frac{Hl/AE}{2\cos^2\theta}; \qquad v = \frac{Wl/AE + \alpha Tl}{2\sin^2\theta + 1/\sin\theta} \tag{1.16}$$

We have now completed a clockwise circuit of Fig. 1.9, and we back-track to find the stresses. Substituting (1.16) back into the stress–displacement equations (1.14), we find

$$N_{12} = \frac{H}{2\cos\theta} + \frac{(W + AE\alpha T)\sin^2\theta}{1 + 2\sin^3\theta}$$

$$N_{13} = -\frac{H}{2\cos\theta} + \frac{(W + AE\alpha T)\sin^2\theta}{1 + 2\sin^3\theta}$$

$$N_{14} = \frac{W - 2AE\alpha T\sin^3\theta}{1 + 2\sin^3\theta}$$

The displacement method has here been used for a statically indeterminate framework because the previous equilibrium approach will only suffice to find stresses directly if the structure is statically determinate. This former method is usually known as the *force method* simply because the unknowns which are first solved are the internal stresses or forces. However, there is absolutely no reason why the displacement method cannot be applied to statically determinate structures also. In the above example the bar (1–4) could be omitted and the same solution obtained as the first example's force solution, except that now the displacements will be found first and the stresses last. We leave this exercise to the reader.

We may now summarize the force method and the displacement method in the following fashion:

(a) The *force method* (for *statically* determinate frames). The equations of joint *equilibrium* are used first to obtain the bar *stresses* in terms of joint *forces*. The *stresses* are then converted to *strains* and the equations of joint *compatibility* are used. Thus the bar *forces* are solved first and then the *displacements*.

(b) The *displacement method* (for *kinematically* determinate frames). The equations of joint *compatibility* are used first to obtain the bar *strains* in terms of the joint *displacements*. The *strains* are then converted to *stresses* and the equations of joint *equilibrium* are used. Thus the bar *displacements* are solved first and then the *forces*.

Now we have some explaining to do. The phrase 'kinematically determinate' simply means the 'dual' of 'statically determinate', where the latter means a solution in which *statics* is sufficient to obtain bar *stresses* in terms of joint *forces*. Thus the former means that *kinematics* (geometry of movement) is sufficient to obtain bar *strains* in terms of joint *displacements*. Thus all frames are kinematically determinate provided we have labelled all joints with sufficient displacement components.

The next explanation is for the plethora of italics and why the descriptions of the force and displacement methods are so similar. In fact the two descriptions are identical if we switch the following pairs of 'dual' relationships

Statics ↔ Kinematics
Equilibrium ↔ Compatibility
Stress ↔ Strain
Force ↔ Displacement

The link between all these dual quantities is *work*, which is done by force over displacement or stress over strain. We shall see how work arguments enable us to switch from equilibrium and force variables to their dual compatability and displacement variables. We say that these duals *correspond*. Thus force H and displacement u correspond: W and u do not.

Before looking at work, let us briefly extend the direct force and displacement solutions to more complicated frameworks which may be statically or kinematically determinate or indeterminate.

1.3 Direct solution of complex frameworks – the force method

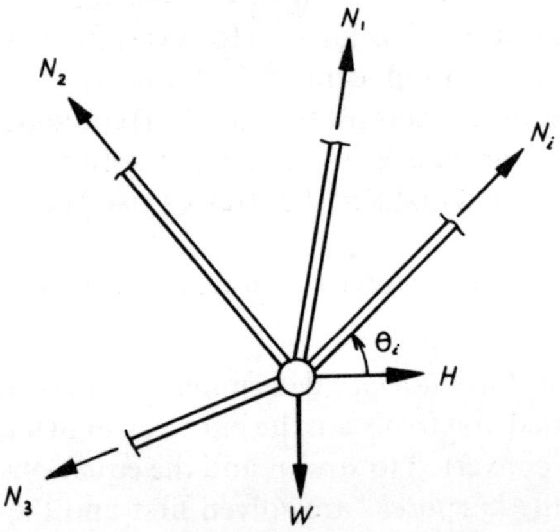

The figure shows a typical joint in a two-dimensional framework, loaded by forces H and W; a typical bar force is N_i. The equations of equilibrium are simply

$$\sum_i N_i \cos\theta_i + H = 0; \qquad \sum_i N_i \sin\theta_i - W = 0 \tag{1.17}$$

If the framework is statically determinate we expect to be able to obtain all the bar forces N_i from a set of equations like (1.17). A necessary – but not sufficient – condition for this to be possible is that the number of joint equilibrium equations shall match the number of unknowns. So if these are n bar forces N_i, r unknown supporting reactions, and j joints, we must have

$$n + r = 2j \tag{1.18}$$

If $n + r > 2j$ we expect the frame to be statically indeterminate. If $n + r < 2j$ we expect this deficiency in bars or supports to result in a mechanism.

In some frameworks the set of equations like (1.17) may be successively solved at each joint if the number of unknown bar forces at a joint never exceeds two. For example, consider the frame in Fig. 1.11. At the loaded joint 1 we have just two unknowns:

$$N_{13} + N_{12}/\sqrt{2} = 0; \qquad N_{12}/\sqrt{2} + W = 0$$

Therefore

$$N_{12} = -\sqrt{2}W; \qquad N_{13} = W$$

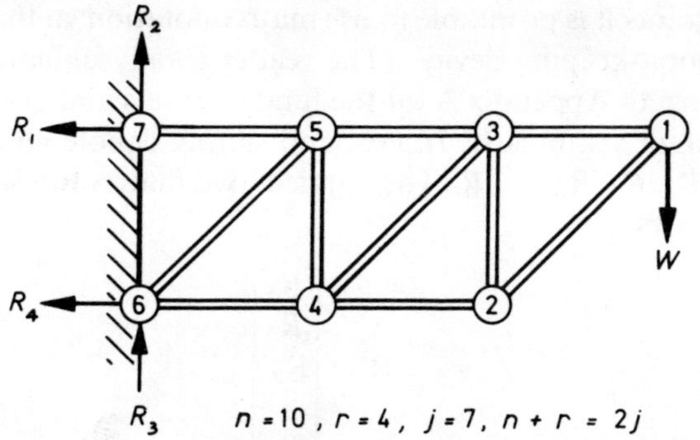

Fig. 1.11 All bars are horizontal, vertical, or inclined at 45°

We then proceed to joint 2 and knowing N_{13} solve immediately N_{23} and N_{24}. Joints 3, 4, 5 then follow and finally we may solve for the supporting reactions at joints 6 and 7. (See Problem 1.1 at the end of the chapter.)

The framework in Fig. 1.12 is not so straightforward. We may systematically start solving bar forces, two at a time, starting at joint 1 then 2 then 4, but then we stop: joints 6 and 7 have three unknown bar forces and 5 has four. However, if we make a temporary diversion we can use *overall equilibrium* of the total structure to

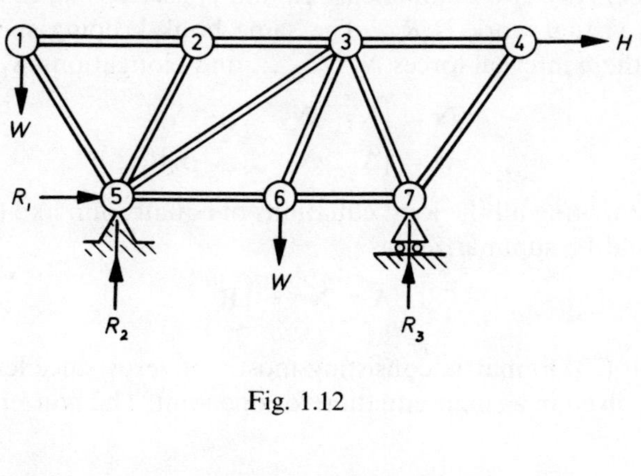

Fig. 1.12

obtain the three reactions R_1, R_2, and R_3 in terms of applied loads. Having solved these, joint 4 then yields N_{73} and N_{76}, then 6 gives N_{63} and N_{65}. There only remains N_{35} and there are four further equations of equilibrium available at joints 5 and 3 – consequently we can use three of these equations as a check. These three surplus equations are only to be expected as we have already used three overall equilibrium equations and these cannot conflict with the sum of all separate joint equations.

It should be fairly clear that in complex three-dimensional frameworks consisting of perhaps hundreds of bars, we may have to be a little more systematic. To be honest, a complex framework is unlikely to be statically determinate, but there might be special reasons for constructing such a structure – they are immune from thermal stresses, for example. In considering many bars and joints it is profitable to use matrix notation, at this stage purely as a labelling or book-keeping device. (The reader totally unfamiliar with matrix notation may turn to Appendix A for the fundamentals and examples.) Suppose we forsake different symbols W, H, etc., and simply denote all the applied load components as $R_1, R_2, R_3, \ldots, R_l$. The applied load matrix $\mathbf{R}$ is written as a single column with l rows:

$$\mathbf{R} = \begin{bmatrix} R_1 \\ R_2 \\ R_3 \\ \vdots \\ R_l \end{bmatrix}$$

In future, to please the printer, we will denote a column matrix as a row inside curly brackets, so the column $\mathbf{R}$ is written as

$$\mathbf{R} = \{R_1 \quad R_2 \quad \ldots \quad R_l\} \tag{1.19}$$

Later we shall need the joint displacement components, and again we drop u, v, and write a corresponding displacement matrix,

$$\mathbf{r} = \{r_1 \quad r_2 \quad \ldots \quad r_l\} \tag{1.20}$$

(It is common practice to use upper case, R, for forces, lower case, r for displacements.) Any two components R_i and r_i *correspond*, in both sense and direction, so virtual work is $R_i r_i$. The same book-keeping shorthand can be adopted for the n internal forces $N_1, N_2, \ldots$ and elongations $\Delta_1, \Delta_2, \ldots$. Thus,

$$\mathbf{N} = \{N_1 \quad N_2 \quad \ldots \quad N_n\} \tag{1.21}$$

$$\mathbf{\Delta} = \{\Delta_1 \quad \Delta_2 \quad \ldots \quad \Delta_n\} \tag{1.22}$$

If we imagine writing all the joint equations of equilibrium like (1.17), it is clear that they could be summarized as

$$\underset{\langle l \times n \rangle}{\mathbf{A}} \; \underset{\langle n \times 1 \rangle}{\mathbf{N}} = \underset{\langle l \times 1 \rangle}{\mathbf{R}} \tag{1.23}$$

where $\mathbf{A}$ is an $(l \times n)$ matrix consisting mostly of zeros since few of all the bar forces are involved in a single equation for one joint. The nonzero terms, noting

(1.17), must be a collection of direction cosines $\cos\theta_i$, $\sin\theta_i$, etc. Now if the structure is statically determinate we expect to be able to solve (1.23), say:

$$\mathbf{N} = \mathbf{bR} \tag{1.24}$$

where

$$\mathbf{b} = \mathbf{A}^{-1} \tag{1.25}$$

Two prerequisites for the inversion of (1.23) to give (1.24) are necessary:

(a) **A** must be square, i.e. $l = n$. Now the number of *known* joint forces is equal to $2j - r$, since reaction forces are unknowns. Therefore $l = 2j - r = n$, which agrees with (1.18).

(b) Even if $l = n$, **A** must not be singular, that is we must be able to invert it. This can happen in various ways. For example, the structure may not be supported adequately. The framework shown for example in Fig. 1.13 has $n = 14$, $r = 6$, $j = 10$, so $r + n = 2j$.

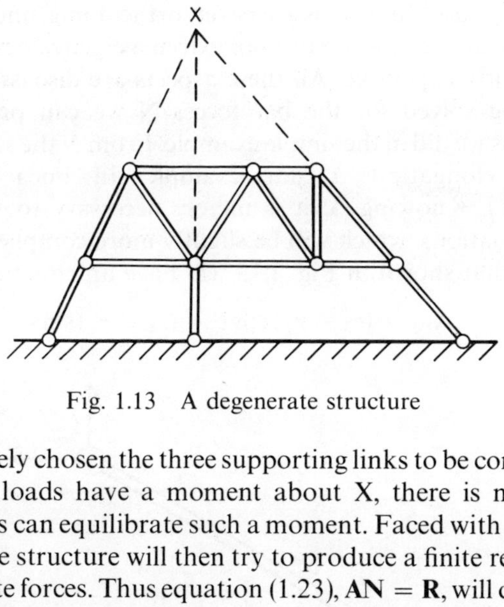

Fig. 1.13 A degenerate structure

But we have perversely chosen the three supporting links to be concurrent at X, so that if the applied loads have a moment about X, there is no way that the supporting reactions can equilibrate such a moment. Faced with a zero lever arm on which to lean, the structure will then try to produce a finite resisting moment by developing infinite forces. Thus equation (1.23), $\mathbf{AN} = \mathbf{R}$, will deliver infinite **N** for finite **R**, i.e. **A** is singular, or if we write

$$\mathbf{N} = \frac{\mathbf{A}_{\mathrm{adj}}\mathbf{R}}{|\mathbf{A}|}, \qquad \text{then } |\mathbf{A}| = 0$$

Therefore one test for singular behaviour (not the best) would be to see if the determinant of **A** was zero or very small.

Alternatively, a frame with the right number of bars may have them in the wrong places (Fig. 1.14). Figure 1.14(a) is a stable shape but (b) clearly is a mechanism. In fact part of (b) is statically indeterminate and it is possible to induce self-straining forces (like temperature changes) which require no applied

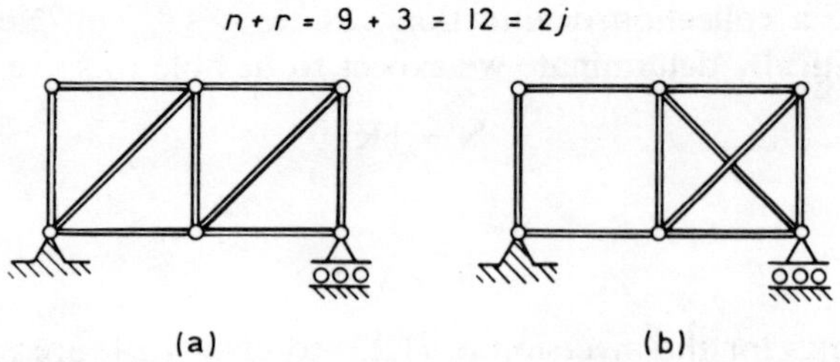

Fig. 1.14 (a) Bars in the right places. (b) Bars in the wrong places

forces **R** at all. There is possibly a solution therefore to $\mathbf{AN} = \mathbf{0}$, confirming that **A** is singular, $|\mathbf{A}| = 0$. In complex frameworks, it will probably be quite impossible to spot mechanisms visually. The singularity of **A**, however, can be tested numerically once the equations of equilibrium are formed. If we were to use techniques like Gauss or Jordan elimination to convert **A** to diagonal form, we would find complete rows of zeros so that with a finite right-hand side **R** there is no possible solution. We could also see if $|\mathbf{A}| = 0$, but because computers have finite word-lengths, absolute zeros are never forthcoming, and some sort of norm is needed. One such check is the ratio of extreme eigenvalues of **A** to see if all is well; it is also fairly expensive. All these aspects are discussed in Appendix A.

Finally, having solved for the bar forces **N** we can proceed to the joint displacements **r** as we did in the simple example. From N the stress–strain law will yield the bar elongation Δ; for example, in linear elastic materials $\Delta = Nl/AE + \alpha Tl$. Knowing Δ, it is merely necessary to write down all the compatibility equations, which will be slightly more complicated than (1.13). A typical bar like that shown in Fig. 1.15 will have an equation like

$$\Delta_{12} = (r_3 - r_1)\cos\theta + (r_4 - r_2)\sin\theta \tag{1.26}$$

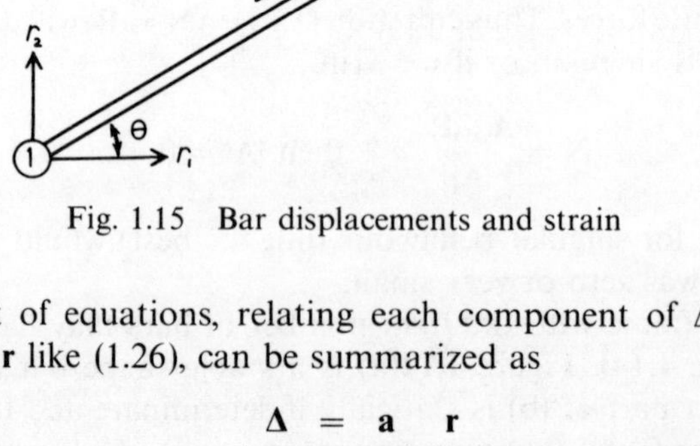

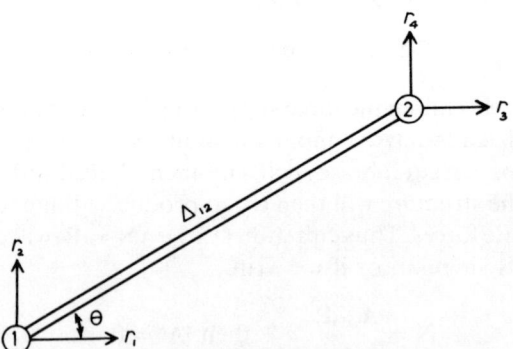

Fig. 1.15 Bar displacements and strain

A complete set of equations, relating each component of $\mathbf{\Delta}$ to four or less components of **r** like (1.26), can be summarized as

$$\underset{\langle n \times 1\rangle}{\mathbf{\Delta}} = \underset{\langle n \times l\rangle}{\mathbf{a}} \; \underset{\langle l \times 1\rangle}{\mathbf{r}} \tag{1.27}$$

where **a** is another matrix like **A** consisting mostly of zeros and the direction cosines. (In section 1.6 we show that $\mathbf{a}^t = \mathbf{A}$.) If the structure is statically determinate then we know that this $(n \times l)$ matrix is square and hence (1.27) should be solvable for **r**, knowing the bar strains $\mathbf{\Delta}$. But one must also be wary of the possibility of a singular **a**, which would imply the existence of solutions of the form $\mathbf{ar} = \mathbf{0}$. Any set of such displacements which can arise without any bars straining must mean that the 'structure' is either a mechanism or perhaps the supports are inadequate in preventing rigid-body motion. Figure 1.13 can be viewed as this type of deficiency; in this case rotation about the point X is not prevented by the three supports. (Such a floppy structure would deform grossly and *then* perhaps recover its stiffness when the deformed supporting links were no longer concurrent.)

1.4 Direct solution of complex frameworks – the displacement method

The procedure is an extension of our solution of the three-bar truss of Fig. 1.10. We first establish the compatibility equations (1.27). Next we write the stress–strain law, like $N/A = E\,\Delta/l$, as

$$N_{12} = k_{12}\Delta_{12} \tag{1.28}$$

where $k_{12} = AE/l$ is the stiffness of bar (1–2). We now assemble – purely for illumination – all the bar stiffnesses as a diagonal matrix.

$$\mathbf{k} = \begin{bmatrix} k_{12} & & & \\ & k_{23} & & \\ & & \cdot & \\ & & & \cdot \end{bmatrix} = \lceil k_{12} \quad k_{23} \quad k_{34} \quad \ldots \rfloor$$

so we may write schematically, as a summary of all equations like (1.28),

$$\mathbf{N} = \mathbf{k\Delta} \tag{1.29}$$

Using (1.27),

$$\mathbf{N} = \mathbf{kar}$$

and substituting into the equilibrium equations

$$\mathbf{AN} = \mathbf{R} \tag{1.23}$$

we find

$$\underset{\langle l\times n\rangle}{\mathbf{A}}\ \underset{\langle n\times n\rangle}{\mathbf{k}}\ \underset{\langle n\times l\rangle}{\mathbf{a}}\ \underset{\langle l\times 1\rangle}{\mathbf{r}} = \underset{\langle l\times 1\rangle}{\mathbf{R}} \tag{1.30}$$

Equations (1.30) are the final 'stiffness' equations from which **r** is solved. The form here is purely symbolic; it would be grossly inefficient to construct a set of equations by multiplication of three matrices **A**, **k**, and **a** consisting of so many zeros. Moreover, in the next section we use virtual work to set up equations like (1.30) where their structure is shown to be rather simpler.

In conclusion, it must be honestly admitted that two-dimensional pinjointed frameworks are very simple structures to use as an introduction to fundamental concepts. They are simple in that the stress or strain in each component bar is summarized by a single scalar N or Δ. Other structures – including beams and stiff-jointed frameworks – have a varying stress field and it is convenient to express the internal stresses or strains in terms of (say) end forces or displacements. These 'forces' could be tensions, shears, bending moments, etc., together with corresponding displacements or rotations. But we leave such details to Chapter 2.

1.5 Virtual work and simple pinjointed frameworks

We have already discussed the idea of virtual work applied to rigid bodies, in which we argued that the virtual work done should be zero. We now extend this idea to deformable structures by stating that *the virtual work done by the applied forces is equal to that done by the internal stresses.*

We remind ourselves that we are seeking an alternative to equilibrium or compatibility arguments and that we wish to avoid the stress–strain or force–displacement law which can be linear or nonlinear. This we do by simply defining virtual work as product of force and displacement (Fig. 1.16).

As an introduction consider again the simple problem of Fig. 1.17.

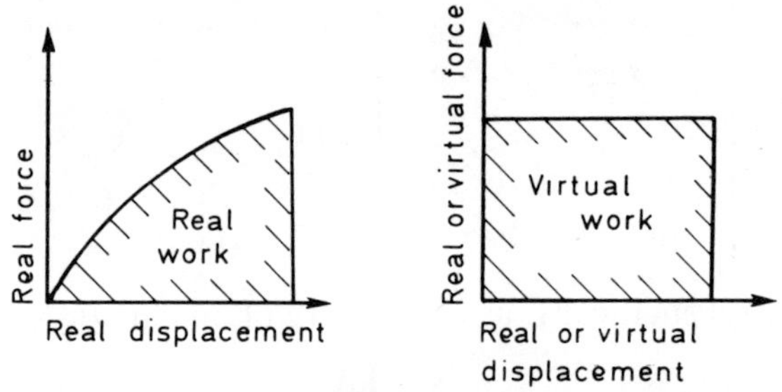

Fig. 1.16

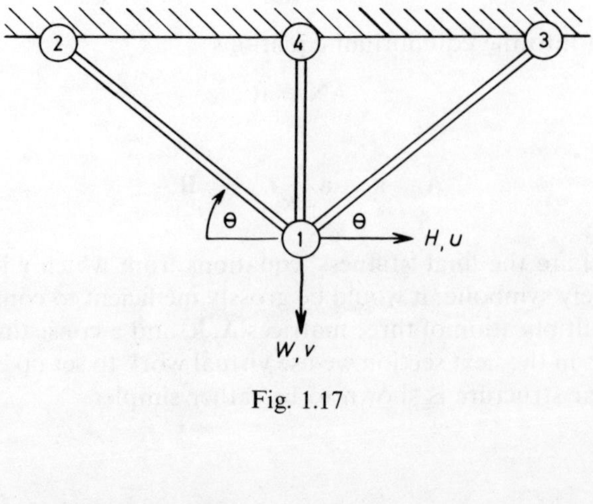

Fig. 1.17

The principle of virtual displacements (PVD)

This allows virtual displacement and bar elongations $\bar{u}, \bar{v}, \bar{\Delta}$; and equates internal to external work, thus:

$$N_{12}\bar{\Delta}_{12} + N_{13}\bar{\Delta}_{13} + N_{14}\bar{\Delta}_{14} = H\bar{u} + W\bar{v} \tag{1.31}$$

This is not a promising beginning. Equation (1.31) provides no information at all until we insist that *the virtual displacements must satisfy compatibility*, and then the equation is transformed. Using then (1.13), $\Delta_{12}, \Delta_{13} = \pm u\cos\theta + v\sin\theta$ and $\Delta_{14} = v$, the PVD (1.31) becomes, on collecting terms,

$$\begin{aligned} &\bar{u}(N_{12}\cos\theta - N_{13}\cos\theta - H) \\ &+ \bar{v}(N_{12}\sin\theta + N_{13}\sin\theta + N_{14} - W) = 0 \end{aligned} \tag{1.32}$$

Now $\bar{u}$ and $\bar{v}$ are unrelated so their individual coefficients must vanish. This gives us the equations of equilibrium as before:

$$\begin{aligned} N_{12}\cos\theta - \mathrm{N}_{13}\cos\theta - H &= 0 \\ N_{12}\sin\theta + N_{13}\sin\theta + N_{14} - W &= 0 \end{aligned} \tag{1.15}$$

We repeat that the PVD was not usable until we made the virtual displacements compatible; further we were able to extract the equilibrium equations when the total virtual work was written as coefficients of the separate virtual displacements. These two points are a common theme throughout this text. We have not proved the PVD in general yet, but we have started.

The principle of virtual forces (PVF)

It is natural to now attempt the dual of the PVD and switch stress ↔ strain, force ↔ displacement, and equilibrium ↔ compatibility. Thus, we now suppose virtual forces $\bar{W}$, $\bar{H}$, and $\bar{N}$ to exist and write our virtual work equality as

$$\bar{N}_{12}\Delta_{12} + \bar{N}_{13}\Delta_{13} + \bar{N}_{14}\Delta_{14} = \bar{H}u + \bar{W}v \tag{1.33}$$

Like (1.31), this equation is totally unproductive until this time we make the virtual force system satisfy equilibrium – equations (1.15). Substituting these into the PVF (1.33) for $\bar{H}$ and $\bar{W}$ and collecting terms, we find

$$\begin{aligned} &\bar{N}_{12}(\Delta_{12} - u\cos\theta - v\sin\theta) \\ &+ \bar{N}_{13}(\Delta_{13} + u\cos\theta - v\sin\theta) \\ &+ \bar{N}_{14}(\Delta_{14} - v) = 0 \end{aligned} \tag{1.34}$$

Again we could select any $\bar{N}$ to be zero or unity, so their coefficients in (1.34) must be zero:

$$\begin{aligned} &\Delta_{12} - u\cos\theta - v\sin\theta = 0 \\ &\Delta_{13} + u\cos\theta - v\sin\theta = 0 \\ &\Delta_{14} - v = 0 \end{aligned} \tag{1.35}$$

Equations (1.35) are the expected compatibility equations (cf (1.13)). It is straightforward to extend this simple example to a completely general proof for frameworks, but there seems little point since we will prove both the PVD and PVF for a general arbitrary continuum. Summarizing, then:

The principle of virtual displacements will indirectly satisfy equilibrium conditions provided the virtual displacements are made to satisfy compatibility.

The principle of virtual forces will indirectly satisfy compatibility conditions provided the virtual forces are made to satisfy equilibrium.

1.6 Virtual work and complex pinjointed frameworks

The extension to complex frameworks is straightforward using our matrix notation. The PVD becomes

$$N_1\bar{\Delta}_1 + N_2\bar{\Delta}_2 + N_3\bar{\Delta}_3 + \cdots + N_g\bar{\Delta}_g = R_1\bar{r}_1 + R_2\bar{r}_2 + \cdots$$

which can be summarized as

$$\mathbf{N}^t\bar{\mathbf{\Delta}} = \mathbf{R}^t\bar{\mathbf{r}} \tag{1.36}$$

Note that work is a simple scalar, so (1.36) could equally be transposed:

$$\bar{\mathbf{\Delta}}^t\mathbf{N} = \bar{\mathbf{r}}^t\mathbf{R}$$

The PVD (1.36) remains idle until we enforce compatibility $\mathbf{\Delta} = \mathbf{ar}$, whence

$$\mathbf{N}^t\mathbf{a}\bar{\mathbf{r}} = \mathbf{R}^t\bar{\mathbf{r}} \qquad \text{or} \qquad (\mathbf{N}^t\mathbf{a} - \mathbf{R}^t)\bar{\mathbf{r}} = 0 \tag{1.37}$$

Now, since *all* components of $\bar{\mathbf{r}}$ are virtual, all equations in (1.37) must be zero, that is

$$\mathbf{N}^t\mathbf{a} - \mathbf{R}^t = 0 \qquad \text{or} \qquad \mathbf{a}^t\mathbf{N} = \mathbf{R} \tag{1.38}$$

We know (1.38) must be the equations of equilibrium, and comparing these with the original (1.23), $\mathbf{AN} = \mathbf{R}$, we see that

$$\mathbf{a}^t = \mathbf{A} \tag{1.39}$$

This very important relationship forces a link between kinematic compatibility (1.27) and static compatibility (1.23). If the structure is statically determinate then $\mathbf{A}$ is square and we now see formally why a mechanism was found to be predictable by testing for singular behaviour of either $\mathbf{a}$ or $\mathbf{A}$. The structure of course may be redundant, in which case we cannot solve (1.38) immediately. We go to the next stage and construct the 'stress–strain' law, which is $\mathbf{N} = \mathbf{k\Delta}$ if linear, and hence (1.38) becomes

$$\mathbf{a}^t\mathbf{k\Delta} = \mathbf{R}$$

or using $\mathbf{\Delta} = \mathbf{ar}$,

$$(\mathbf{a}^t\mathbf{ka})\mathbf{r} = \mathbf{R} \tag{1.40}$$

which is usually written as,

$$\mathbf{Kr} = \mathbf{R} \tag{1.41}$$

where the structure's *stiffness matrix*

$$\mathbf{K} = \mathbf{a}^t\mathbf{k}\mathbf{a} \tag{1.42}$$

We recall that **k** is diagonal, so

$$\mathbf{K}^t = \mathbf{a}^t\mathbf{k}^t\mathbf{a} = \mathbf{a}^t\mathbf{k}\mathbf{a} = \mathbf{K}$$

that is, the *stiffness matrix is symmetrical*, a fact not apparent previously (1.30). The solution, or inverse, of **K** in (1.41) we later see is not too much of a problem even for several thousand unknowns. Again we must be vigilant in case **K** is singular or *rank deficient* (see Appendix A). The rank of the product (1.42) is equal at the most to that of its factors, so again we see that rank deficiency of **a** will be the culprit if it happens.

The dual PVF can also be used for large systems in an analogous fashion. The PVF is simply

$$\overline{\mathbf{N}}^t\mathbf{\Delta} = \overline{\mathbf{R}}^t\mathbf{r}$$

which is useful only when we enforce equilibrium

$$\mathbf{A}\mathbf{N} = \mathbf{R} \tag{1.23}$$

and the PVF then becomes

$$\overline{\mathbf{N}}^t\mathbf{\Delta} = \overline{\mathbf{N}}^t\mathbf{A}^t\mathbf{r}$$

or

$$\overline{\mathbf{N}}^t(\mathbf{\Delta} - \mathbf{A}^t\mathbf{r}) = \mathbf{0}$$

As all components of $\overline{\mathbf{N}}$ are virtual, we deduce the equations of compatibility

$$\mathbf{\Delta} = \mathbf{A}^t\mathbf{r}$$

To be consistent in our dual approach, we should now write all bar *flexibility* strain-stress laws like $\Delta = fN$ ($f = l/AE$) as $\mathbf{\Delta} = \mathbf{f}\mathbf{N}$, where $\mathbf{f} = \lceil f_1 \quad f_2 \quad f_3 \quad \ldots \rfloor$. Thus

$$\mathbf{\Delta} = \mathbf{f}\mathbf{N} = \mathbf{A}^t\mathbf{r} \tag{1.43}$$

This can be inverted, $\mathbf{N} = \mathbf{f}^{-1}\mathbf{A}^t\mathbf{r}$,
where

$$\mathbf{f}^{-1} = \lceil f_1^{-1} \quad f_2^{-1} \quad f_3^{-1} \quad \ldots \rfloor = \lceil k_1 \quad k_2 \quad \ldots \rfloor = \mathbf{k} \tag{1.44}$$

Therefore $\mathbf{N} = \mathbf{k}\mathbf{A}^t\mathbf{r}$ and substituting into $\mathbf{A}\mathbf{N} = \mathbf{R}$ we find

$$\mathbf{A}\mathbf{k}\mathbf{A}^t\mathbf{r} = \mathbf{R} \tag{1.45}$$

Thus we have set up a stiffness solution again, but we have this time used equilibrium equations and the PVF. In simple frameworks there is absolutely no difference between (1.45) and (1.40), but when we come to *approximate* finite-element methods there is a real difference which we can exploit.

Enough has been said of simple frameworks, and we now turn to more ambitious things. We shall look at a completely arbitrary continuum and attempt to answer two general questions:

1. Why exactly did the PVD deliver the equations of equilibrium only when compatibility was satisfied?
2. Why did the PVF deliver compatibility only when the virtual forces satisfied equilibrium?

1.7 Direct analysis of a general continuum

Before appealing to indirect arguments it is necessary to establish the equations directly. To describe the stresses and strains in a body we will use the very convenient cartesian coordinate system x, y, z, but the interested reader can turn to Appendix C for equivalent statements in polar coordinates.

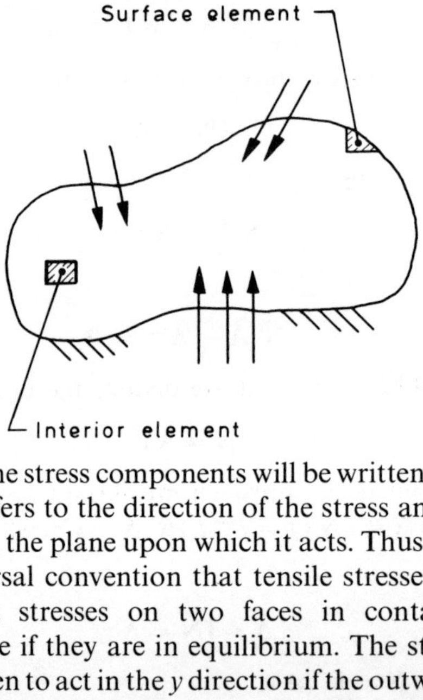

Referring to Fig. 1.18, the stress components will be written as $\sigma_{xx}\,\sigma_{yy}\,\sigma_{zz}\,\sigma_{xy}\,\sigma_{yz}$ σ_{zx}. The first subscript refers to the direction of the stress and the second to the direction of the normal to the plane upon which it acts. Thus σ_{xx} is a *direct* stress and we accept the universal convention that tensile stresses are positive; then automatically the direct stresses on two faces in contact are equal and opposite – as they must be if they are in equilibrium. The stress σ_{yx} is clearly a *shear* stress and this is taken to act in the y direction if the outward normal is in the positive x direction; if not then it is drawn in the opposite sense. Figure 1.19 shows that the shears will then also be equal and opposite on two faces in contact. This

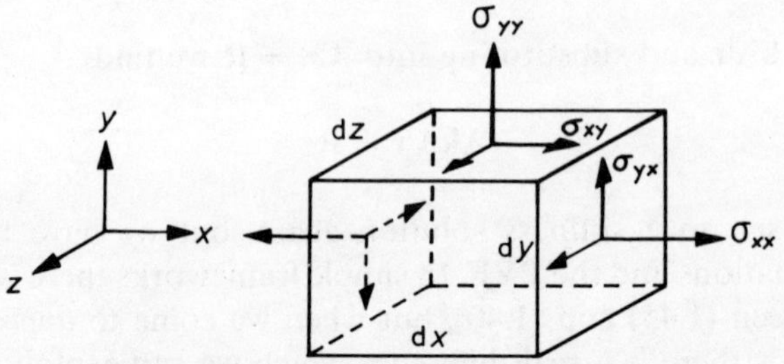

Fig. 1.18 Stresses in a continuum

convention ensures equilibrium across a single face: now we go further. If the stresses vary continuously throughout the body, we must ensure equilibrium of the varying stress field with the applied body forces such as gravity or inertia. Denote the body force *per unit volume* as having components p_{vx}, p_{vy}, p_{vz}; for example, due to gravity $p_{vz} = -g$. Assume our element is so small that variations across it can be approximated as linear, for example, σ_{xx} increasing to

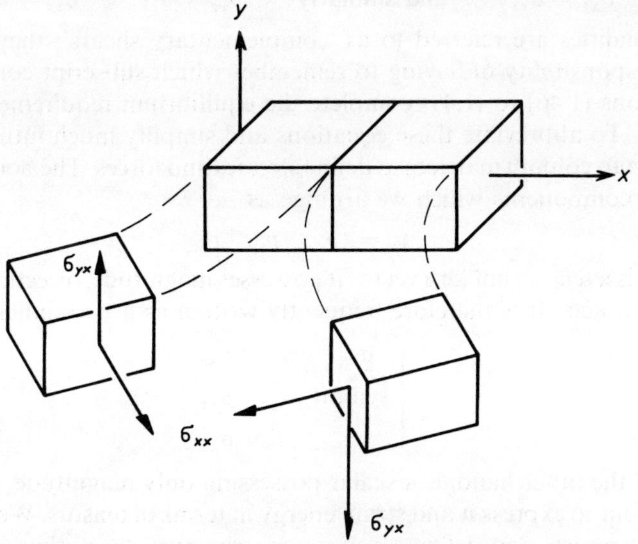

Fig. 1.19 Shear stress convention

$\sigma_{xx} + (\partial\sigma_{xx}/\partial x)\,\mathrm{d}x$. The pattern of stress components – restricted to x components for clarity – is shown in Fig. 1.20. Summing all the forces in the x direction, and cancelling $\mathrm{d}x\,\mathrm{d}y\,\mathrm{d}z$, we find

$$\frac{\partial\sigma_{xx}}{\partial x} + \frac{\partial\sigma_{xy}}{\partial y} + \frac{\partial\sigma_{xz}}{\partial z} + p_{vx} = 0 \tag{1.46}$$

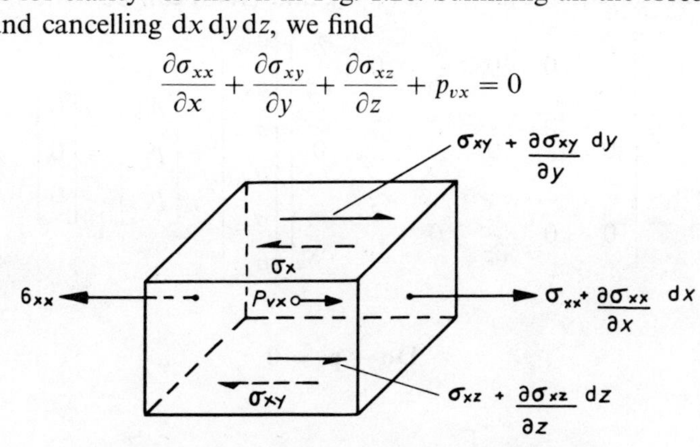

Fig. 1.20 Stresses and forces in x direction

Cycling subscripts, we likewise obtain for equilibrium in the y and z directions

$$\frac{\partial\sigma_{yx}}{\partial x} + \frac{\partial\sigma_{yy}}{\partial y} + \frac{\partial\sigma_{yz}}{\partial z} + p_{vy} = 0 \tag{1.47}$$

$$\frac{\partial\sigma_{zx}}{\partial x} + \frac{\partial\sigma_{zy}}{\partial y} + \frac{\partial\sigma_{zz}}{\partial z} + p_{vz} = 0 \tag{1.48}$$

There remains moment equilibrium and, if we take moments about axes through the centre of the cube, the forces on the faces will be mean values of stress × elemental area. Thus, taking moments about the z axis,

$$\sigma_{xy}(\mathrm{d}x\,\mathrm{d}z)\,\mathrm{d}y - \sigma_{yx}(\mathrm{d}y\,\mathrm{d}z)\,\mathrm{d}x = 0$$

or

$$\sigma_{xy} = \sigma_{yx}; \qquad \text{and similarly} \qquad \sigma_{yz} = \sigma_{zy}; \qquad \sigma_{zx} = \sigma_{xz} \tag{1.49}$$

These equalities are referred to as 'complementary shears': they relieve us of further responsibility of having to remember which subscript comes first.

Equations (1.46) to (1.49) complete the equilibrium requirements inside the volume V. To abbreviate these equations and simplify much future algebra we now resort to column matrices to define stresses and forces. The body force *vector* has three components which we arrange as

$$\mathbf{p}_v = \{p_{vx} \quad p_{vy} \quad p_{vz}\} \tag{1.50}$$

But stress is a *tensor*: unlike a vector it possesses magnitude, direction, *and* a plane on which it acts. It is therefore frequently written as a two-dimensional array:

$$\begin{bmatrix} \sigma_{xx} & \sigma_{xy} & \sigma_{xz} \\ \text{symm} & \sigma_{yy} & \sigma_{yz} \\ & & \sigma_{zz} \end{bmatrix}$$

Work, on the other hand, is a scalar possessing only magnitude, and it is most inconvenient to express it and strain energy in terms of tensors. We will therefore ignore the purists and define a column matrix stress 'vector' as

$$\boldsymbol{\sigma} = \{\sigma_{xx} \quad \sigma_{yy} \quad \sigma_{zz} \quad \sigma_{xy} \quad \sigma_{yz} \quad \sigma_{zx}\} \tag{1.51}$$

The equilibrium equations (1.46), (1.47), and (1.48) can be written as

$$\begin{bmatrix} \dfrac{\partial}{\partial x} & 0 & 0 & \dfrac{\partial}{\partial y} & 0 & \dfrac{\partial}{\partial z} \\ 0 & \dfrac{\partial}{\partial y} & 0 & \dfrac{\partial}{\partial x} & \dfrac{\partial}{\partial z} & 0 \\ 0 & 0 & \dfrac{\partial}{\partial z} & 0 & \dfrac{\partial}{\partial y} & \dfrac{\partial}{\partial x} \end{bmatrix} \begin{bmatrix} \sigma_{xx} \\ \sigma_{yy} \\ \sigma_{zz} \\ \sigma_{xy} \\ \sigma_{yz} \\ \sigma_{zx} \end{bmatrix} + \begin{bmatrix} p_{vx} \\ p_{vy} \\ p_{vz} \end{bmatrix} = \begin{bmatrix} 0 \\ 0 \\ 0 \end{bmatrix}$$

or simply

$$\mathbf{D}\boldsymbol{\sigma} + \mathbf{p}_v = \mathbf{0} \tag{1.52}$$

where

$$\mathbf{D} = \begin{bmatrix} \dfrac{\partial}{\partial x} & 0 & 0 & \dfrac{\partial}{\partial y} & 0 & \dfrac{\partial}{\partial z} \\ 0 & \dfrac{\partial}{\partial y} & 0 & \dfrac{\partial}{\partial x} & \dfrac{\partial}{\partial z} & 0 \\ 0 & 0 & \dfrac{\partial}{\partial z} & 0 & \dfrac{\partial}{\partial y} & \dfrac{\partial}{\partial x} \end{bmatrix} \tag{1.53}$$

Equation (1.52), although merely a shorthand form, is very useful in investigating precisely what it is that our work statements are doing, particularly when using approximate methods such as finite-element methods. We note that the equations have been derived, like our frameworks, using the undistorted geometry of an elemental cube. The equations are not valid for gross deformation in which the element dx, dy, dz alters shape; in this case the array of linear operators in **D** does not arise. Most structural materials do not accommodate strains in excess of 1 per cent so the linear equations (1.52) are not too restrictive.

Suppose we have two separate loading systems $\mathbf{p}_1$ and $\mathbf{p}_2$ which produce stress systems $\boldsymbol{\sigma}_1$ and $\boldsymbol{\sigma}_2$. Then by (1.52),

$$\mathbf{D}\boldsymbol{\sigma}_1 + \mathbf{p}_1 = \mathbf{0} \qquad \text{and} \qquad \mathbf{D}\boldsymbol{\sigma}_2 + \mathbf{p}_2 = \mathbf{0}$$

Adding,

$$\mathbf{D}\boldsymbol{\sigma}_1 + \mathbf{D}\boldsymbol{\sigma}_2 + \mathbf{p}_1 + \mathbf{p}_2 = \mathbf{0}$$

But because **D** is linear we may group this equation as

$$\mathbf{D}(\boldsymbol{\sigma}_1 + \boldsymbol{\sigma}_2) + (\mathbf{p}_1 + \mathbf{p}_2) = \mathbf{0}$$

which is in the same form as (1.52). Thus two loading systems, when added, will produce simply the sum of the two separate stress systems. This *principle of superposition* is useful, for example, in constructing complex solutions from simpler ones.

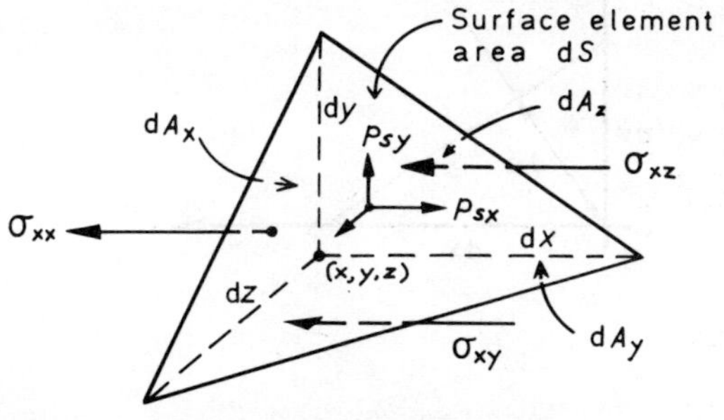

Fig. 1.21 A surface element

On the surface of a body, the simple cube element will not suffice and the surface element has to be treated as a special case. Let the surface traction – force per unit area – have components $p_{sx} p_{sy} p_{sz}$, and take tensile applied traction to be positive. Consider the triangular surface element of area dS, in Fig. 1.21, produced by the three sides dx, dy, dz, all drawn as positive increments from a point (x, y, z) close to the surface. For clarity, only the x components of stress are shown acting over the sides dA_x, dA_y, and dA_z. The elemental area of surface dS has a normal n to it, passing through (x, y, z). Resolving in the x direction,

$$\sigma_{xx}\,\mathrm{d}A_x + \sigma_{xy}\,\mathrm{d}A_y + \sigma_{xz}\,\mathrm{d}A_z = p_{sx}\,\mathrm{d}S$$

To simplify this we note that the volume of the tetrahedron may be written in four ways:

$$\mathrm{d}V = \tfrac{1}{6}\mathrm{d}A_x\,\mathrm{d}x = \tfrac{1}{6}\mathrm{d}A_y\,\mathrm{d}y = \tfrac{1}{6}\mathrm{d}A_z\,\mathrm{d}z = \tfrac{1}{6}\mathrm{d}S\,\mathrm{d}n$$

Thus,

$$\frac{\partial A_x}{\partial S} = \frac{\partial n}{\partial x}, \qquad \frac{\partial A_y}{\partial S} = \frac{\partial n}{\partial y}, \qquad \frac{\partial A_z}{\partial S} = \frac{\partial n}{\partial z}$$

and the equation of equilibrium becomes

$$\sigma_{xx}\frac{\partial n}{\partial x} + \sigma_{xy}\frac{\partial n}{\partial y} + \sigma_{xz}\frac{\partial n}{\partial z} = p_{sx} \tag{1.54}$$

Obviously we may again cycle x, y, z to produce the other two equations,

$$\sigma_{yx}\frac{\partial n}{\partial x} + \sigma_{yy}\frac{\partial n}{\partial y} + \sigma_{yz}\frac{\partial n}{\partial z} = p_{sy} \tag{1.55}$$

$$\sigma_{zx}\frac{\partial n}{\partial x} + \sigma_{zy}\frac{\partial n}{\partial y} + \sigma_{zz}\frac{\partial n}{\partial z} = p_{sz} \tag{1.56}$$

The terms $\partial n/\partial x$, $\partial n/\partial y$, $\partial n/\partial z$ are simply the direction cosines of the local normal to the surface.

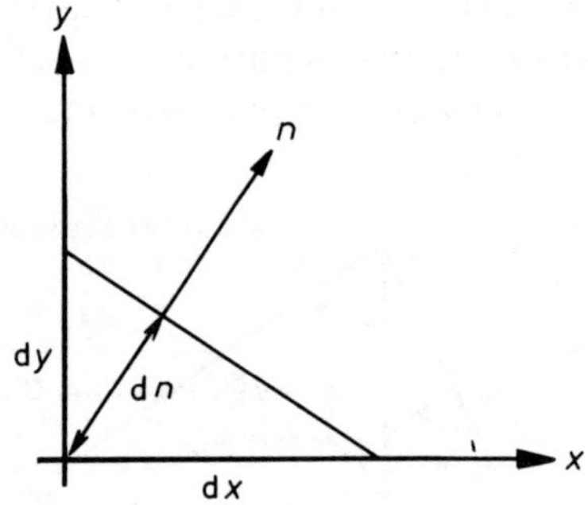

The surface equilibrium equations (1.54, 1.55, and 1.56) can be summarized as the single expression

$$(\mathbf{D}n)\boldsymbol{\sigma} = \mathbf{p}_s \tag{1.57}$$

where $\mathbf{p}_s = \{p_{sx} \quad p_{sy} \quad p_{sz}\}$ is the surface force vector, and $\mathbf{D}$ is as before, (1.53). $(\mathbf{D}n)$ implies that all the operators in $\mathbf{D}$ operate upon the single scalar variable n. The brackets are necessary because $\mathbf{D}(n\boldsymbol{\sigma})$ means something quite different, as the curious reader may verify. Like the internal equations, the surface ones are also linear.

We now turn to compatibility, assuming as we did for frameworks, that the displacements are continuous everywhere as the structure deforms. This merely means that dislocations or ruptures do not take place, joints do not break, and cracks of an indeterminate length do not occur. (A known crack which does not propagate is permissible since it is just another surface.) If then the displacements are continuous, their derivatives must exist and we may use simple geometrical arguments to relate strains to displacement derivatives. This is all that kinematic compatibility means. (The reader may find other 'equations of compatibility' such as Timoshenko's[16] from which the displacements have been eliminated.)

Consider the point (x, y, z) which is displaced by components u, v, w. It is simpler to first imagine (and easier to draw) just the components u and v in the x–y plane. In Fig. 1.22 an element $\mathrm{d}x\,\mathrm{d}y$ is shown (exaggeratedly) displaced, and, as the variations of displacement are approximated as *linear* over a small element, the sides of the distorted element are still straight. In the absence of components v and w there would be a simple 'stretch' in the x-direction which, by analogy with the direct tensile stress σ_{xx}, we denote as ε_{xx}. It is universal to denote this strain as a fractional change in length, and since strains are assumed small we are spared the embarrassment of having to distinguish between the distorted side AB and its projection AC – that is, $\cos\beta \simeq 1$.

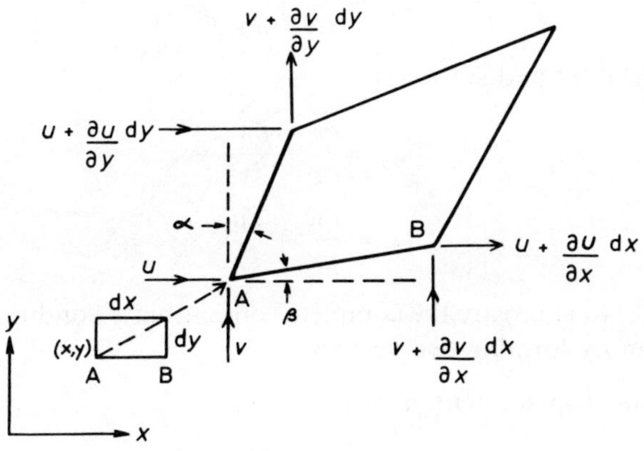

Fig. 1.22 Deformation of an element

Thus,

$$\varepsilon_{xx} = \frac{(\mathrm{d}x + \partial u/\partial x - \mathrm{d}x)}{\mathrm{d}x} = \frac{\partial u}{\partial x} \tag{1.58}$$

similarly,

$$\varepsilon_{yy} = \frac{\partial v}{\partial y} \tag{1.59}$$

$$\varepsilon_{zz} = \frac{\partial w}{\partial z} \tag{1.60}$$

An angular distortion can clearly take place without any of the element sides stretching, and this is referred to as 'shear strain' since it occurs in isotropic materials under a pure shear stress (Fig. 1.23). It is most common in engineering practice to measure this type of strain by the total angle $(\alpha + \beta)$ in Fig. 1.22.

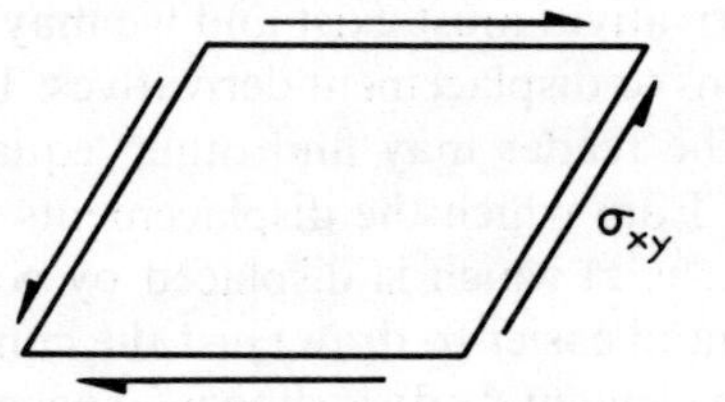

Fig. 1.23 Positive shear stress and strain

Thus

$$\varepsilon_{xy} = \frac{(\partial v/\partial x)\,dx}{dx + (\partial u/\partial x)\,dx} + \frac{(\partial u/\partial y)\,dy}{dy + (\partial v/\partial y)\,dy}$$

or, as we have already stated that strains like $\partial u/\partial x$ are $\ll 1$,

$$\varepsilon_{xy} = \frac{\partial v}{\partial x} + \frac{\partial u}{\partial y} \tag{1.61}$$

Cycling the variables and subscripts,

$$\varepsilon_{yz} = \frac{\partial v}{\partial z} + \frac{\partial w}{\partial y} \tag{1.62}$$

$$\varepsilon_{zx} = \frac{\partial w}{\partial x} + \frac{\partial u}{\partial z} \tag{1.63}$$

Equations (1.58) to (1.63) are the complete compatibility conditions and we can abbreviate them by forming two vectors:

for displacement $\mathbf{u} = \{u \quad v \quad w\}$ (1.64)

and,

for strains $\boldsymbol{\varepsilon} = \{\varepsilon_{xx} \quad \varepsilon_{yy} \quad \varepsilon_{zz} \quad \varepsilon_{xy} \quad \varepsilon_{yz} \quad \varepsilon_{zx}\}$ (1.65)

Writing compatibility as

$$\begin{bmatrix} \varepsilon_{xx} \\ \varepsilon_{yy} \\ \varepsilon_{zz} \\ \varepsilon_{xy} \\ \varepsilon_{yz} \\ \varepsilon_{zx} \end{bmatrix} = \begin{bmatrix} \frac{\partial}{\partial x} & 0 & 0 \\ 0 & \frac{\partial}{\partial y} & 0 \\ 0 & 0 & \frac{\partial}{\partial z} \\ \frac{\partial}{\partial y} & \frac{\partial}{\partial x} & 0 \\ 0 & \frac{\partial}{\partial z} & \frac{\partial}{\partial y} \\ \frac{\partial}{\partial z} & 0 & \frac{\partial}{\partial x} \end{bmatrix} \begin{bmatrix} u \\ v \\ w \end{bmatrix}$$

this is summarized as

$$\varepsilon = \mathbf{D}^t\mathbf{u} \tag{1.66}$$

where **D** is identical to definition (1.53) previously used. (It may appear fortuitous that **D** appears, but virtual work will reveal all.)

Two points are worth emphasizing in the compatibility equation:

(a) Like the equilibrium equations, (1.66) is *linear*, and the principle of superposition also applies to strain-displacements.
(b) *The cause of the strains has not been specified.* They may be due to stress, temperature, creep, fabrication processes, etc. This seemingly self-evident statement is often forgotten by students, perhaps due to the confusing approach used in some texts to solve thermal problems as a two-part process – first a thermal expansion and then an induced 'thermal stress'. As a digression, a simple example illustrates the point.

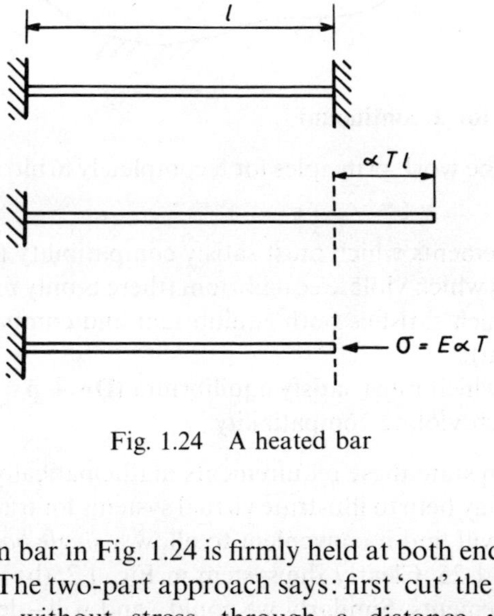

Fig. 1.24 A heated bar

Suppose the uniform bar in Fig. 1.24 is firmly held at both ends and then heated through T degrees. The two-part approach says: first 'cut' the bar and allow free thermal expansion, without stress, through a distance αTl. Now restore compatibility by applying a compressive stress σ which, if the material modulus is E, must be given by

$$\sigma = -E\frac{\alpha Tl}{l} = -\mathrm{E}\alpha T$$

We prefer to follow our standard scheme (Fig. 1.9) for redundant structures.

Step 1 Compatibility:

No displacements occur at all, therefore by (1.66)

the *total* strain $\varepsilon = \dfrac{\partial u}{\partial x} = 0$

Step 2 Stress–strain law:

$$\text{total strain } \varepsilon = \frac{\sigma}{E} + \alpha T = 0$$

$$\therefore \quad \sigma = -E\alpha T$$

Summarizing:

$$\text{Equilibrium} \quad \mathbf{D}\boldsymbol{\sigma} + \mathbf{p}_v = \mathbf{0} \quad \text{in } V \qquad (1.52)$$

$$(\mathbf{Dn})\boldsymbol{\sigma} = \mathbf{p}_s \quad \text{on } S \qquad (1.57)$$

$$\text{Compatibility } \boldsymbol{\varepsilon} = \mathbf{D}^t\mathbf{u} \quad \text{in } V \qquad (1.66)$$

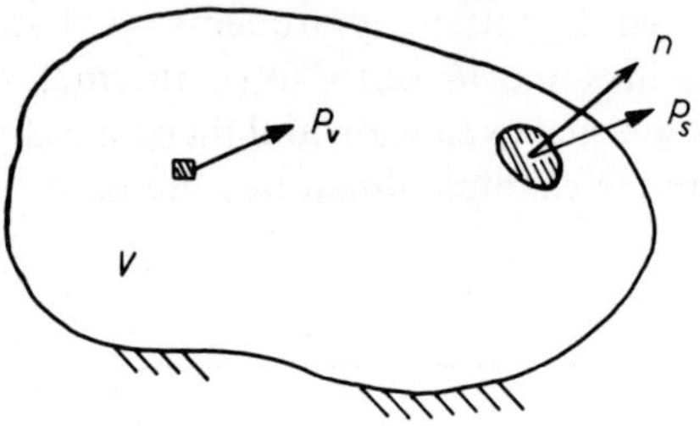

1.8 Virtual work for a continuum

In demonstrating the work principles for a completely arbitrary sort of structure we will be using:

(a) virtual displacements which must satisfy compatibility ($\bar{\boldsymbol{\varepsilon}} = \mathbf{D}^t\bar{\mathbf{u}}$) but would lead to stresses which violate equilibrium (there is only *one* system of stresses and strains which satisfies both equilibrium and compatibility, and that is the true system);
(b) virtual forces which must satisfy equilibrium ($\mathbf{D}\bar{\boldsymbol{\sigma}} + \bar{\mathbf{p}}_v = \mathbf{0}$) but would lead to strains which violate compatibility.

It is easy enough to state these requirements mathematically, but a little odd to visualize them. It may help to illustrate virtual systems for frameworks again. For example, we later will find it convenient to allow a *single bar* in a framework to distort like in Fig. 1.25. Clearly the system in Fig. 1.25(b) is unlikely to satisfy equilibrium requirements. Similarly we could – and will – destroy compatibility in a redundant structure, and yet still be able to keep a virtual stress system in equilibrium with virtual forces – like Fig. 1.26.

The virtual work of the applied forces is again the product of force and displacement, that is,

$$\int_V (p_{vx}u + p_{vy}v + p_{vz}w)\mathrm{d}v + \int_S (p_{sx}u + p_{sy}v + p_{sz}w)\,\mathrm{d}s$$

$$= \int_V \mathbf{p}_v^t\,\mathbf{u}\,\mathrm{d}v + \int_S \mathbf{p}_s^t\,\mathbf{u}\,\mathrm{d}s \qquad (1.67)$$

$$= \int_V \mathbf{u}^t\mathbf{p}_v\,\mathrm{d}v + \int_S \mathbf{u}^t\,\mathbf{p}_s\,\mathrm{d}s \qquad \text{also}$$

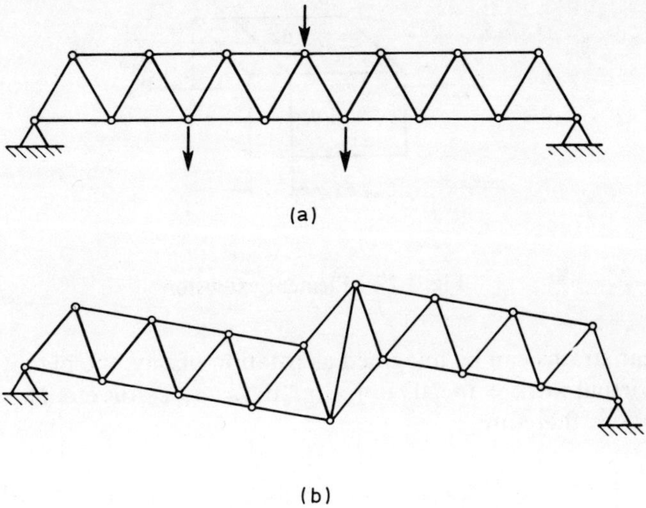

Fig. 1.25 (a) Actual loads and stresses $(\mathbf{D}n)\ \boldsymbol{\sigma} = \mathbf{p}_s$; $\mathbf{D}\boldsymbol{\sigma} + \mathbf{p}_v = \mathbf{0}$. (b) Virtual displacement $\bar{\boldsymbol{\epsilon}} = \mathbf{D}^t\bar{\mathbf{u}}$; $(\mathbf{D}n)\ \bar{\boldsymbol{\sigma}} \neq \mathbf{p}_s$

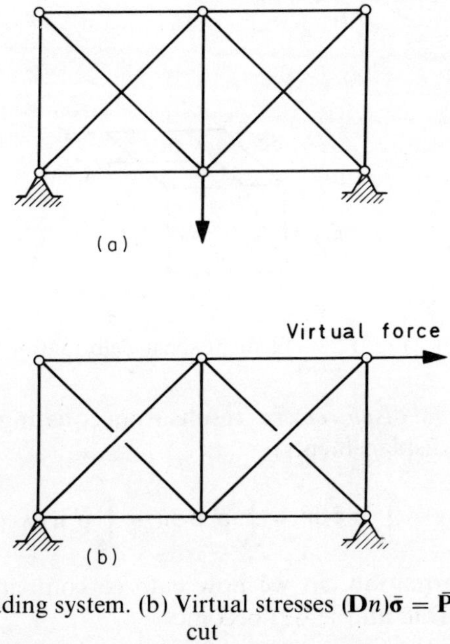

Fig. 1.26 (a) Actual loading system. (b) Virtual stresses $(\mathbf{D}n)\bar{\boldsymbol{\sigma}} = \bar{\mathbf{P}}_s$; $\bar{\boldsymbol{\epsilon}} \neq \mathbf{D}^t\bar{\mathbf{u}}$ since bars cut

The internal virtual work is a similar product obtained by examining the stresses (Fig. 1.27). Thus, for the direct stress σ_{xx} we have:

$$\text{virtual work} = (\sigma_{xx}\,\mathrm{d}y\,\mathrm{d}z) \times (\varepsilon_{xx}\,\mathrm{d}x) = \sigma_{xx}\,\varepsilon_{xx}\,\mathrm{d}v$$

and similarly $\sigma_{yy}\varepsilon_{yy}\,\mathrm{d}v$ and $\sigma_{zz}\,\varepsilon_{zz}\,\mathrm{d}v$.

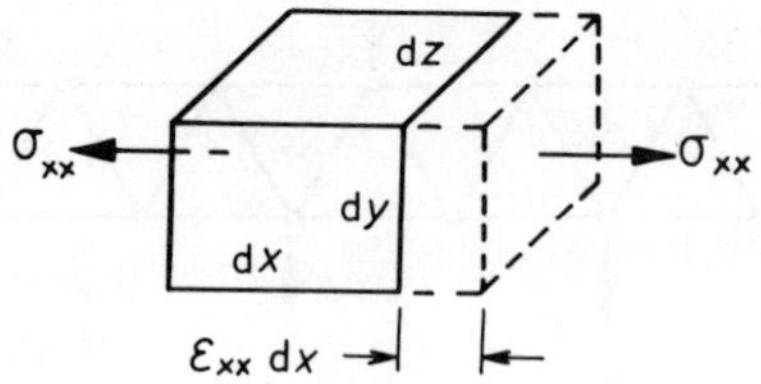

Fig. 1.27 Element extension

The shear strains can be imagined as rotation of, say, the planes $\mathrm{d}y\,\mathrm{d}z$ in Fig. 1.28. The virtual work $= (\sigma_{xy}\,\mathrm{d}x\,\mathrm{d}z) \times (\varepsilon_{xy}\,\mathrm{d}y) = \sigma_{xy}\varepsilon_{xy}\,\mathrm{d}v$, etc. The total internal virtual work is therefore

$$\int_V (\sigma_{xx}\varepsilon_{xx} + \sigma_{yy}\varepsilon_{yy} + \sigma_{zz}\varepsilon_{zz} + \sigma_{xy}\varepsilon_{xy} + \sigma_{yz}\varepsilon_{yz} + \sigma_{zx}\varepsilon_{zx})\,\mathrm{d}v$$

$$= \int_V \boldsymbol{\sigma}^{\mathrm{t}}\boldsymbol{\varepsilon}\,\mathrm{d}v \qquad (1.68)$$

$$= \int_V \boldsymbol{\varepsilon}^{\mathrm{t}}\boldsymbol{\sigma}\,\mathrm{d}v$$

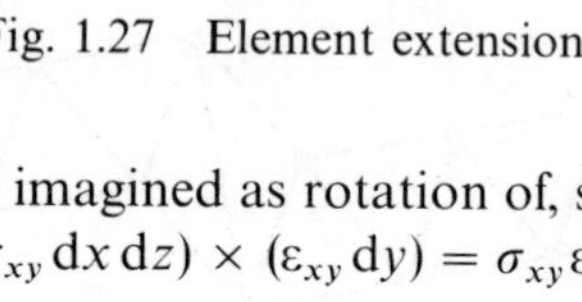

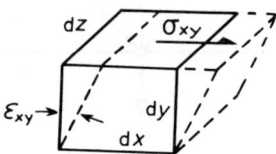

Fig. 1.28 Element shear deformation

The principle of virtual displacements results from equating (1.67) and (1.68) with virtual strains and displacements:

$$\int_V \boldsymbol{\sigma}^{\mathrm{t}}\bar{\boldsymbol{\varepsilon}}\,\mathrm{d}v = \int_V \mathbf{p}_v^{\mathrm{t}}\,\bar{\mathbf{u}}\,\mathrm{d}v + \int_S \mathbf{p}_s^{\mathrm{t}}\,\bar{\mathbf{u}}\,\mathrm{d}s \qquad (1.69)$$

This yields no information, so we now enforce compatibility on the virtual displacements, $\bar{\boldsymbol{\varepsilon}} = \mathbf{D}^{\mathrm{t}}\bar{\mathbf{u}}$, and (1.69) becomes

$$\int_V (\boldsymbol{\sigma}^{\mathrm{t}}\mathbf{D}^{\mathrm{t}}\bar{\mathbf{u}} - \mathbf{p}_v^{\mathrm{t}}\,\bar{\mathbf{u}})\,\mathrm{d}v - \int_S \mathbf{p}_s^{\mathrm{t}}\,\bar{\mathbf{u}}\,\mathrm{d}s = 0 \qquad (1.70)$$

But (1.70) does not yield equilibrium as we have been led to expect. Our optimism is based on the over-simple frameworks where it was possible to express both internal and external work in terms of joint displacements. To be able to use

(1.70) we must convert the term $\mathbf{D}^t\bar{\mathbf{u}}$ to displacements $\bar{\mathbf{u}}$, and then we can examine the whole expression as a coefficient of $\bar{\mathbf{u}}$, and deduce as before that this coefficient must be zero. To convert $\mathbf{D}^t\bar{\mathbf{u}}$ to $\bar{\mathbf{u}}$ we need to use a mathematical identity which is a novel form of Gauss's theorem, or the Gauss–Ostrogradski identity.[17] This theorem is a valuable and quite general identity used extensively in practically all forms of continuum physics where a 'flux' is created inside a volume V and we wish to preserve continuity by matching this creation to the 'emergence' through the surface S. In our case we can shorten the proof by considering specifically 'work' flux, say initially in one stress component only, σ_{xx}. The creation of work flux inside V is

$$\int_V \frac{\partial}{\partial x}[(\sigma_{xx}\,\mathrm{d}y\,\mathrm{d}z)\,u]\,\mathrm{d}x$$

and the emergence through S

$$\int_S \sigma_{xx}\left(\mathrm{d}s\frac{\partial n}{\partial x}\right)u$$

where n is the outward normal again.

Thus

$$\int_V \frac{\partial}{\partial x}(\sigma_{xx}u)\,\mathrm{d}v = \int_S \sigma_{xx}\frac{\partial n}{\partial x}u\,\mathrm{d}s$$

If we repeat this statement for all stress components, and then sum, the resulting identity can be summarized as

$$\int_V (\mathbf{u}^t\mathbf{D}\boldsymbol{\sigma} + \boldsymbol{\sigma}^t\mathbf{D}^t\mathbf{u})\,\mathrm{d}v = \int_S \mathbf{u}^t(\mathbf{D}n)\boldsymbol{\sigma}\,\mathrm{d}s \tag{1.71}$$

In fact, (1.71) is nothing more than a three-dimensional vectorial version of 'integration by parts'; so if we take a one-dimensional rod of length $x = 0$ to l, we obtain

$$\int_0^l \left(u\frac{\partial\sigma}{\partial x} + \sigma\frac{\partial u}{\partial x}\right)\mathrm{d}x = [u\sigma]_{x=l} - [u\sigma]_{x=0}$$

(noting $\partial n/\partial x = +1$ at $x = l$, and -1 at $x = 0$).

Thus

$$\int_0^l u\frac{\partial\sigma}{\partial x}\mathrm{d}x = [u\sigma]_0^l - \int_0^l \sigma\frac{\partial u}{\partial x}\mathrm{d}x$$

Now substitute for $\int_V \boldsymbol{\sigma}^t\mathbf{D}^t\bar{\mathbf{u}}\,\mathrm{d}v$ from (1.71) into (1.70) and the PVD becomes

$$\int_V \bar{\mathbf{u}}^t(\mathbf{D}\boldsymbol{\sigma} + \mathbf{p}_v)\,\mathrm{d}v - \int_S \bar{\mathbf{u}}^t\,[(\mathbf{D}n)\boldsymbol{\sigma} - \mathbf{p}_s]\,\mathrm{d}s = 0 \tag{1.72}$$

Hence, since $\bar{\mathbf{u}}$ is arbitrary everywhere

$$\mathbf{D}\boldsymbol{\sigma} + \mathbf{p}_v = \mathbf{0} \text{ inside } V, \qquad \text{and} \qquad (\mathbf{D}n)\boldsymbol{\sigma} = \mathbf{p}_s \text{ on } S.$$

These are the expected equilibrium equations (1.52) and (1.57) both inside and on the surface of the body. It is worth while noting the manner in which these have been extracted:

(a) apply the PVD (1.69);
(b) use compatibility for the virtual displacements (1.66);
(c) integrate by parts and isolate coefficients of $\bar{\mathbf{u}}$.

This is the procedure we will follow in solving any problem – even beams – where the variables are continuous functions. The process will be shown to be virtually automatic, and often very powerful since expressing work for a particular structure is usually much easier than deriving differential equations directly. As a bonus we do not have to worry unduly about signs in the product $\boldsymbol{\sigma}^t\boldsymbol{\varepsilon}$ (which must always be positive) in contrast to equations of equilibrium. Further, it is frequently necessary to simplify the behaviour of a structure and neglect certain terms which are thought to be unimportant. The advantage of a work formulation lies in the simple scalar $\int_V \boldsymbol{\sigma}^t\boldsymbol{\varepsilon}\, dv$, which contains products of all stress and strain components. If we doubt the significance of any term then we just include it and examine the difference.

If we use the PVD for a *statically determinate* structure the equilibrium equations are sufficient to solve for the stresses. Mostly we are not able to do this, and it is then usual to obtain the *displacement solution* in which not only do we enforce $\bar{\boldsymbol{\varepsilon}} = \mathbf{D}^t\bar{\mathbf{u}}$, but we also convert the real stress $\boldsymbol{\sigma}$ to $\boldsymbol{\varepsilon}$ and then use $\boldsymbol{\varepsilon} = \mathbf{D}^t\mathbf{u}$ for the real strains as well. The PVD will then deliver, giving the displacements in terms of the applied loads.

In most problems the structure is likely to be held in some fashion, that is some portion of the surface S has *prescribed displacements* $\tilde{\mathbf{u}}$ – denote this portion as S_u – while the remainder, S_p, is where the tractions $\mathbf{p}_s$ are prescribed (Fig. 1.29). In most cases $\tilde{\mathbf{u}}$ is zero. Now the surface integral in (1.69) is unrestricted, and we may include both S_u and S_p. If we do then the PVD will deliver surface equations of equilibrium in both the known forces $\mathbf{p}_s$ and the *unknown* support reactions on S_u. In seeking a displacement solution it is *convenient* to avoid solving these unknown reactions at the same time, and we may do this simply by stipulating that $\bar{\mathbf{u}}$ *are taken to be zero on* S_u. (In Chapter 4 on approximate methods we show how $\bar{\mathbf{u}}$ on S_u may be allowed with advantage.)

The principle of virtual forces is the dual of the PVD and is simply (1.69) with the 'bars' switched.

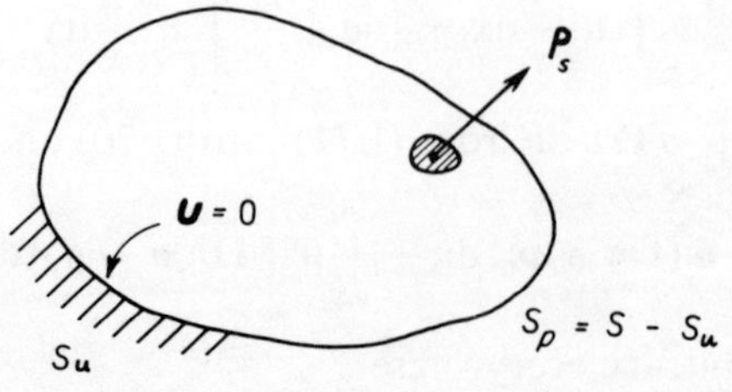

Fig. 1.29 Prescribed forces or prescribed displacements

Thus

$$\int_V \bar{\boldsymbol{\sigma}}^t \boldsymbol{\varepsilon}\,dv = \int_V \bar{\mathbf{p}}_v^t \mathbf{u}\,dv + \int_S \bar{\mathbf{p}}_s^t \mathbf{u}\,ds \tag{1.73}$$

In an analogous fashion it is here permissible to allow virtual forces $\bar{\mathbf{p}}_s$ on both S_p and S_u; indeed we will later deliberately use virtual surface forces on S_p to obtain the displacements there. However, if we intend to use the PVF as a way of obtaining a *force solution*, it is convenient to eliminate unknown surface displacements on S_p by stipulating that $\bar{\mathbf{p}}_s$ vanish there. For the same reason, to exclude real displacements in the interior V, we will stipulate that $\bar{\mathbf{p}}_v = \mathbf{0}$ also; consequently there are no virtual applied forces at all except on the surface S_u where the displacements are prescribed. The PVF (1.73) then takes the special form

$$\int_V \bar{\boldsymbol{\sigma}}_i^t \boldsymbol{\varepsilon}\,dv = \int_{S_u} \bar{\mathbf{p}}_s^t \tilde{\mathbf{u}}\,ds \tag{1.74}$$

The subscript i is usually used to denote that $\bar{\boldsymbol{\sigma}}_i$ is an 'internal' self-equilibrating stress field satisfying

$$\mathbf{D}\bar{\boldsymbol{\sigma}}_i = \mathbf{0} \text{ in } V \qquad \text{and} \qquad (\mathbf{D}n)\bar{\boldsymbol{\sigma}}_i = \mathbf{0} \text{ on } S_p$$

The choice of $\bar{\boldsymbol{\sigma}}_i$ is absolutely straightforward as can be seen in Chapter 3, devoted exclusively to the applications of the PVF.

To demonstrate that (1.74) will deliver compatibility, we must now enforce equilibrium on $\bar{\boldsymbol{\sigma}}$. This is not immediately possible because the equations of equilibrium (1.52) contain $\mathbf{D}\bar{\boldsymbol{\sigma}}$ which has to be integrated, and we do this again by using the Gauss equation (1.71) to convert (1.74) to

$$\int_V \bar{\boldsymbol{\sigma}}^t (\boldsymbol{\varepsilon} - \mathbf{D}^t\mathbf{u})\,dv - \int_V \mathbf{u}^t(\mathbf{D}\bar{\boldsymbol{\sigma}})\,dv + \int_S \mathbf{u}^t (\mathbf{D}n)\bar{\boldsymbol{\sigma}}\,ds - \int_{S_u} \bar{\mathbf{p}}_s^t \tilde{\mathbf{u}}\,ds = 0$$

Now enforce $\mathbf{D}\bar{\boldsymbol{\sigma}} = \mathbf{0}$ in V and $(\mathbf{D}n)\bar{\boldsymbol{\sigma}} = \bar{\mathbf{p}}_s$ on S_u but zero on S_p. The above equation then becomes

$$\int_V \bar{\boldsymbol{\sigma}}^t (\boldsymbol{\varepsilon} - \mathbf{D}^t\mathbf{u})\,dv + \int_{S_u} [(\mathbf{D}n)\bar{\boldsymbol{\sigma}}]^t (\mathbf{u} - \tilde{\mathbf{u}})\,ds = 0 \tag{1.75}$$

We see (for the last time) that because $\bar{\boldsymbol{\sigma}}$ is virtual everywhere in V and on S_u, its coefficients must vanish to ensure equation (1.75) holds.

Thus

$$\boldsymbol{\varepsilon} = \mathbf{D}^t\mathbf{u} \text{ in } V$$

and

$$\mathbf{u} = \tilde{\mathbf{u}} \text{ on } S_u, \quad \text{as expected.}$$

Equation (1.75) is clearly the virtual force dual to the virtual displacement equations (1.72). In later chapters we may use (1.72) and (1.75) in preference to the simpler PVD and PVF, but not yet.

1.9 Summary

The principle of virtual work

PVD

$$\int_V \boldsymbol{\sigma}^t \bar{\boldsymbol{\varepsilon}} \, dv = \int_V \mathbf{p}_v^t \bar{\mathbf{u}} \, dv + \int_{S_p} \mathbf{p}_s^t \bar{\mathbf{u}} \, ds$$

Satisfy (kinematic) compatibility directly,

$$\bar{\boldsymbol{\varepsilon}} = \mathbf{D}^t \bar{\mathbf{u}}$$

The PVD will provide indirectly the equations of equilibrium

$$\mathbf{D}\boldsymbol{\sigma} + \mathbf{p}_v = \mathbf{0} \text{ in } V$$
$$(\mathbf{D}_n)\boldsymbol{\sigma} = \mathbf{p}_s \text{ on } S_p$$

PVF

$$\int_V \bar{\boldsymbol{\sigma}}_i^t \boldsymbol{\varepsilon} \, dv = \int_{S_u} \bar{\mathbf{p}}_s^t \tilde{\mathbf{u}} \, ds$$

Satisfy equilibrium (static compatibility) directly,

$$\mathbf{D}\bar{\boldsymbol{\sigma}}_i = \mathbf{0} \text{ in } V, \quad (\mathbf{D}n)\bar{\boldsymbol{\sigma}}_i = \mathbf{0} \text{ on } S_p$$

The PVF will then provide indirectly the equations of compatibility

$$\boldsymbol{\varepsilon} = \mathbf{D}^t \mathbf{u} \text{ in } V$$
$$\mathbf{u} = \tilde{\mathbf{u}} \text{ on } S_u$$

Problems

(Asterisks indicate problems with worked solutions in the section before the Appendixes.)

P 1.1

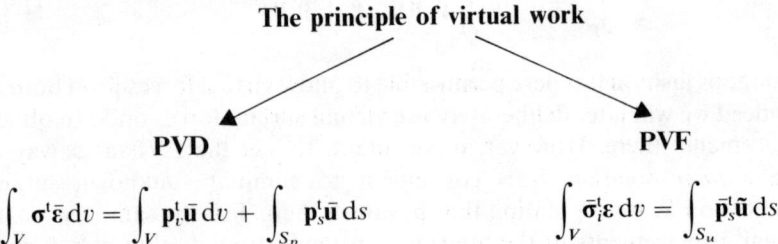

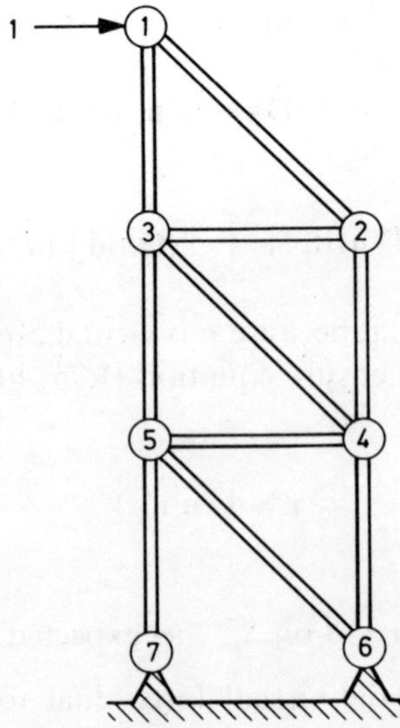

Fig. P. 1.1

All bars are vertical, horizontal or inclined at 45°. Check that the structure is statically determinate and the bar forces are:

$$N_{12} = -\sqrt{2};\quad N_{13} = 1;\quad N_{23} = 1;\quad N_{24} = -1;\quad N_{34} = -\sqrt{2};$$
$$N_{35} = 2;\quad N_{45} = 1;\quad N_{46} = -2;\quad N_{57} = 3;\quad N_{56} = -\sqrt{2}.$$

P 1.2

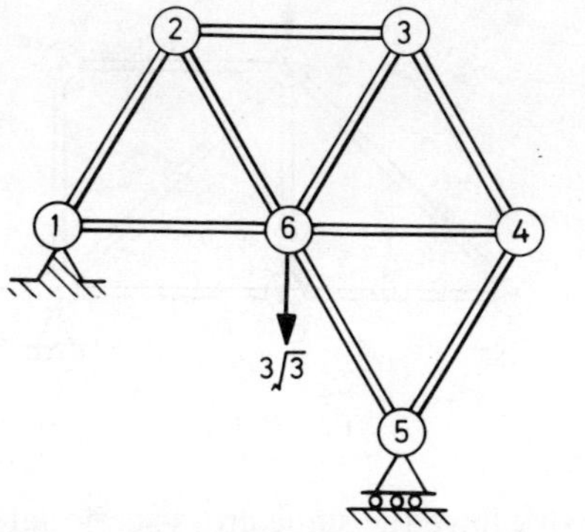

Fig. P. 1.2

All bars are the same length. Find the bar forces induced by the single vertical load $3\sqrt{3}$.

Ans: $N_{26} = N_{36} = N_{64} = -N_{12} = -N_{23} = -N_{34} = -N_{45} = -N_{56} = 2;$
$N_{16} = 1.$

P 1.3

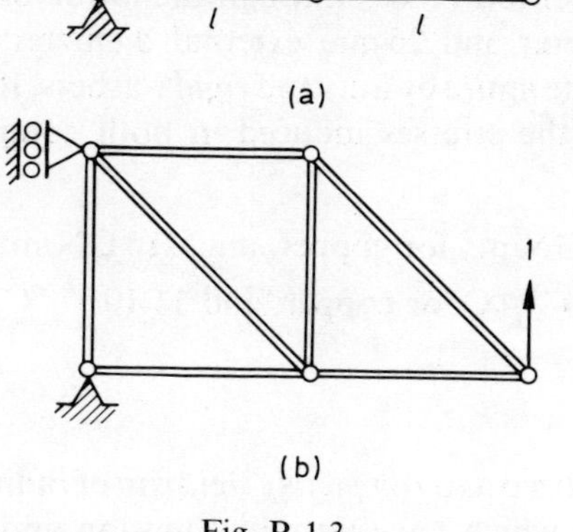

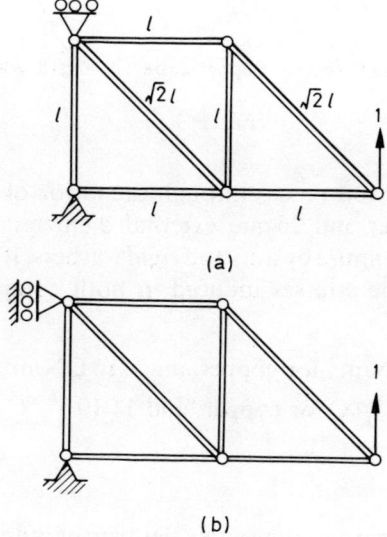

Fig. P. 1.3

Without finding the support reactions first (which gives the game away) show that finite bar forces cannot be determined in the first configuration, but can in the second.

*P 1.4**

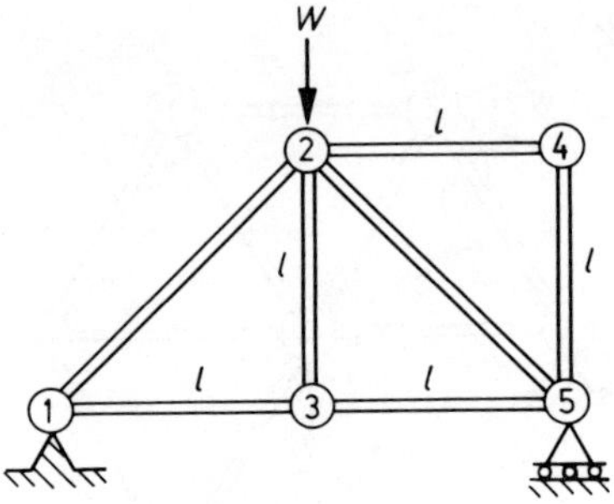

Fig. P. 1.4

All bars in this framework have the same cross-sectional area A and modulus E. Find the statically determinate bar forces directly, and then solve for the joint displacements. Sketch the deflected shape.

*P 1.5**

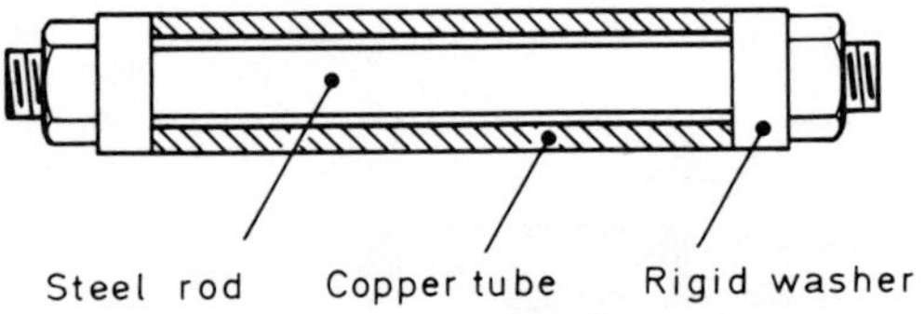

Fig. P. 1.5

A 12 mm diameter steel rod passes through the inside of a copper tube having 15 mm internal diameter and 25 mm external diameter. The two are secured together as shown in the figure by nuts and *rigid* washers. If the assembly is heated through 100 °C, find the stresses induced in both components. The material properties are:

$$E = 105\,\mathrm{GN/m^2} \text{ for copper, and } 210\,\mathrm{GN/m^2} \text{ for steel.}$$

$$\alpha = 18.10^{-6}/°\mathrm{C} \text{ for copper, and } 11.10^{-6}/°\mathrm{C} \text{ for steel.}$$

P 1.6

A bicycle wheel is idealized as a *completely rigid* rim of radius r connected to a hub by thin radial spokes which have no initial tension and are also incapable of

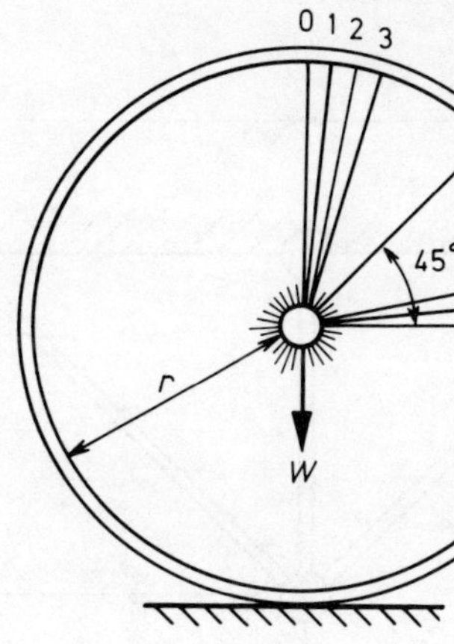

Fig. P. 1.6

resisting compression. The equally spaced spokes all have the same area A and modulus E, and there are $8n$ of them as shown in Fig. P 1.6. Show that a weight W at the hub will deflect a distance $Wr/2AEn$.

P1.7

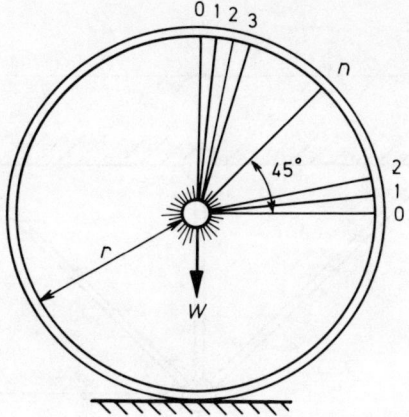

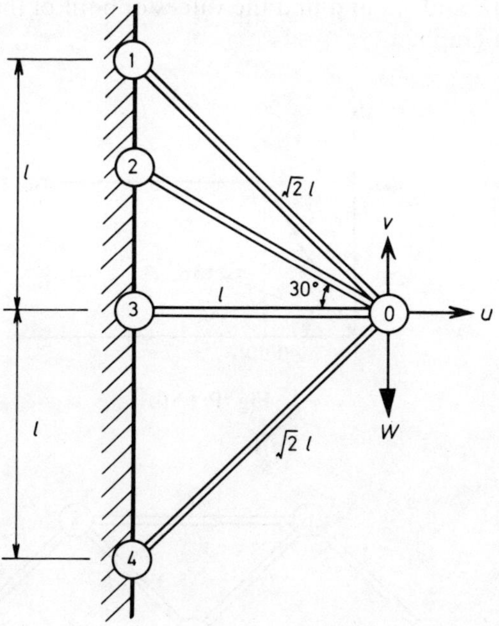

Fig. P. 1.7

All bars again have the same AE. Show that the joint displacements are $u = -0.184\,Wl/AE$ and $v = -1.158\,Wl/AE$; and the bar forces are $N_{10} = 0.487\,W$, $N_{20} = 0.363\,W$, $N_{30} = -0.184\,W$, $N_{40} = -0.670\,W$.

*P 1.8**

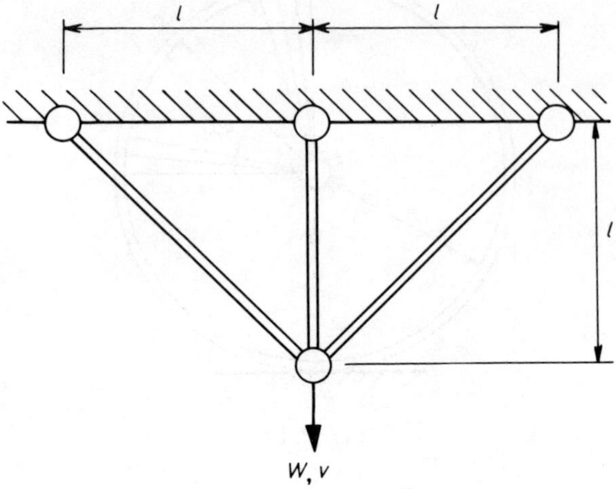

Fig. P. 1.8(a)

The cross-sectional area of each bar is the same, and the material has an elastic-plastic stress-strain law of the form shown in Fig. 1.8(b). If the deflection of the joint, loaded by the weight *W*, is *v*, sketch the graph showing the relationship between W/AE and v/l, and find the values of both of these variables at salient points on the graph.

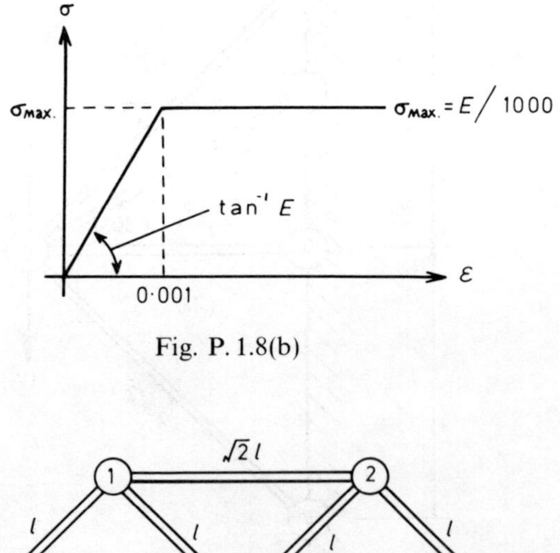

Fig. P. 1.8(b)

P 1.9

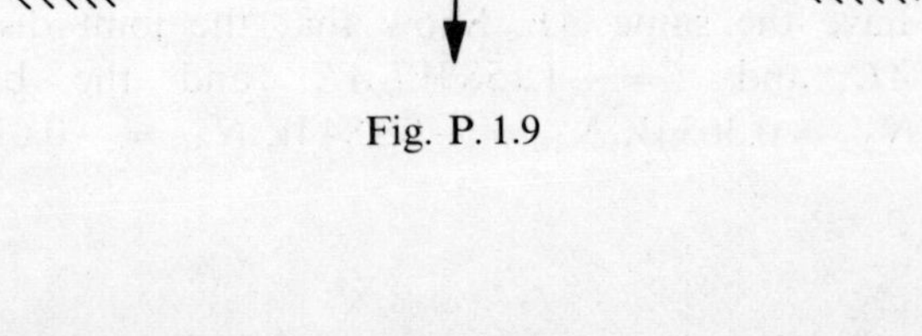

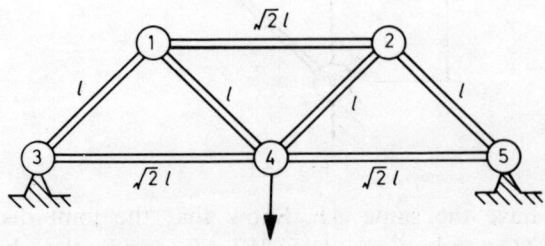

Fig. P. 1.9

Taking advantage of symmetry to reduce the algebra ($u_2 = -u_1$, $v_2 = v_1$, $u_4 = 0$), use the PVD to generate the equations of equilibrium and show that they can be solved. Did you think that the framework was statically indeterminate?

$$Ans: N_{12} = -W; \quad N_{24} = N_{14} = -N_{25} = -N_{13} = W/\sqrt{2};$$
$$N_{45} = N_{43} = 0.$$

*P1.10**

The framework of Problem 1.8 has bars of cross-sectional area 100 mm^2 and the length $l = 300$ mm. The material is a light aluminium alloy whose stress–strain relationship can be approximated as

$$\sigma = 72\varepsilon[1 - (100\varepsilon)^2]\,\text{kN/mm}^2, \quad \text{up to } (100\varepsilon) = \pm 1/\sqrt{3}$$

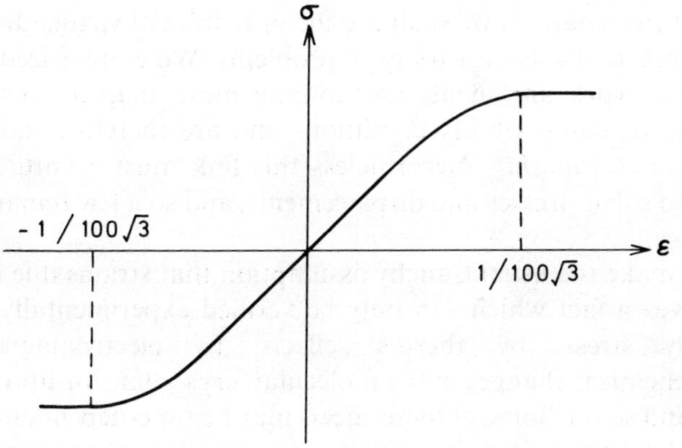

Fig. P. 1.10

and beyond this strain the stress is constant. Use the PVD to show that the load–deflection relation is:

$$v[1 - 0.0766\,v^2] = 0.0244\,W$$

Show that at $W = 30$ kN, the central bar carries 57 per cent of the load while at $W = 50$ kN this proportion falls to 53 per cent.

2

THE PRINCIPLE OF VIRTUAL DISPLACEMENTS

2.1 Stress–strain relationships

In the next two chapters we shall use the principles of virtual displacements and virtual forces, and solve a variety of problems. We emphasized in the previous chapter that work arguments are nothing more than a substitute for direct equilibrium or compatibility conditions and are therefore independent of the *stress–strain relationship.* Nevertheless this link must eventually be forged to enable us to solve stresses and displacements, and so a few remarks are pertinent at this stage.

We will make the usual Cauchy assumption that strains due to several effects are additive – a fact which can only be verified experimentally. Strain may be caused by stress, by thermal effects, by electromagnetic fields, by physical–chemical changes in the molecular, crystalline, or fibrous nature of the material, and so on. Some of these effects may be time-dependent, but we will not consider such 'strain-rate' phenomena as 'creep'. It is convenient to separate the strains due to stress and the strains due to all other effects ($\boldsymbol{\eta}$) which are present when the stress is zero. That is, the total strain is equal to the 'elastic strain' plus the 'initial strain'. The strains $\boldsymbol{\eta}$ may literally be an initial strain in the case of redundant structures whose components in the unstressed state do not fit together. In Fig. 2.1 for example, there is an initial strain $\eta = -\Delta/l$ in the short member prior to attaching it to the pinjoint. (Alternatively we could consider the short member to be unstrained while all the other members meeting at the joint

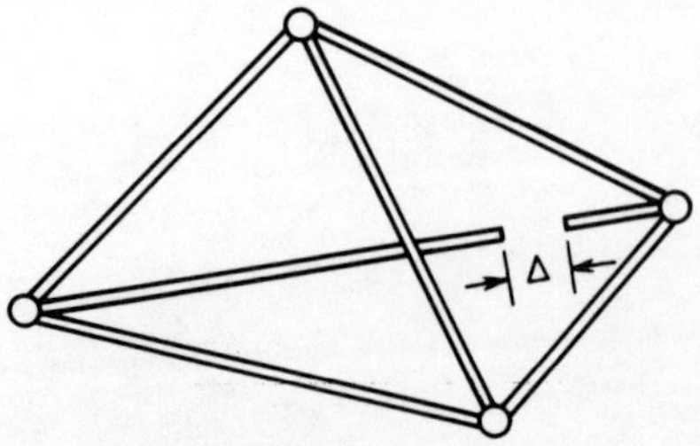

Fig. 2.1 A lack of fit

had appropriate positive initial strains: for $\Delta \ll l$ there is no difference in either definition.)

This particular form of initial strain is best described simply in terms of the 'lack of fit' Δ, as we shall see when using the PVF to solve pinned and stiff-jointed frameworks. However, in the case of thermal strains the values of $\boldsymbol{\eta}$ are given directly. For example, if the material is *isotropic* (having identical properties in all directions), and a moderate temperature increase T occurs, it is found for most materials that $\eta_{xx} = \eta_{yy} = \eta_{zz} = \alpha T$, where α is the linear coefficient of expansion. $\eta_{xy} = \eta_{yz} = \eta_{zx} = 0$, since no angular distortions are induced in an isotropic material. For large variations of temperature, α may vary so we have to evaluate

$$\eta = \int \alpha(T)\, \mathrm{d}T$$

The relationship between $\boldsymbol{\varepsilon}$ and the stress $\boldsymbol{\sigma}$ may be complex, especially for ductile materials, and for certain materials it may not even be describable in functional form. Fortunately, for most structural materials it is found that working strains are linearly proportionate to stress, that is,

$$\boldsymbol{\varepsilon} = \boldsymbol{\phi\sigma}$$

where $\boldsymbol{\phi}$ is a measure of the *material flexibility* and is a symmetric matrix with 21 constants. In the common isotropic case, for instance,

$$\begin{bmatrix} \varepsilon_{xx} \\ \varepsilon_{yy} \\ \varepsilon_{zz} \\ \varepsilon_{xy} \\ \varepsilon_{yz} \\ \varepsilon_{zx} \end{bmatrix} = \frac{1}{E} \begin{bmatrix} 1 & -\nu & -\nu & 0 & 0 & 0 \\ & 1 & -\nu & 0 & 0 & 0 \\ & & 1 & 0 & 0 & 0 \\ & \text{Symmetric} & & 2(1+\nu) & 0 & 0 \\ & & & & 2(1+\nu) & 0 \\ & & & & & 2(1+\nu) \end{bmatrix} \begin{bmatrix} \sigma_{xx} \\ \sigma_{yy} \\ \sigma_{zz} \\ \sigma_{xy} \\ \sigma_{yz} \\ \sigma_{zx} \end{bmatrix} \tag{2.1}$$

E is Young's modulus, and ν Poisson's ratio. The *total strain* is therefore

$$\boldsymbol{\varepsilon} = \boldsymbol{\phi\sigma} + \boldsymbol{\eta} \tag{2.2}$$

The inverse of the flexibility $\boldsymbol{\phi}$ is the *material stiffness* $\boldsymbol{\kappa}$, so in the above isotropic case:

$$\boldsymbol{\kappa} = \frac{E}{(1-2\nu)(1+\nu)} \begin{bmatrix} 1-\nu & \nu & \nu & 0 & 0 & 0 \\ & 1-\nu & \nu & 0 & 0 & 0 \\ & & 1-\nu & 0 & 0 & 0 \\ & & & \tfrac{1}{2}-\nu & 0 & 0 \\ & & & & \tfrac{1}{2}-\nu & 0 \\ & & & & & \tfrac{1}{2}-\nu \end{bmatrix} \tag{2.3}$$

2.2 The unit displacement method

There are two 'unit' methods or theorems involving either *unit displacement* or *unit load*, and both are simply devices for isolating a solitary equation of equilibrium or a solitary displacement. Now invariably we have to use *all* the available equations of equilibrium, so isolating a solitary one is not terribly useful, except as a book-keeping aid. Isolating a solitary displacement in statically determinate structures can be much more convenient as it may not be necessary to know all displacements in assessing whether the structure is excessively floppy. Thus the unit load method for finding a displacement will be used extensively in the next chapter on the PVF.

The analogous unit displacement method is not popularly used for another reason. It is easier to picture a structure loaded by a solitary force, but not so easy to imagine a solitary displacement with all other displacement freedoms zero. And yet this is a prejudiced view, for such a deformed state is a very convenient one in which to find $\bar{\boldsymbol{\varepsilon}} = \mathbf{D}^t\bar{\mathbf{u}}$ since the displacement pattern $\bar{\mathbf{u}}$ is confined to a fairly small region as we shall see.

We recall from the previous chapter that we used all virtual displacements simultaneously in setting up equations of equilibrium for frameworks. So if compatibility for all bars was $\boldsymbol{\Delta} = \mathbf{ar}$ (equation (1.27)), the PVD using *all* displacements was $\mathbf{N}^t\bar{\boldsymbol{\Delta}} = \mathbf{R}^t\bar{\mathbf{r}}$ (equation (1.36)) and hence

$$(\mathbf{N}^t\mathbf{a} - \mathbf{R}^t)\bar{\mathbf{r}} = 0$$

giving $\mathbf{a}^t\mathbf{N} = \mathbf{R}$ as equilibrium equations (1.38).

If we select a single virtual displacement, say $r_i = 1$, then equation (1.36) becomes simply

$$\mathbf{N}^t\bar{\boldsymbol{\Delta}} = R_i \tag{2.4}$$

which is a single equation of equilibrium relating an applied force R_i to the internal forces $\mathbf{N}$, and $\bar{\boldsymbol{\Delta}}$ are the virtual strains $\bar{\boldsymbol{\Delta}} = \mathbf{a}\bar{\mathbf{r}}$ due to

$$\bar{\mathbf{r}} = \{0 \quad 0 \quad 0 \quad 0 \quad 1 \quad 0 \quad 0 \quad 0 \quad \ldots\}$$

In general, for an arbitrary structure the framework form (2.4) may be more complicated: so if a virtual displacement $\bar{\mathbf{u}}$ has a value unity at load R_i, and $\bar{\boldsymbol{\varepsilon}}$ are the strains $\bar{\boldsymbol{\varepsilon}} = \mathbf{D}^t\bar{\mathbf{u}}$, compatible with $\bar{\mathbf{u}}$, then the PVD is simply:

$$\int_V \boldsymbol{\sigma}^t\bar{\boldsymbol{\varepsilon}}\,\mathrm{d}v = R_i \tag{2.5}$$

We have in fact already used the above forms in examples where only one degree of freedom was involved anyway, like the three-bar framework of Fig. 2.2. Suppose we look at this example again, but for interest (see also Problem 1.9) we take a nonlinear stress–strain law typical of light aluminium alloys or some high-tensile steels:

$$\sigma = E\varepsilon\left[1 - \frac{1}{3}\left(\frac{\varepsilon}{\varepsilon_p}\right)^2\right] \tag{2.6}$$

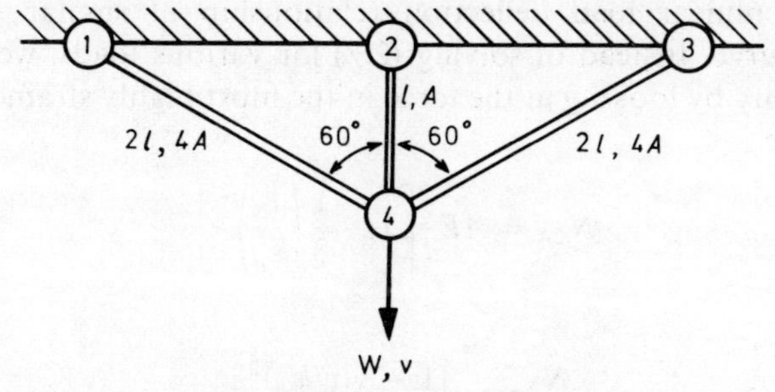

Fig. 2.2 Inelastic three-bar frame

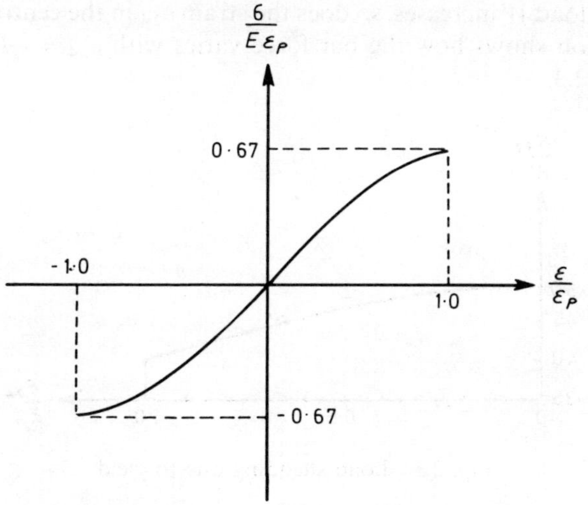

Curve 2.6 Stress–strain law of equation (2.6)

The *tangent modulus* is given by

$$E_T = \frac{\partial \sigma}{\partial \varepsilon} = E\left[1 - \left(\frac{\varepsilon}{\varepsilon_p}\right)^2\right]$$

so the material is linear ($E_T = E$) for small strains, but becomes fully plastic ($E_T = 0$) as ε approaches ε_p. The curve (2.6) is ignored beyond $\varepsilon = \varepsilon_p$.

The actual elongation of bar (2–4) is v, and of bars (1–4) and (3–4) is $v/2$, so using equation (2.4) and the stress given by (2.6) the unit-displacement version of the PVD becomes

$$2.4AE\frac{v}{4l}\left[1 - \frac{1}{3}\left(\frac{v}{4l\varepsilon_p}\right)^2\right]\left(\frac{1}{2}\right) + AE\frac{v}{l}\left[1 - \frac{1}{3}\left(\frac{v}{l\varepsilon_p}\right)^2\right](1) = W$$

Collecting terms,

$$\frac{v}{l}\left[2 - \frac{17}{48}\left(\frac{v}{l\varepsilon_p}\right)^2\right] = \frac{W}{AE} \qquad (2.7)$$

which is a nonlinear load–deflection relationship of similar shape to the stress–strain curve. Instead of solving (2.7) for various loads, we examine the effect of plasticity by looking at the force in the most highly strained central bar (2–4). We have

$$N_{24} = AE\frac{v}{l}\left[1 - \frac{1}{3}\left(\frac{v}{l\varepsilon_p}\right)^2\right]$$

so using (2.7),

$$\frac{N_{24}}{W} = \frac{[1 - \frac{1}{3}(v/l\varepsilon_p)^2]}{[2 - \frac{17}{48}(v/l\varepsilon_p)^2]}$$

As the applied load W increases, so does the strain ε_{24} in the central bar, and the above expression shows how the bar force varies with $\varepsilon_{24} = v/l$. The result is shown in Fig. 2.3.

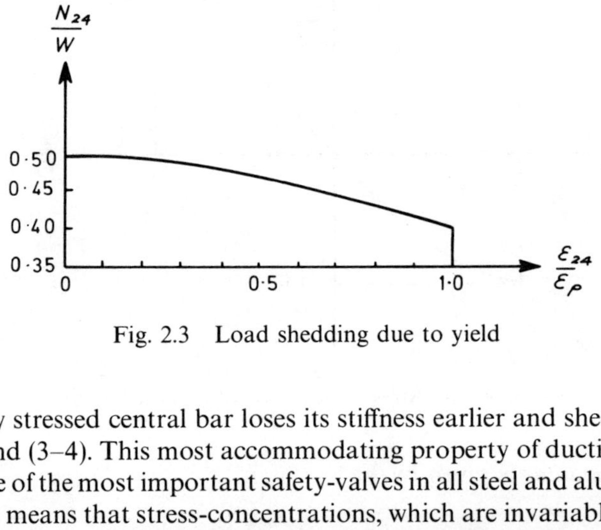

Fig. 2.3 Load shedding due to yield

The highly stressed central bar loses its stiffness earlier and sheds the load to bars (1–4) and (3–4). This most accommodating property of ductile materials is probably one of the most important safety-valves in all steel and aluminium alloy structures. It means that stress-concentrations, which are invariably present, but often difficult to estimate, do not necessarily cause failure if plasticity allows the material to redistribute the load. Stress concentrations may, of course, cause fatigue, but even in a cyclically loaded structure a good dose of plasticity due to a large load can produce beneficial stress-reversals which limit future stress amplitudes.

2.3 Elastic stiffness matrix of a large framework

We have already looked at a framework's stiffness matrix, and shown that if

$$\mathbf{Kr} = \mathbf{R} \tag{1.41}$$

then

$$\mathbf{K} = \mathbf{a}^t\mathbf{k}\mathbf{a} \tag{1.42}$$

But this way to assemble **K** is expensive, and so we will put some flesh on the bones of equation (1.42) and show both physically and mathematically that the assembly of the structure's stiffness matrix is both cheap and simple. Consider the relationship (1.41) in detail:

$$\begin{bmatrix} R_1 \\ R_2 \\ \vdots \\ R_i \\ \vdots \\ R_l \end{bmatrix} = \begin{bmatrix} k_{11} & \ldots & k_{1j} & \ldots & k_{1l} \\ k_{21} & & \vdots & & \vdots \\ \vdots & & \vdots & & \vdots \\ k_{i1} & \ldots & k_{ij} & \ldots & k_{il} \\ \vdots & & \vdots & & \vdots \\ k_{l1} & \ldots & k_{lj} & \ldots & k_{ll} \end{bmatrix} \begin{bmatrix} r_1 \\ \vdots \\ r_j \\ \vdots \\ r_i \\ \vdots \\ r_l \end{bmatrix} \tag{2.8}$$

We can interpret physically the nature of the elements k_{ij}; for suppose we impose a *single displacement* $r_j = 1$, keeping all the others zero, and examine the forces **R** needed to maintain this highly improbable deformed state. From (2.8) this special pattern must be

$$\mathbf{R} = \{R_1 \quad R_2 \quad \ldots \quad R_i \quad \ldots \quad R_l\} = \{k_{1j} \quad k_{2j} \quad \ldots \quad k_{ij} \quad \ldots \quad k_{lj}\}$$

In particular the ith force must be $R_i = k_{ij}$. We can find R_i, and all the others, by direct equilibrium arguments. From $r_j = 1$ we can immediately find the strains in all affected bars, thence the bar forces, and so using joint equilibrium equations the values of **R** are found (the reverse of the usual process, finding **N** in terms of **R**). We leave this technique as an exercise for the reader (see Problems 2.1 and 2.2) since it is a feasible method for only the simplest structures. However, without actually using the method to generate numbers, we can anticipate the form of the stiffness matrix by considering a typical joint displacement $r_j = 1$ in Figure 2.4,

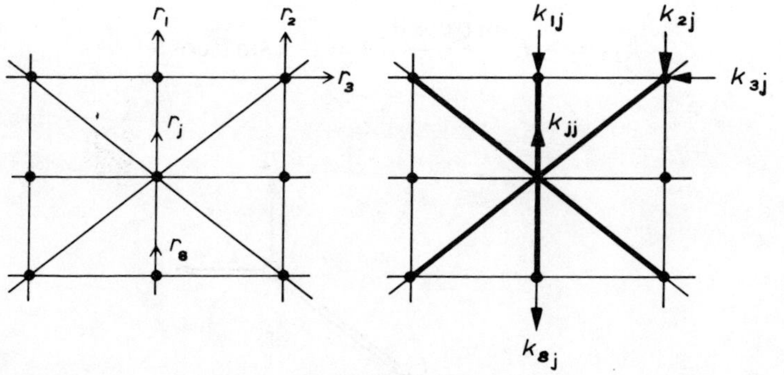

Fig. 2.4 Localized nature of k_{ij}

where the surrounding joints have displacement components $r_1, r_2, r_3, r_4, \ldots$. The displacement $r_j = 1$ induces known strains and bar forces in the members shown bold in Fig. 2.4. From these forces we can deduce the set of joint forces needed to equilibrate them and which we know to be $k_{jj}, k_{1j}, k_{2j}, k_{3j}, \ldots$. It is obvious that k_{jj} will be greater than all the other joint forces, and that for regions in the frame away from the shown sample there will be no forces at all. Thus the elements of a global stiffness matrix will be clustered around the leading diagonal (unless we use a curious numbering system for **r**). This property can be seen another way: suppose we isolate the solitary force $R_i = k_{ij}$ by using our unit displacement method. Thus we impose a unit displacement $r_i = 1$ and produce strains ε_i. If the strains (and stresses) due to $r_j = 1$ are $\sigma_j = E\varepsilon_j$, the unit displacement method (2.5) produces

$$k_{ij} = \int_V \sigma_j \varepsilon_i \, dv \tag{2.9}$$

$$= E \int_V \varepsilon_j \varepsilon_i \, dv = \int_V \varepsilon_j \sigma_i \, dv = k_{ji}$$

Most of the elements k_{ij} can now be seen to be zero since a solitary displacement $r_i = 1$ or $r_j = 1$ must strain only those bars having an end with r_i or r_j. Only bars strained in *both* systems contribute to (2.9) and hence the large number of zeros for all elements k_{ij} whenever r_i and r_j refer to the ends of two bars which do not have another joint in common. The solution of a large set of equations $\mathbf{Kr} = \mathbf{R}$ (1.41) is greatly facilitated if **K** possesses this diagonal banded nature; it can be both cheap and accurate to obtain solutions for perhaps thousands of degrees of freedom. The numerical aspects of solving such equations are discussed in Appendix A.

As (2.9) is an integral, we can sum the contributions to it, bar by bar (and computers are well equipped to perform repetitive summations). Consider the bar in Fig. 2.5. If we select, say, $r_2 = 1$ and $r_3 = 1$, we produce strains

$$\varepsilon_2 = -\sin\theta/l \qquad \text{and} \qquad \varepsilon_3 = \cos\theta/l$$

so the contribution to (2.9) for k_{23} is

$$k_{23} = -E\frac{\sin\theta}{l}\frac{\cos\theta}{l}lA = -k\sin\theta\cos\theta$$

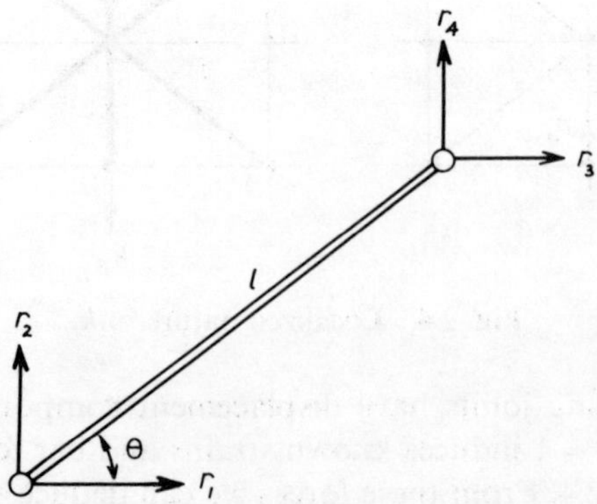

Fig. 2.5 Bar displacements again

where $k = AE/l$. Similarly, we get contributions to $k_{22} = k\sin^2\theta$, $k_{33} = k\cos^2\theta$, and so on. The assembly of a stiffness matrix is therefore seen to be a straightforward process involving only bar stiffnesses, k, and direction cosines, $\cos\theta$, $\sin\theta$. We therefore have to make this process systematic, having in mind that the technique should be extendable to more complicated structures. We must also bridge the gap between the summation concept and the apparently quite different expression for $\mathbf{K} = \mathbf{a}^t\mathbf{k}\mathbf{a}$ used before (equation 1.42).

It will be helpful, now and in future structures, to recognize that simple pin-ended bars are really a special and simple case. For instance, we are able to summarize their stiffness completely by the single expression $k = AE/l$. This stiffness is a property only of the bar and does not depend on where the bar is located or inclined in space. The term '*natural stiffness*' was first coined by Argyris[14] for this property, and we will denote this stiffness in future as k_N for simple bars or $\mathbf{k}_N$ for complex elements. We now examine the compatibility statement for a bar, $\boldsymbol{\Delta} = \mathbf{ar}$ (equation (1.27)) where $\mathbf{a}$ is a collection of direction cosines. In future we will label an individual bar's displacements as $\boldsymbol{\rho}$ in Fig. 2.5 instead of $\mathbf{r}$, and reserve $\mathbf{r}$ for the global list of all displacements. To identify a particular bar, say the gth bar, in a framework we simply write

$$\boldsymbol{\rho}_g = \mathbf{a}_g\mathbf{r} \tag{2.10}$$

where $\mathbf{a}_g$ is now a (Boolean) matrix of zeros mostly and four unit values to select the $\boldsymbol{\rho}_g$ from all the $\mathbf{r}$. It $(\mathbf{a}_g)$ is sometimes referred to as a connectivity matrix as it ensures that all bars meeting at a joint share the same joint displacement. The gth bar's elongation Δ_g is then selected from $\boldsymbol{\rho}_g$ by the second selection matrix $\mathbf{a}_N$ (cf. equation (1.26)):

$$\Delta_g = \mathbf{a}_N\boldsymbol{\rho}_g \tag{2.11}$$

where

$$\mathbf{a}_N = [-\cos\theta \quad -\sin\theta \quad \cos\theta \quad \sin\theta] \tag{2.12}$$

Now we enforce equilibrium for the single bar by using the PVD

$$N_g\bar{\Delta}_g = \mathbf{P}^t_g\bar{\boldsymbol{\rho}}_g$$

where $\mathbf{P}_g$ must be the set of joint forces corresponding to the displacements $\boldsymbol{\rho}_g$. Putting (2.11) into the above,

$$N_g\mathbf{a}_N\bar{\boldsymbol{\rho}}_g = \mathbf{P}^t_g\bar{\boldsymbol{\rho}}_g$$

$$\text{or} \quad \mathbf{P}_g = \mathbf{a}^t_N N_g \tag{2.13}$$

And using (2.12) for $\mathbf{a}_N$,

$$\mathbf{P}_g = \begin{bmatrix} P_1 \\ P_2 \\ P_3 \\ P_4 \end{bmatrix} = \begin{bmatrix} -\cos\theta \\ -\sin\theta \\ \cos\theta \\ \sin\theta \end{bmatrix} N_g$$

which is clearly a set of bar end forces in equilibrium with N (Figure 2.6). Writing $N = k_N\Delta$, equation (2.13) becomes

$$\mathbf{P}_g = \mathbf{a}_N^t k_N \mathbf{a}_N \boldsymbol{\rho}_g$$

$$\text{or} \quad \mathbf{P}_g = \mathbf{k}_g \boldsymbol{\rho}_g \tag{2.14}$$

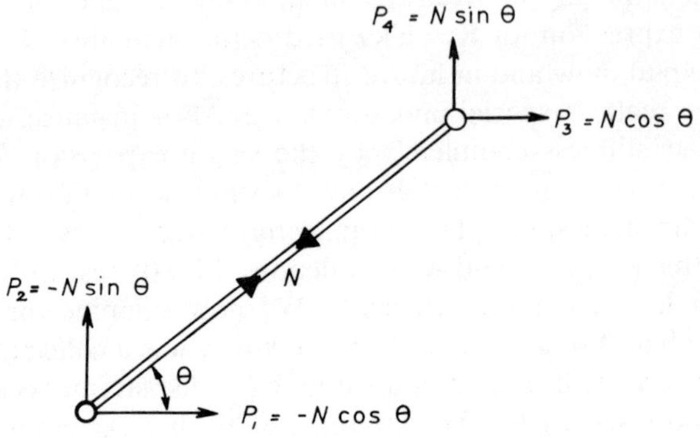

Fig. 2.6 Bar forces

The 4 × 4 matrix $\mathbf{k}_g$ is a new stiffness matrix relating the four forces $\mathbf{P}_g$ to the four displacements $\boldsymbol{\rho}_g$, and

$$\mathbf{k}_g = \mathbf{a}_N^t \mathbf{k}_N \mathbf{a}_N \tag{2.15}$$

On substituting for $\mathbf{a}_N$ from (2.12) and $k_N = AE/l$, (2.15) delivers

$$\mathbf{k}_g = \frac{AE}{l}\begin{bmatrix} \cos^2\theta & \cos\theta\sin\theta & -\cos^2\theta & -\cos\theta\sin\theta \\ & \sin^2\theta & -\cos\theta\sin\theta & -\sin^2\theta \\ & & \cos^2\theta & \cos\theta\sin\theta \\ & & & \sin^2\theta \end{bmatrix} \tag{2.16}$$

The terms in (2.16) can be recognized as those we were producing using the unit displacement method. This bar stiffness is for a single element unsupported in space and it cannot be inverted to obtain $\boldsymbol{\rho}_g$ from $\mathbf{P}_g$. $\mathbf{k}_g$ is singular since there are three possible independent rigid-body movements which require no forces to produce them, that is, there are solutions to $\mathbf{k}_g\boldsymbol{\rho}_g = \mathbf{0}$.

Consider now the PVD *applied to the whole structure*, that is,

$$\mathbf{R}^t\bar{\mathbf{r}} = \sum_{g=1,2,\ldots} N_g\bar{\Delta}_g = \sum \mathbf{P}_g^t\bar{\boldsymbol{\rho}}_g = \sum \boldsymbol{\rho}_g^t\mathbf{k}_g\bar{\boldsymbol{\rho}}_g \qquad \text{using (2.14)}$$

We now relate bar displacements $\boldsymbol{\rho}_g$ to $\mathbf{r}$ using $\boldsymbol{\rho}_g = \mathbf{a}_g\mathbf{r}$, equation (2.10), to obtain

$$[\mathbf{r}^t \textstyle\sum \mathbf{a}_g^t\mathbf{k}_g\mathbf{a}_g - \mathbf{R}^t]\bar{\mathbf{r}} = \mathbf{0}$$

$$\text{or} \quad \mathbf{r}^t\mathbf{K} - \mathbf{R}^t = \mathbf{0}, \quad \text{or} \quad \mathbf{R} = \mathbf{K}\mathbf{r}$$

where

$$\mathbf{K} = \sum_g \mathbf{a}_g^t\mathbf{k}_g\mathbf{a}_g, \qquad \text{in a similar fashion to (1.42).}$$

But the two expressions $\sum \boldsymbol{\rho}_g^t \mathbf{k}_g \bar{\boldsymbol{\rho}}_g$ and $\mathbf{r}^t \mathbf{K} \bar{\mathbf{r}}$ both represent the same thing, the work done by forces $\mathbf{P}_g = \mathbf{k}_g \boldsymbol{\rho}_g$ over the displacements $\bar{\boldsymbol{\rho}}_g$. Moreover, the displacements $\boldsymbol{\rho}_g$ are identical to $\mathbf{r}$, only the labelling differs. Consequently, if two local bar displacements ρ_i and ρ_j correspond to global displacements r_I and r_J, then the stiffness k_{ij} has merely to be inserted at K_{IJ}, and if it occurs more than once in the above summation, then K_{IJ} is the sum of all such sources. Contributions k_{ij} from different bars need not, of course, be the same.

The transformation $\mathbf{a}_g^t \mathbf{k}_g \mathbf{a}_g$ – called a congruent transformation – is therefore merely a relocation of elements of $\mathbf{k}_g$ into another, differently labelled $\mathbf{K}$. The reader is invited to generate a matrix $\mathbf{a}_g$ containing unit elements which equate two displacements ρ_i and ρ_j to r_I and r_J, and to confirm that the congruent transformation takes k_{ij} and places it at (I, J) in $\mathbf{K}$.

The free-body element stiffness $\mathbf{k}_g$ is singular as we said, but the assembled stiffness $\mathbf{K}$ is not, provided that the structure is supported adequately and is not a mechanism.

Summary of displacement solution $\mathbf{Kr} = \mathbf{R}$

(1) $\mathbf{r}$: Label all joint displacements.
(2) $\mathbf{a}_g$: Label all bars and establish the equivalence between local $\boldsymbol{\rho}_g$ and global $\mathbf{r}$ lists.
(3) $\mathbf{k}_g$: From the coordinates of the gth bar's ends, find θ and l and hence the elements of $\mathbf{k}_g$, (2.16). Repeat for all bars, reducing the $\mathbf{k}$ size for any bar which has one or more joint displacements held at a support.
(4) $\sum \mathbf{a}_g^t \mathbf{k}_g \mathbf{a}_g$: Using (2), insert elements of $\mathbf{k}_g$ into appropriate locations in $\mathbf{K}$ and repeat for all elements.
(5) $\mathbf{Kr} = \mathbf{R}$: Solve for joint dispacements $\mathbf{r}$.
(6) For every bar extract $\boldsymbol{\rho}_g = \mathbf{a}_g \mathbf{r}$, and form $\Delta = \mathbf{a}_N \boldsymbol{\rho}_g$ (2.11 and 2.12) and find $N = k_N \Delta$.

As a very simple illustrative example consider the statically determinate framework in Fig. 2.7. All bars have the same A, E and l. The element stiffnesses $\mathbf{k}_g$, using (2.16), are:

$$g = 1;\ \theta = 60^\circ. \quad \mathbf{k}_1 = \frac{AE}{l}\begin{bmatrix} \frac{1}{4} & \frac{\sqrt{3}}{4} & -\frac{1}{4} & -\frac{\sqrt{3}}{4} \\ & \frac{3}{4} & -\frac{\sqrt{3}}{4} & -\frac{3}{4} \\ & \text{symm.} & \frac{1}{4} & \frac{\sqrt{3}}{4} \\ & & & \frac{3}{4} \end{bmatrix} \begin{matrix} \rho_1 = 0 \\ \rho_2 = 0 \\ \rho_3 = r_1 \\ \rho_4 = r_2 \end{matrix} \quad \therefore \quad 3,4 \rightarrow 1,2$$

$$g = 2;\ \theta = 120°. \quad \mathbf{k}_2 = \frac{AE}{l}\begin{bmatrix} \frac{1}{4} & -\frac{\sqrt{3}}{4} & -\frac{1}{4} & \frac{\sqrt{3}}{4} \\ & \frac{3}{4} & \frac{\sqrt{3}}{4} & -\frac{3}{4} \\ & & \frac{1}{4} & -\frac{\sqrt{3}}{4} \\ & & & \frac{3}{4} \end{bmatrix} \begin{matrix} \rho_1 = r_3 \\ \rho_2 = 0 \\ \rho_3 = r_1 \\ \rho_4 = r_2 \end{matrix} \quad \therefore \quad 1, 3, 4 \rightarrow 3, 1, 2$$

$$g = 3;\ \theta = 0°. \quad \mathbf{k}_3 = \frac{AE}{l}\begin{bmatrix} 1 & 0 & -1 & 0 \\ & 0 & 0 & 0 \\ & & 1 & 0 \\ & & & 0 \end{bmatrix} \begin{matrix} \rho_1 = 0 \\ \rho_2 = 0 \\ \rho_3 = r_3 \\ \rho_4 = 0 \end{matrix} \quad \therefore \quad 3 \rightarrow 3$$

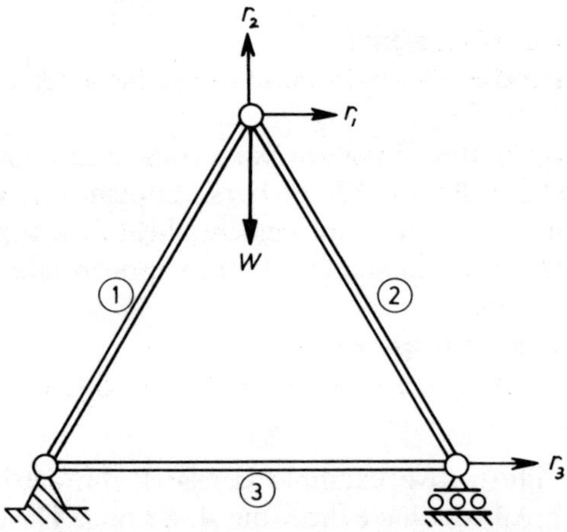

Fig. 2.7 Example

Now add $\mathbf{k}_g$ into $\mathbf{K}$. The details of contributions from $\mathbf{k}_1, \mathbf{k}_2, \mathbf{k}_3$ are shown here before summing.

$$\mathbf{K} = \frac{AE}{l}\begin{bmatrix} \frac{1}{4} + \frac{1}{4} + 0 = \frac{1}{2} & \frac{\sqrt{3}}{4} - \frac{\sqrt{3}}{4} + 0 = 0 & 0 - \frac{1}{4} + 0 = -\frac{1}{4} \\ & \frac{3}{4} + \frac{3}{4} + 0 = \frac{3}{2} & 0 + \frac{\sqrt{3}}{4} + 0 = \frac{\sqrt{3}}{4} \\ & & 0 + \frac{1}{4} + 1 = \frac{5}{4} \end{bmatrix}$$

In this simple structure it is easy to see that the unit displacement method gives precisely the same information. For example, putting $r_3 = 1$ produces $N_3 = AE/l$, $N_2 = (AE/l)\frac{1}{2}$, $N_1 = 0$. Thus,

$$\begin{aligned}\{k_{31} \quad k_{32} \quad k_{33}\} &= \{R_1 \quad R_2 \quad R_3\} \\ &= \left\{\frac{1}{2}N_1 - \frac{1}{2}N_2 \quad \frac{\sqrt{3}}{2}N_1 + \frac{\sqrt{3}}{2}N_2 \quad N_3 + \frac{1}{2}N_2\right\} \\ &= \frac{AE}{l}\left\{-\frac{1}{4} \quad \frac{\sqrt{3}}{4} \quad \frac{5}{4}\right\}\end{aligned}$$

Solving $\mathbf{Kr} = \mathbf{R} = \{0 \quad -W \quad 0\}$ we find

$$\mathbf{r} = \{r_1 \quad r_2 \quad r_3\} = \frac{Wl}{AE}\left\{\frac{1}{4\sqrt{3}} \quad -\frac{3}{4} \quad \frac{1}{2\sqrt{3}}\right\}$$

Back-substituting using (2.11) and (2.12),

$$N_1 = \frac{AE}{l}\left(\frac{1}{2}r_1 + \frac{\sqrt{3}}{2}r_2\right) = -\frac{W}{\sqrt{3}}$$

$$N_2 = \frac{AE}{l}\left(-\frac{1}{2}r_1 + \frac{\sqrt{3}}{2}r_2 + \frac{1}{2}r_3\right) = -\frac{W}{\sqrt{3}}$$

$$N_3 = \frac{AE}{l}r_3 = \frac{W}{2\sqrt{3}}$$

These results of course could in this case have been found immediately by statics.

2.4 Influence lines for statically determinate frames works

As a final example, and using an amusing extreme form of virtual unit displacement, we consider the preparation of *influence lines* for statically determinate structures. Influence lines, which nowadays have become somewhat old-hat with the advent of large and accessible digital computers, are a useful device for assessing the worst effects of a moving load or loads across a structure such as a bridge. They are simply a graph showing the variation in the magnitude of the stress at a particular point, as a unit load moves across the structure, for all positions of that unit load. If the structure is statically determinate, then obviously this information can be found using direct equilibrium arguments, albeit tedious ones, if many unit-load positions have to be examined. The PVD is a much quicker alternative.

Consider, for example, the bridge truss in Fig. 2.8; all bars are identical, and suppose we require the influence line for the tension N in AD as a unit load moves across the deck. As we are interested only in the tension N, and in no other internal stresses, we seek a virtual-displacement pattern for which compatible strains occur only in the member AD. In other words, we cause member AD to

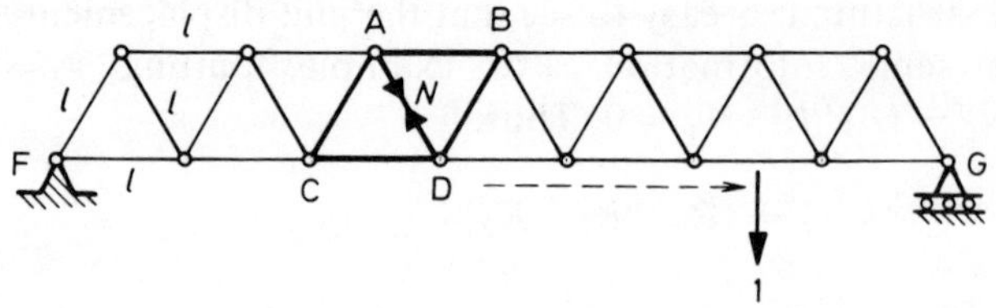

Fig. 2.8 Virtual strain of single bar AD

stretch a (unit) amount and equate this sole internal work to that done by the applied unit load. The deformation of the truss consists simply of a distortion of the parallelogram ABCD shown bold in Fig. 2.8, with the structure to the left and right undergoing displacements as rigid bodies. ABCD still remains a parallelogram so FC remains parallel to DG. If the displacements are small, then the sketch, Fig. 2.9, shows that D must displace a small distance $\sqrt{3}/2$ normal to the deck. Thus, DG is displaced a distance $\sqrt{3}/2$ parallel to FC and the distorted deck looks like Fig. 2.10. We see that the unit load has moved through a virtual displacement $\bar{h}$.

By the PVD $$N1 = 1\bar{h}$$

The value of N is therefore equal to $\bar{h}$ irrespective of where the unit load happens to be, and consequently the shaded graph in Fig. 2.10 is the variation of N with the position of the unit load. The displaced structure draws its own influence line! We can immediately see that the maximum force in AD occurs when the travelling load is at D. When at C, the value of N is a compressive $-\sqrt{3}/7$ since now $\bar{h}$ is in the opposite sense to the unit load. If the effect of a series of loads – like a train – is

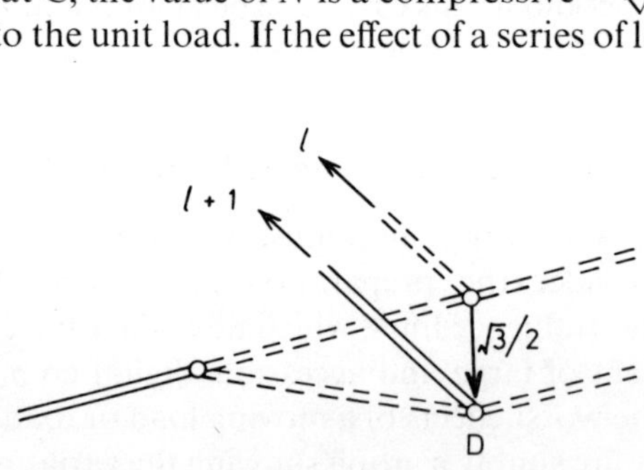

Fig. 2.9 Displacement of bridge deck

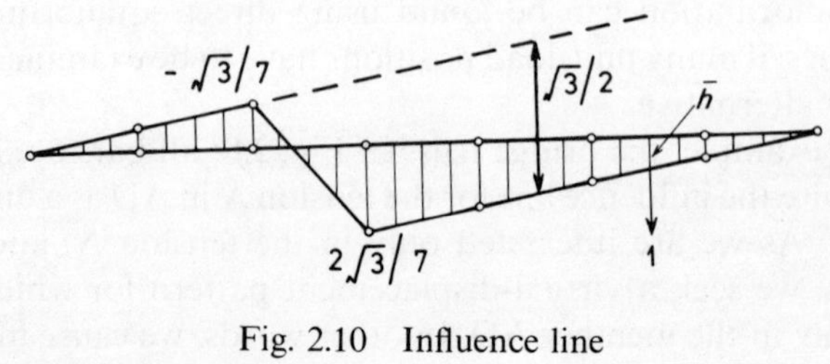

Fig. 2.10 Influence line

desired, then we may superpose several influence lines suitably shifted with respect to each other. This ability of a structure to draw its own influence line can of course be extended to shear forces, torques, and moments, etc., although a little ingenuity is sometimes required to pick a unit displacement such that the internal virtual work is confined to the desired internal force. For example, to isolate a moment in a beam it is necessary to confine the deformation to an infinitesimal element as shown in Fig. 2.11(a). The internal virtual work equals $M \times 1$. The influence line for the bending moment at the point A in Fig. 2.11(b) is constructed immediately as shown.

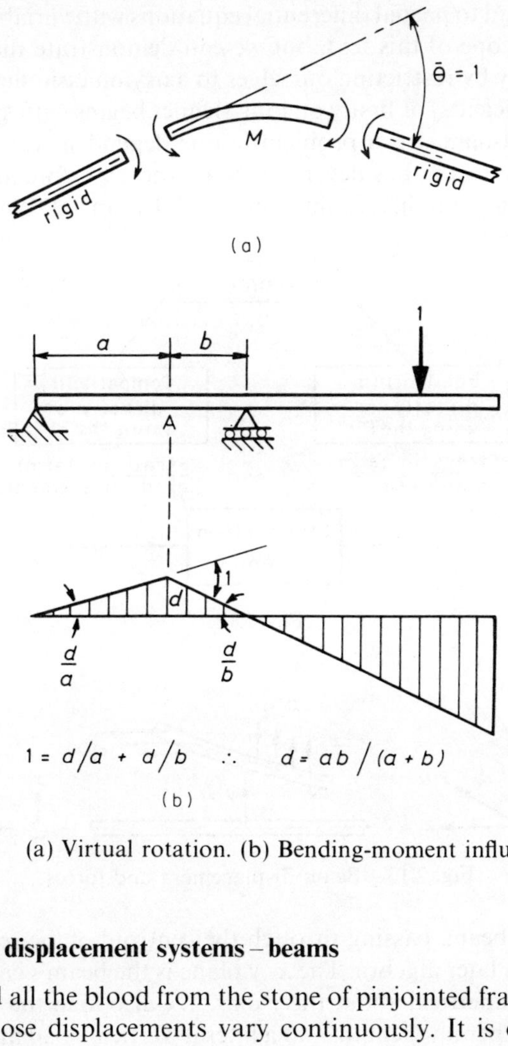

Fig. 2.11 (a) Virtual rotation. (b) Bending-moment influence line

2.5 Continuous displacement systems – beams

Having extracted all the blood from the stone of pinjointed frameworks, we turn to structures whose displacements vary continuously. It is opportune to ask under what circumstances would we prefer to use the PVD for continuous systems, rather than the direct formulation. Obviously we will not use it for a general continuum since we would succeed only in producing the general equations (1.52) again. Although the real power of virtual-work formulations undoubtedly lies in the use of approximate methods, they are also useful in deriving for *specific structures* awkward equations of equilibrium or compatibility which might otherwise be elusive. This invariably happens when for a particular type of structure, some assumptions have to be made to make it tractable. For example, the equations governing the behaviour of thin shells of arbitrary shape are complex and depend crucially on whether we ignore or retain certain strains. The work approach is here fairly systematic since we merely toss into the product $\boldsymbol{\sigma}^t\boldsymbol{\varepsilon}$ those components which we think are significant – we can always check our judgement after we have obtained a solution. Naturally, arbitrary shells lead to partial differential equations with variable coefficients and are beyond the scope of this text, but we can demonstrate the PVD technique quite satisfactorily by restricting ourselves to axisymmetric shells and flat plates (no variable coefficients) or firstly on long slender beams with symmetrical cross-sections. This last-named is a problem in one dependent variable only.

We suppose the reader has never met beam theory before and so develop the PVD solution from scratch, recalling our Fig. 1.9 as an aid. We choose the z-axis

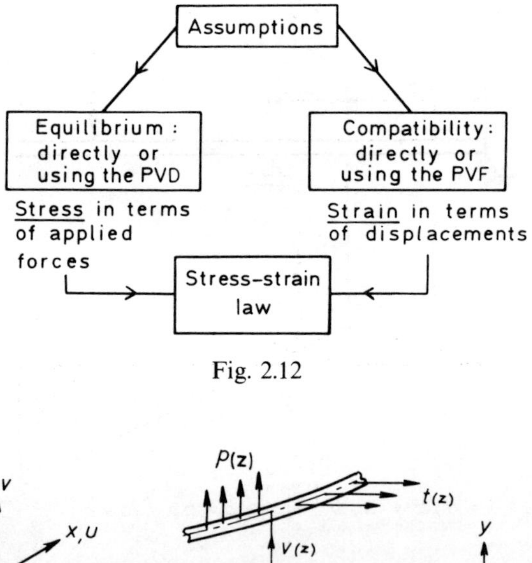

Fig. 2.12

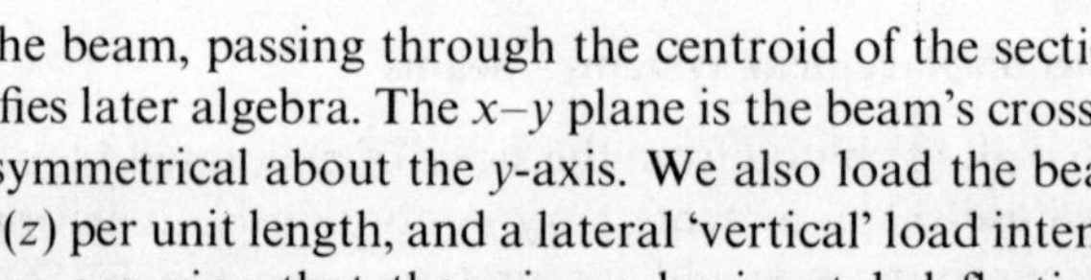

Fig. 2.13 Beam displacement and forces

as the axis of the beam, passing through the centroid of the section, as we shall find this simplifies later algebra. The x–y plane is the beam's cross-section which we take to be symmetrical about the y-axis. We also load the beam by an axial load intensity $t(z)$ per unit length, and a lateral 'vertical' load intensity of $p(z)$ per unit length, thus ensuring that there is no horizontal deflection, u. This last restriction is also simply to shorten the algebra. We now make assumptions about the likely displacement field.

Assumption 1 The cross-section is stocky and does not undergo changes in shape; so the deflections v and w are functions of z only. Our theory will not therefore be valid for beams with delicate flanges, for example.

Assumption 2 Plane cross-sections remain plane and normal to the axis after bending and stretching occur – the classical beam-theory assumption. This assumption is only strictly valid for beams of uniform section subjected to a pure moment, but for slender beams of moderate taper, the out-of-plane shear deformations can be safely ignored. Figure 2.14 shows the displaced position of a section whose centroid has moved $w_0(z)$ and $v(z)$ as shown, and the plane section has rotated through the same angle as the centroidal axis. For small deflections this rotation is $\partial v/\partial z = v'(z)$. Thus the axial displacement $w(z)$ of any point distance y from the centroidal axis is given by

$$w(z) = w_0(z) - yv'(z) \tag{2.17}$$

This completes all the assumptions and physical arguments necessary to define the displacement field. No further structural thought is necessary; the rest is automatic. We are required to satisfy $\bar{\boldsymbol{\varepsilon}} = \mathbf{D}^t\bar{\mathbf{u}}$ directly, and having obtained this relationship it is most convenient to use it also for stress $\boldsymbol{\sigma} = \boldsymbol{\kappa}\boldsymbol{\varepsilon}$, and allow the PVD to deliver a solution for the displacements.

The displacement vector is

$$\mathbf{u} = \{u \quad v \quad w\} = \{0 \quad v(z) \quad w_0 - yv'\}$$

whence $\boldsymbol{\varepsilon} = \mathbf{D}^t\mathbf{u}$ (equation (1.66)) gives

$$\begin{bmatrix} \varepsilon_{xx} \\ \varepsilon_{yy} \\ \varepsilon_{zz} \\ \varepsilon_{xy} \\ \varepsilon_{yz} \\ \varepsilon_{zx} \end{bmatrix} = \begin{bmatrix} \dfrac{\partial}{\partial x} & 0 & 0 \\ 0 & \dfrac{\partial}{\partial y} & 0 \\ 0 & 0 & \dfrac{\partial}{\partial z} \\ \dfrac{\partial}{\partial y} & \dfrac{\partial}{\partial x} & 0 \\ 0 & \dfrac{\partial}{\partial z} & \dfrac{\partial}{\partial y} \\ \dfrac{\partial}{\partial z} & 0 & \dfrac{\partial}{\partial x} \end{bmatrix} \begin{bmatrix} 0 \\ v \\ w_0 - yv' \end{bmatrix} = \begin{bmatrix} 0 \\ 0 \\ w_0' - yv'' \\ 0 \\ 0 \\ 0 \end{bmatrix}$$

All the zeros, bar one, are due to both w_0 and v being functions of z only, and ε_{yz} is zero because

$$\varepsilon_{yz} = \frac{\partial v}{\partial z} + \frac{\partial}{\partial y}(-yv') = 0$$

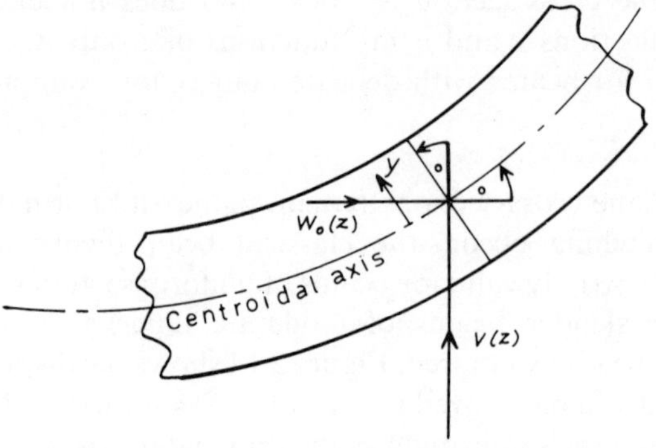

Fig. 2.14 Beam theory—plane sections remain plane

confirming our original assumption that such shear strains are zero. The PVD (1.69) now becomes

$$\int_V \sigma_{zz}\,\bar{\varepsilon}_{zz}\,\mathrm{d}v = \int_S p_s \bar{u}\,\mathrm{d}s$$

or
$$\int_0^l \int_A E(w_0' - yv'')(\bar{w}_0' - y\bar{v}'')\,\mathrm{d}A\,\mathrm{d}z = \int_0^l p(z)\bar{v}\,\mathrm{d}z + \int_0^l t(z)\bar{w}_0\,\mathrm{d}z$$

Evaluating the integrals over the cross-section, and noting that $\int_A y\,\mathrm{d}A = 0$ because the centroid is $y = 0$, and putting $\int_A y^2\,\mathrm{d}A = I$, the second moment of area, we find

$$\int_0^l (EAw_0'\bar{w}_0' + EIv''\bar{v}'' - p\bar{v} - t\bar{w}_0)\,\mathrm{d}z = 0$$

In deriving this expression we have simply followed standard procedure: assume a displacement field $\mathbf{u}$, apply $\boldsymbol{\varepsilon} = \mathbf{D}^t\mathbf{u}$, and evaluate $\int \boldsymbol{\sigma}^t\bar{\boldsymbol{\varepsilon}}\,\mathrm{d}v$. However, in this case, had we wished, we could have deduced the virtual work directly. Thus the work done by an axial tension $A\sigma\,(= AEw_0')$ over an element stretching by $\bar{\varepsilon}_{zz}\,\mathrm{d}z\,(= \bar{w}_0'\,\mathrm{d}z)$ is $(EAw_0')\bar{w}_0'\,\mathrm{d}z$. Similarly, one end of an element rotates by $\bar{v}''\,\mathrm{d}z$ with respect to the other end, so the virtual work due to the moment is $(EIv'')\bar{v}''\,\mathrm{d}z$.

Now as usual we take advantage of the virtual nature of $\bar{w}_0(z)$ and $\bar{v}(z)$. Although $\bar{v}(z)$ is arbitrary, as is $\bar{v}''(z)$, they are not separately arbitrary; that is, once $\bar{v}(z)$ is chosen, then so is $\bar{v}''(z)$, and we cannot insist that their separate coefficients vanish. We must therefore convert $\bar{v}''(z)$ to $\bar{v}(z)$, integrating by parts twice. This is equivalent to using Gauss's theorem in equation (1.71), where we had to convert $\mathbf{D}^t\bar{\mathbf{u}}$ to $\bar{\mathbf{u}}$. The two displacements $\bar{v}(z)$ and $\bar{w}_0(z)$, however, are unrelated and their coefficients may be separately treated. Thus the above equation becomes

$$\int_0^l [(-EAw_0'' - t)\bar{w}_0 + (EIv'''' - p)\bar{v}]\,\mathrm{d}z + [EAw_0'\bar{w}_0 + EIv''\bar{v}' - EIv'''\bar{v}]_0^l = 0 \tag{2.18}$$

We can now appeal to the virtual nature of $\bar{v}(z)$ and $\bar{w}_0(z)$. Firstly, $\bar{w}_0(z)$:

$$EAw_0'' + t = 0$$

This is an equation of axial equilibrium, for if we recognize that the axial tension $N = AEw_0'$, it is simply saying $\partial N/\partial z + t = 0$. If there is no applied force $t(z)$, then $w_0(z) = 0$, confirming that under lateral forces alone the centroidal axis is the unstrained neutral axis. Secondly, $\bar{v}(z)$:

$$EIv'''' - p = 0 \tag{2.19}$$

This equation is the classical fourth-order differential equation for beams. It is valid whether the beam is statically determinate or indeterminate since it does not depend on boundary conditions at all. For a fourth-order differential equation we expect to provide four boundary conditions, two at each end $z = 0$ and $z = l$. We now note that only the integrals in (2.18) have vanished; there still remain the terms in square parenthesis and these must vanish as a consequence of the boundary conditions. There are two possibilities:

(a) The boundary conditions are kinematic. If the end of the beam is held in some way, we stipulated earlier that the unknown support reactions can be excluded from the PVD by specifying zero corresponding virtual displacements. Thus, if the beam is prevented from displacing ($w_0 = v = 0$) or rotating ($v' = 0$), we also specify that $\bar{v} = \bar{v}' = 0$. This ensures vanishing of these terms in (2.18) and we then use $v = v' = 0$ as boundary conditions.
(b) The boundary conditions may be equilibrium. We expect the PVD to provide this type of information, and it duly does. If the beam end (say $z = l$) is *free*, then $\bar{v}(l)$ and $\bar{v}'(l)$ can exist and we must equate their coefficients to zero:

$$EIv'''(l) = 0; \quad EIv''(l) = 0$$

These are recognizable as equilibrium statements. There is no shear force or bending moment at a free end. Had we never met beam theory before, we would have accepted that the PVD delivers equilibrium statements and that the coefficient of rotation $\bar{v}'$ must be a moment (EIv'') and of displacement $\bar{v}$ must be a shear force (EIv''').

Combinations of kinematic and equilibrium boundary conditions are clearly possible but there will always be four boundary conditions depending on whether v, $\bar{v}$, v', and $\bar{v}'$ are free or zero.

The same information is provided by the other term in (2.18), $[EAw_0'\bar{w}_0]_0^l$. Thus if the end is axially unconstrained, $\bar{w}_0$ exists and so EAw_0' must be zero, that is no axial force.

To show both the versatility and automatic nature of the PVD, consider another beam problem. The axial displacements will now be ignored since it is clear from the previous example that $w_0(z)$ and $v(z)$ are uncoupled problems. The beam shown in Fig. 2.15 is firmly clamped at one end and supported at the other by a vertical spring of stiffness k_1 (force per unit deflection) and rotational spring

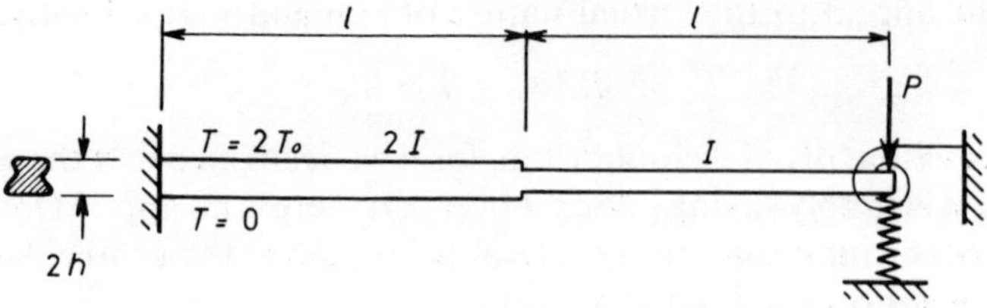

Fig. 2.15 Example

of stiffness k_2 (moment per rotation). The beam stiffness is $2EI$ from $z = 0$ to l, and EI from $z = l$ to $2l$. The addition of the two springs presents no problem: they are merely further elastic structures which are included in $\int_V \boldsymbol{\sigma}^t \bar{\boldsymbol{\varepsilon}}\,dv$ which must be evaluated over the *whole* structure. (Endless confusion will arise if springs are treated as applied forces – which they are not.) By quoting spring 'stiffnesses' we avoid having to evaluate actual integrals, since virtual work is force × displacement, that is, $k_1 v(2l) \cdot \bar{v}(2l)$ or $k_2 v'(2l) \cdot \bar{v}'(2l)$. The beam is loaded by a single force P at $z = 2l$, and from $z = 0$ to l the upper surface is heated to a temperature $2T_0$, the variation through the depth being assumed linear. To simplify algebra we assume a doubly symmetrical cross-section of depth $2h$ so that the centroid is in the middle, and the temperature variation is

$$T(y) = T_0(1 + y/h)$$

The thermal strains are quite straightforward to handle, they are just included in the total strain (2.2), so that

$$\varepsilon = \frac{\sigma}{E} + \alpha T; \qquad \text{or} \qquad \sigma = E(\varepsilon - \alpha T)$$

With $\varepsilon_{zz} = -yv''$ as before, the PVD becomes

$$\int_0^l \int_A E[-yv'' - \alpha T_0(1 + y/h)](-y\bar{v}'')\,dA\,dz + \int_l^{2l} \int_A E(-yv'')(-y\bar{v}'')\,dA\,dz$$
$$+ k_1 v(2l)\bar{v}(2l) + k_2 v'(2l)\bar{v}'(2l) = -P\bar{v}(2l)$$

The negative sign on the right-hand side is due to the fact that P and $v(2l)$ are in opposite senses. There is never any ambiguity at all about signs on the left-hand side: all internal virtual work must be positive. Evaluating integrals over the cross-section A, and recalling that $\int y\,dA = 0$, the PVD becomes

$$\int_0^l (2EIv''\bar{v}'' + 2EI\alpha T_0\bar{v}''/h)\,dz + \int_l^{2l} EIv''\bar{v}''\,dz$$
$$+ k_1 v(2l)\bar{v}(2l) + k_2 v'(2l)\bar{v}'(2l) + P\bar{v}(2l) = 0$$

Integrating by parts and collecting terms,

$$\int_0^l 2EI(v'''' + \alpha T_0''/h)\bar{v}\,dz + [2EI(v'' + \alpha T_0/h)\bar{v}' - 2EI(v''' + \alpha T_0'/h)\bar{v}]_0^l$$

$$+\int_{l}^{2l} EIv''''\bar{v}\,dz + [EIv''\bar{v}' - EIv'''\bar{v}]_{l}^{2l}$$

$$+ k_1 v(2l)\bar{v}(2l) + k_2 v'(2l)\bar{v}'(2l) + P\bar{v}(2l) = 0 \quad (2.20)$$

Equation (2.20) is a fair mouthful, but then a fair amount of information has been delivered. We use it in stages.

(1) $\bar{v}(z)$ from $z = 0$ to l: $v'''' + \alpha T_0''/h = 0$

$$\therefore \quad v_L = C_0 + C_1 z + C_2 z^2 + C_3 z^3 - \iint \alpha T_0/h\, dz\, dz$$

(2) $\bar{v}(z)$ from $z = l$ to $2l$: $v'''' = 0$,

$$\therefore \quad v_R = C_4 + C_5 z + C_6 z^2 + C_7 z^3$$

(Subscripts in v_L and v_R denote 'left' and 'right' of $z = l$)

(3) At $z = 0$, $v(0) = 0$ and $v'(0) = 0$

(4) At $z = l$, kinematic continuity says that $v_L(l) = v_R(l)$; $v_L'(l) = v_R'(l)$. Also, since $\bar{v}(l)$ and $\bar{v}'(l)$ are not zero, their coefficients in (2.20) must be zero, thus:

$$\bar{v}'(l): \quad 2EI(v_L'' + \alpha T_0/h) - EIv_R'' = 0$$

$$\bar{v}(l): \quad -2EI(v_L''' + \alpha T_0'/h) + EIv_R''' = 0$$

These two equations must be continuity of moment and shear force respectively.

(5) At $z = 2l$, $\bar{v}'(2l)$ and $\bar{v}(2l)$ are not zero, therefore:

$$\bar{v}'(2l): \quad EIv''(2l) + k_2 v'(2l) = 0$$

$$\bar{v}(2l): \quad -EIv'''(2l) + k_1 v(2l) + P = 0$$

These last are also equations of moment and shear equilibrium. There are now eight equations all told for solving C_0 to C_7. The reader who questions whether the PVD provides any information that could not have been found directly is invited to speculate whether he/she would have deduced (4) and (5) correctly without missing any terms or signs.

We have developed solutions for beams with symmetrical cross-sections only, but arbitrary sections subjected to loads with vertical and horizontal components present no difficulties. We assume both $v(z)$ and $u(z)$ simultaneously, put $w = -yv'' - xu''$, and then turn the PVD handle to obtain coupled equations from $\bar{v}$ and $\bar{u}$ (see Problem 2.4).

2.6 Stiff-jointed frameworks

Now that the behaviour of a single beam has been defined, explained, and solved, we are in a position to solve assemblies of such beams. Initially we look at two-dimensional assemblies loaded in their planes and composed of symmetrical-section beams: thus displacements out-of-plane are avoided. Such 'stiff-jointed' or 'rigid-jointed' frameworks are probably the most common form of structure in use today; they form the basis of most buildings since, although nowhere near as efficient as triangulated frameworks, they do have clear spaces for people to move about in. The sketch in Fig. 2.16 indicates an important feature of such

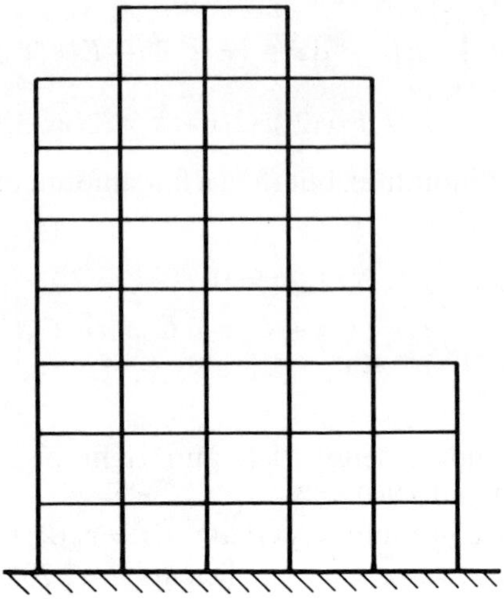

Fig. 2.16 Rigid jointed framework

frameworks – if the joints were 'pinned' then the structure would become a mechanism. We shall concern ourselves with this sort of framework, but we mention in passing that 'pinjointed frameworks' are usually welded and rigid-jointed also, and here there is a difference. A triangulated framework has its joints displace relatively by both extensional deformation of the bars and by bending deformation which produces displacements at right angles to a bar. But no particular direction is the weakest at any joint, therefore extensional *deformation* Nl/AE is of the same order as bending *deformation* which is of order Fl^3/EI, where F is the shear force. Therefore F/N is of order $(EI/l^3)/(l/AE) = k^2/l^2$, where $k = \sqrt{I/A}$ is the radius of gyration and k/l is usually very small for slender beams. Thus $F \ll N$, and we can treat the equilibrium of joints as if the framework were pinjointed. However, for the type of framework in Fig. 2.16 we shall find that both shear and axial *forces* are of the same order as the applied loads. Thus the ratio of extensional to bending deformation is of order $(l/AE)/(l^3/EI) = k^2/l^2$ again. So in stiff-jointed frames it is common to ignore the extensional deformations of component beams (although later we find it is systematically easier to include them in large frameworks).

In describing the behaviour of beam assemblies it is necessary to use not only joint displacements and forces as unknowns, but also joint rotations and moments. The virtual-work arguments can of course still be used provided 'moment' and 'rotation' correspond. In future, the term 'displacement' (or 'force') unknowns may well be a list containing a mixture of translation displacements (or forces) and rotations (or moments).

Consider first the simple portal frame in Fig. 2.17; we choose to label the joint 'displacements' r_1, r_2, and r_3 as shown. Only one displacement r_3 is needed to describe the 'sway' if we ignore stretching of BC. To apply the PVD we need to evaluate $\int_V \sigma\bar{\varepsilon}\,dv = \int EIv''\bar{v}''\,dz$ over all three beams, and this is tedious unless we

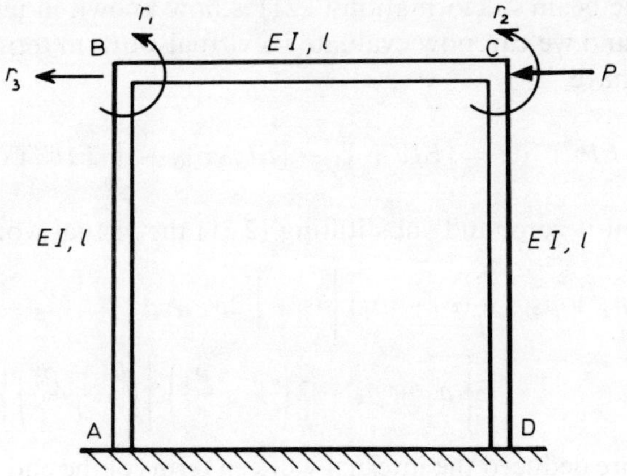

Fig. 2.17 Simple portal frame

take advantage of the fact that the beams are loaded only at the joints. Thus if $p(z) = 0$ in (2.19), $v'''' = 0$, and we may integrate:

$$v'' = C_2 + C_3 z/l$$

and

$$v = C_0 + C_1 z + C_2 z^2/2 + C_3 z^3/6l \tag{2.21}$$

Consider then a typical beam from the frame, displaced at its ends by ρ_1, ρ_2, ρ_3, and ρ_4, as shown in Fig. 2.18. On putting $v(0) = \rho_1$; $v'(0) = \rho_2$; $v(l) = \rho_3$; $v'(l) = \rho_4$; we find

$$C_0 = \rho_1; \qquad C_1 = \rho_2; \qquad C_2 = \frac{2}{l}\left[\frac{3}{l}(\rho_3 - \rho_1) - 2\rho_2 - \rho_4\right]$$

$$C_3 = \frac{6}{l}\left[\rho_4 + \rho_2 - \frac{2}{l}(\rho_3 - \rho_1)\right]$$

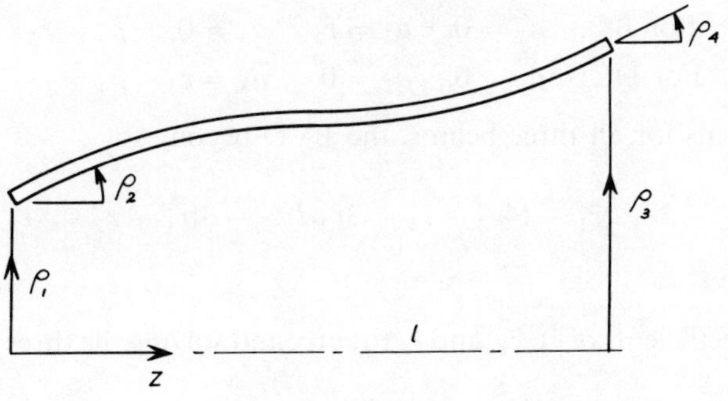

Fig. 2.18 Beam displacement variables

The shape of the beam's deformation (2.21) is now known in terms of its ends' displacements, and we can now evaluate its virtual work in terms of these four quantities. We have

$$\int_0^l EIv''\bar{v}''\,dz = [EIv''\bar{v}']_0^l - [EIv'''\bar{v}]_0^l + \int_0^l EIv''''\bar{v}\,dz \tag{2.22}$$

But the last term is zero, and substituting (2.21) the virtual work becomes

$$\frac{2EI}{l}\left\{\left[2\rho_2 + \rho_4 - \frac{3}{l}(\rho_3 - \rho_1)\right]\bar{\rho}_2 + \left[2\rho_4 + \rho_2 - \frac{3}{l}(\rho_3 - \rho_1)\right]\bar{\rho}_4 - 3\left[\rho_2 + \rho_4 - 2\left(\frac{\rho_3 - \rho_1}{l}\right)\right]\left(\frac{\bar{\rho}_3 - \bar{\rho}_1}{l}\right)\right\} \tag{2.23}$$

We have therefore deduced the internal work in terms of the end values of $\bar{\boldsymbol{\rho}}$ and $\boldsymbol{\rho} = \{\rho_1 \quad \rho_2 \quad \rho_3 \quad \rho_4\}$, and no further integrals have to be performed, just summations. This trick is the basis of the *approximate* finite element method of Chapter 5, where an assumption (like equation (2.21)) is made for the displacements in order to evaluate the integrals. In our case the assumption happens to be *exact*. The nature of (2.23) should also be clear. Since it is virtual work, the coefficients of $\bar{\rho}_2$ and $\bar{\rho}_4$ must be the end moments, and of $\bar{\rho}_3$ and $\bar{\rho}_1$ the end shears. Thus,

$$\begin{aligned} M(0) &= \frac{2EI}{l}\left[2\rho_2 + \rho_4 - \frac{3}{l}(\rho_3 - \rho_1)\right] \\ M(l) &= \frac{2EI}{l}\left[2\rho_4 + \rho_2 - \frac{3}{l}(\rho_3 - \rho_1)\right] \\ F(0) = -F(l) &= \frac{6EI}{l^2}\left[\rho_2 + \rho_4 - \frac{2}{l}(\rho_3 - \rho_1)\right] \end{aligned} \tag{2.24}$$

It is now possible to solve the portal by adding together three contributions like (2.23) from beams AB, BC, and CD. We must, though, continue to enforce compatibility where necessary, so we impose continuity of displacements at joints B and C. Consequently, when using (2.23) we select:

$$\begin{aligned} &\text{For AB:} \quad \rho_1 = 0, \quad \rho_2 = 0, \quad \rho_3 = r_3, \quad \rho_4 = r_1 \\ &\text{For BC:} \quad \rho_1 = 0, \quad \rho_2 = r_1, \quad \rho_3 = 0, \quad \rho_4 = r_2 \\ &\text{For DC:} \quad \rho_1 = 0, \quad \rho_2 = 0, \quad \rho_3 = r_3, \quad \rho_4 = r_2 \end{aligned} \tag{2.25}$$

Collecting terms for all three beams, the PVD becomes

$$\frac{2EI}{l}[(4r_1 + r_2 - 3r_3/l)\bar{r}_1 + (4r_2 + r_1 - 3r_3/l)\bar{r}_2 - 3(r_1 + r_2 - 4r_3/l)\bar{r}_3/l] - P\bar{r}_3 = 0 \tag{2.26}$$

Putting the coefficients of $\bar{r}_1$, $\bar{r}_2$, and $\bar{r}_3$ to zero, and solving the three simultaneous equations, we get

$$r_1 = r_2 = Pl^2/28EI; \qquad r_3 = 5Pl^3/84EI$$

The stresses follow by substituting these displacements into the end moments (2.24): since moments vary linearly, (2.21), the maxima must be at the ends A, B, C, or D. The moment distribution and the deflected shape are both sketched in Fig. 2.19. (The latter sketch is always worth while in simple problems; if any algebraic mistake has been made it is usually found difficult to sketch a shape such that the curvature matches the shape of the bending-moment diagram.)

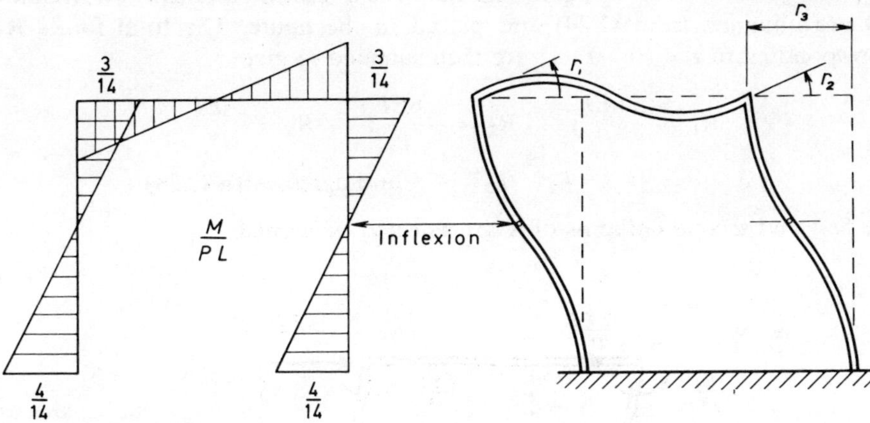

Fig. 2.19 Bending moment and displacement patterns

What we have just done is a fairly simple example where we were able to express $\int_V \boldsymbol{\sigma}^t \bar{\boldsymbol{\varepsilon}}\, dv$ in terms of the three joint displacements r_1, r_2, r_3; and the $\bar{r}_1, \bar{r}_2, \bar{r}_3$ duly delivered three equations. This technique can be made more systematic in exactly the same fashion as we used in pinjointed frameworks, by assembling a global stiffness matrix. The element stiffness matrix $\mathbf{k}_g$ can readily be extracted from the virtual-work expression (2.23) since we recall from (2.9) that the unit displacement method says that $k_{ij} = \int \sigma_j \varepsilon_i\, dv$. But (2.23) is precisely this, already evaluated for all $\sigma_j \bar{\varepsilon}_i$, and so we pick out k_{ij} as the coefficient of $\rho_i \bar{\rho}_j$. On doing this for all $i, j = 1, 2, 3$, we get:

$$\mathbf{k}_g = \frac{2EI}{l^3}\begin{bmatrix} 6 & 3l & -6 & 3l \\ & 2l^2 & -3l & l^2 \\ \text{Symmetric} & & 6 & -3l \\ & & & 2l^2 \end{bmatrix} \tag{2.27}$$

The stiffness elements from $\mathbf{k}_g$ for all three beams AB, BC, and CD (identical here as it happens) are now inserted and summed into the 3×3 global stiffness matrix $\mathbf{K}$ using (2.25) as the connectivity information which gives local $\boldsymbol{\rho}_g$ subscripts in terms of global $\mathbf{r}$ subscripts. Thus

$$\mathbf{K} = \frac{2EI}{l^3}\begin{bmatrix} 4l^2 & l^2 & -3l \\ \text{symm} & 4l^2 & -3l \\ & & 12 \end{bmatrix} \tag{2.28}$$

The elements of **K** are naturally recognizable as the coefficients of the three simultaneous equations (2.26) previously solved.

The reader might be tempted to abandon virtual work and construct the stiffness matrix directly – as we did in pinjointed frames – by imposing a single unit displacement r_i and finding the forces $\{R_1 \quad R_2 \quad R_3\}$ needed to hold this pattern. Indeed, this is quite straightforward in this example, as we demonstrate. Suppose $r_3 = 1, r_2 = r_1 = 0$, as in Fig. 2.20. The forces and moments on AB and CD are obtained from (2.24) and placed in the figure. The total forces **R** corresponding to $\mathbf{r} = \{0 \quad 0 \quad 1\}$ are then summed to give:

$$R_1 = -\frac{6EI}{l^2}; \qquad R_2 = -\frac{6EI}{l^2}; \qquad R_3 = \frac{24EI}{l^3}$$

$$= \{k_{13} \quad k_{23} \quad k_{33}\}, \qquad \text{and agrees with (2.28)}$$

The first and second columns of **K** are as easily generated.

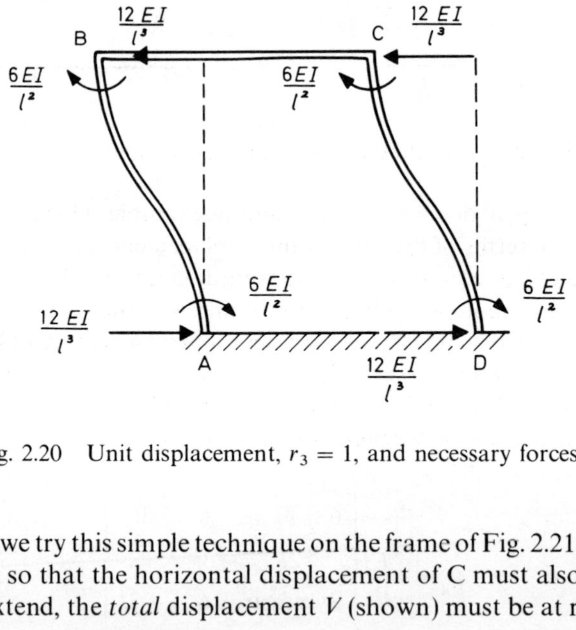

Fig. 2.20 Unit displacement, $r_3 = 1$, and necessary forces

But suppose we try this simple technique on the frame of Fig. 2.21. Again ignore axial extension so that the horizontal displacement of C must also be r_3. But as CD does not extend, the *total* displacement V (shown) must be at right angles to CD, so $v/\sqrt{2} = r_3$. Now apply $r_1 = 1, r_2 = 1, r_3 = 1$, in turn and form the forces and moments according to (2.24). The results are shown in Figs 2.21(a), (b), (c). Support reactions, not included in **R**, are omitted. Suppose we wish to generate $\{k_{13} \quad k_{23} \quad k_{33}\}$, which are the values **R** in Fig. 2.21(c). Here there is a snag: no applied force occurs at the joint C, so how are the necessary forces $6EI/l^3$ and $12EI/l^3$ applied? This quandary is due entirely to the *kinematic* assumption that both BC and CD are inextensional, consequently coupling the displacements of B and C; in which case we should be able to use the unit displacement method (2.9) since this involves *only kinematic* assumptions in evaluating ε_3, σ_1, etc.

Thus $k_{13} = \int_V \sigma_3 \varepsilon_1 \, dv$; but again this integral does not have to be evaluated

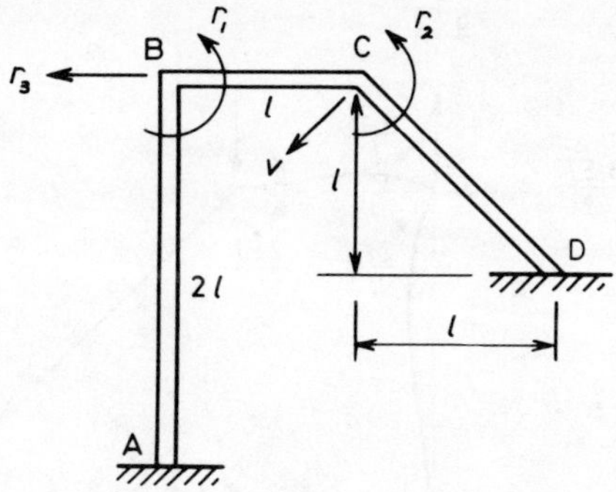

Fig. 2.21 Portal with inclined member

since the beams' end forces and displacements enable us to simply form products for the interior work. Thus,

$$k_{13} = [\text{forces in Fig. (c)}] \times [\text{displacements in Fig. (a)}]$$

$$= \left(\frac{6EI}{l^2} - \frac{3EI}{2l^2}\right) \times 1 = \frac{9EI}{2l^2}$$

$$k_{23} = [\text{Fig. (c)}] \times [\text{Fig. (b)}] = \left(\frac{6EI}{l^2} - \frac{3\sqrt{2}EI}{l^2}\right) \times 1$$

$$= (6 - 3\sqrt{2})\frac{EI}{l^2}$$

$$k_{33} = [\text{Fig. (c)}] \times [\text{Fig. (c)}] = \left(\frac{3EI}{2l^3}\right)1 + \left(\frac{12EI}{l^3}\right)1 + \left(\frac{6EI}{l^3}\right)\sqrt{2}$$

$$= \left(\frac{27}{2} + 6\sqrt{2}\right)\frac{EI}{l^3}$$

Having obtained the answers in this painless fashion, we can now reaffirm using equilibrium (and hindsight) how these values arise. In Fig. 2.21(c) the joint forces can only be provided by R_3 and the ground reaction. But the joint forces at C must be communicated to R_3 and the ground only by internal *axial forces* through BC and CD, since further shear forces would modify the assumed displacement pattern. Figure 2.22 shows the forces resolved into the two *oblique* components along BC and CD, to confirm the previous evaluation for k_{33}. Direct equilibrium arguments do therefore have to be applied with care. After all, to replace a force $12EI/l^3$ by two components, one of which is at *right angles* to it, does make us discard the habits of a lifetime.

2.7 Kinematically equivalent forces

Suppose the previous portal frame had been loaded beween joints, as in Fig. 2.23. One way of coping would be to divide BC into two beam elements by introducing further displacement unknowns at the load W. This is unnecessary, and would not

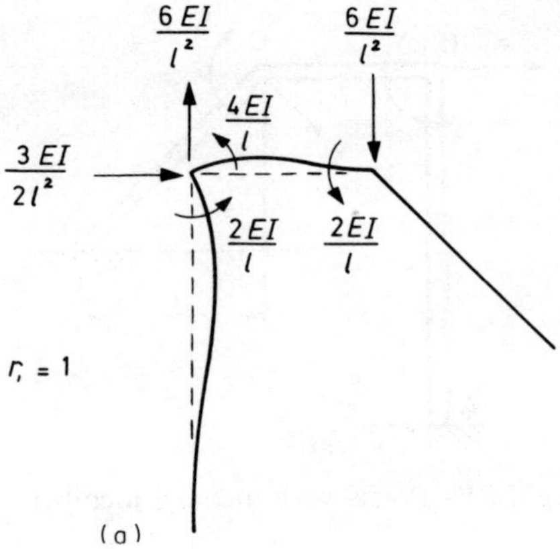
$\frac{6EI}{l^2}$
$\frac{6EI}{l^2}$
$\frac{4EI}{l}$
$\frac{3EI}{2l^2}$
$\frac{2EI}{l}$
$\frac{2EI}{l}$
$r_1 = 1$

(a)

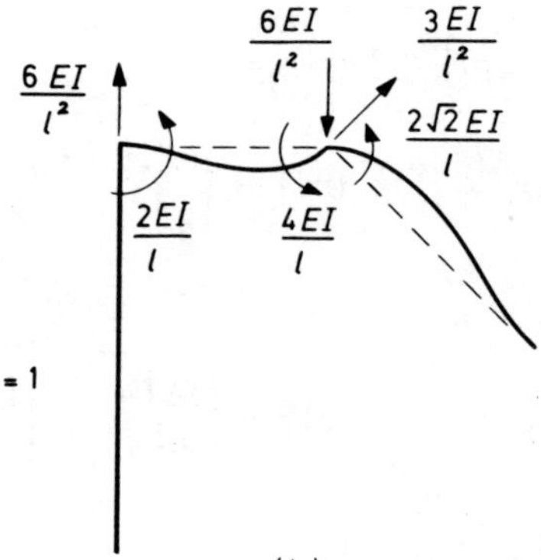
$\frac{6EI}{l^2}$
$\frac{6EI}{l^2}$
$\frac{3EI}{l^2}$
$\frac{2\sqrt{2}EI}{l}$
$\frac{2EI}{l}$
$\frac{4EI}{l}$
$r_2 = 1$

(b)

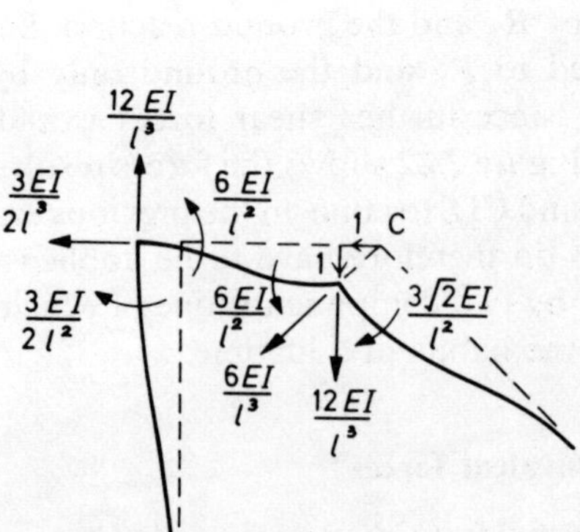
$\frac{12EI}{l^3}$
$\frac{3EI}{2l^3}$
$\frac{6EI}{l^2}$
1 C
1
$\frac{3EI}{2l^2}$
$\frac{6EI}{l^2}$
$\frac{3\sqrt{2}EI}{l^2}$
$\frac{6EI}{l^3}$
$\frac{12EI}{l^3}$
$r_3 = 1$

(c)

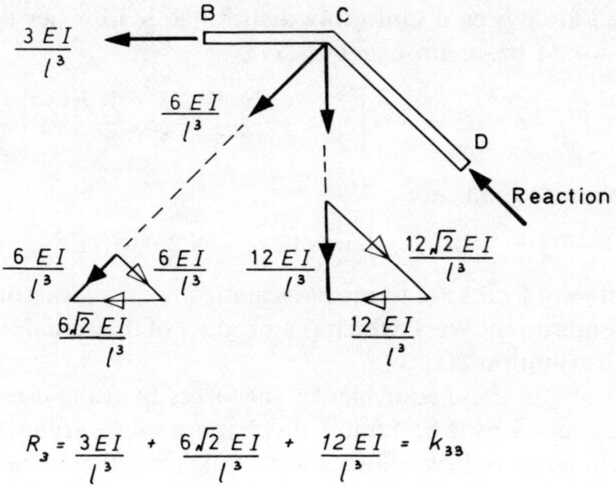

Fig. 2.22 Oblique force resolution

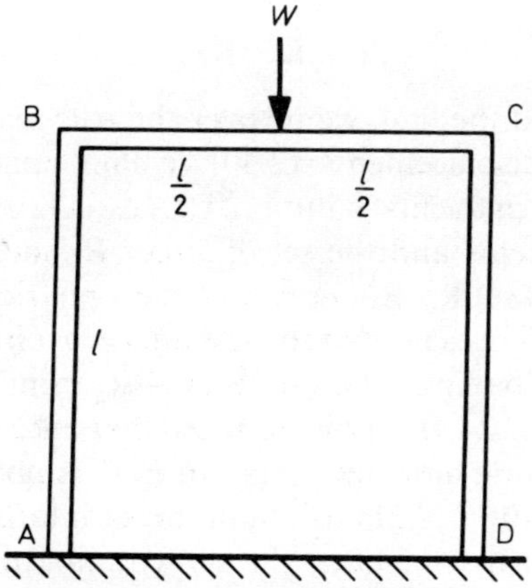

Fig. 2.23 Load between joints

be plausible anyway for, say, a uniformly distributed load. Instead we merely introduce the true load W into the right-hand side of the PVD together with the virtual displacement $\bar{v}(l/2)$, obtained using (2.21), in terms of the end displacements $\{\rho_1 \quad \rho_2 \quad \rho_3 \quad \rho_4\}$. The external virtual work turns out to be

$$-W\bar{v}(l/2) = -W\left[\frac{1}{2}\bar{\rho}_1 + \frac{l}{8}\bar{\rho}_2 + \frac{1}{2}\bar{\rho}_3 - \frac{l}{8}\bar{\rho}_4\right]$$

Consequently, the effect of the central load can be exactly simulated, when solving for joint displacements, by a set of joint forces, deduced from the above, as

$$\mathbf{R}_g = \{-W/2 \quad -Wl/8 \quad -W/2 \quad Wl/8\} \qquad (2.29a)$$

Had the applied force been a uniformly distributed load w per unit length, the external work would be, again using (2.21),

$$\int_0^l (-w)\bar{v}\,dz = -w\left[\frac{l}{2}\bar{\rho}_1 + \frac{l^2}{12}\bar{\rho}_2 + \frac{l}{2}\bar{\rho}_3 - \frac{l^2}{12}\bar{\rho}_4\right]$$

So this time the joint loads are

$$\mathbf{R}_g = \{-wl/2 \quad -wl^2/12 \quad -wl/2 \quad wl^2/12\} \tag{2.29b}$$

Such equivalent joint forces are termed *kinematically equivalent* forces since their evaluation depends on the work which is a product of the actual loading and the displacement distribution $\bar{v}(z)$.

If we wish to obtain these equivalent joint forces by using direct equilibrium arguments, then – not for the first time – much more subtle arguments have to be used. (Simple *statically equivalent* forces will not do since there is no static limit on the magnitude of the two equal and opposite end moments.) Suppose, firstly, that we artificially suppress *all* joint displacements in the loaded structure by applying a series of 'constraint' forces $-\mathbf{R}_0$. Now, secondly, solve the structure subjected to the set of forces $\mathbf{R}_0$ but with no other applied forces, that is, obtain

$$\mathbf{r} = \mathbf{K}^{-1}\mathbf{R}_0 \tag{2.30}$$

If to this solution we add the first, we recover the true loading since $-\mathbf{R}_0 + \mathbf{R}_0$ cancel; moreover the displacements (2.30) remain unaltered since the joint displacements were zero in the first solution. Equation (2.30) therefore represents the true joint displacements and the set of forces $\mathbf{R}_0$ must be the same as our kinematically equivalent set $\mathbf{R}_g$. But now we have a physical interpretation since $-\mathbf{R}_0$ are the 'fixed-end' forces needed to mainain zero joint displacements in the presence of the applied loading. The values of $-\mathbf{R}_g$ from equations (2.29a) and (2.29b) are shown in Fig. 2.24. It should be noted that once we allow applied loads between joints, the bending moment variation EIv'' is no longer linear and it is not possible to use equation (2.21) to obtain exact information between joints. Instead it is permissible to add back the self-equilibrating fixed-end moment distributions like Fig. 2.24 and so obtain exact solutions (see Problem 2.5). This correction is a privileged one and is only possible when an exact solution is

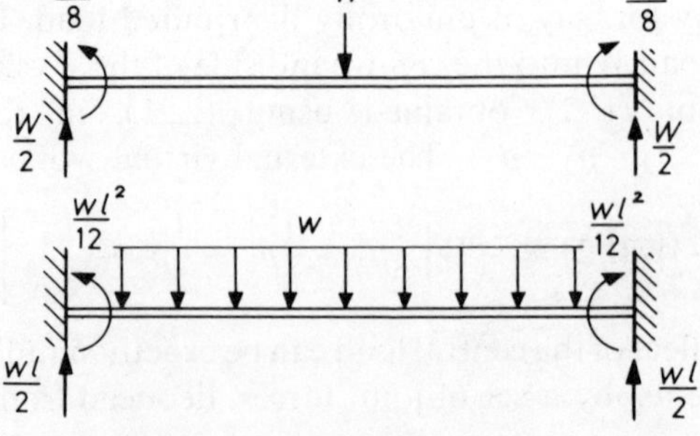

Fig. 2.24 Fixed-end forces

available to the 'fixed-end' problem. In the finite-element method of Chapter 5 we always use kinematically equivalent loads to solve the displacements and then have no alternative but to obtain the stresses and strains from the displacement functions, no matter how inaccurate they are.

2.8 Coordinate transformations

When analysing pinjointed frameworks it was found convenient to use local descriptions of position and displacement before inserting a bar into a framework with a global reference system. The same systematic approach is fruitful also in stiff-jointed frameworks, even though most of the component beams may be horizontal or vertical. Later we look at shells and general finite elements where such simple alignment is never possible. Thus, to place a beam into a framework, and assemble a global stiffness matrix, it is most convenient to evaluate the gth beam element stiffness matrix $\mathbf{k}'_g$ in local coordinates and displacements $\boldsymbol{\rho}'_g$, before rotating the coordinate system to align with the global, and convert $\boldsymbol{\rho}'_g$ to the usual $\boldsymbol{\rho}_g$. Then finally we identify $\boldsymbol{\rho}_g$ with the global list $\mathbf{r}$ in the usual way.

For clarity we resort to cartesian coordinates x, y, z and consider a single displacement $\boldsymbol{\rho} = \{\rho_x \quad \rho_y \quad \rho_z\}$ and corresponding force $\mathbf{P} = \{P_x \quad P_y \quad P_z\}$. If another set of orthoganal axes is x', y', z', then let a displacement be defined also in terms of these components $\boldsymbol{\rho}' = \{\rho'_x \quad \rho'_y \quad \rho'_z\}$ (see Fig. 2.25). Now *any*

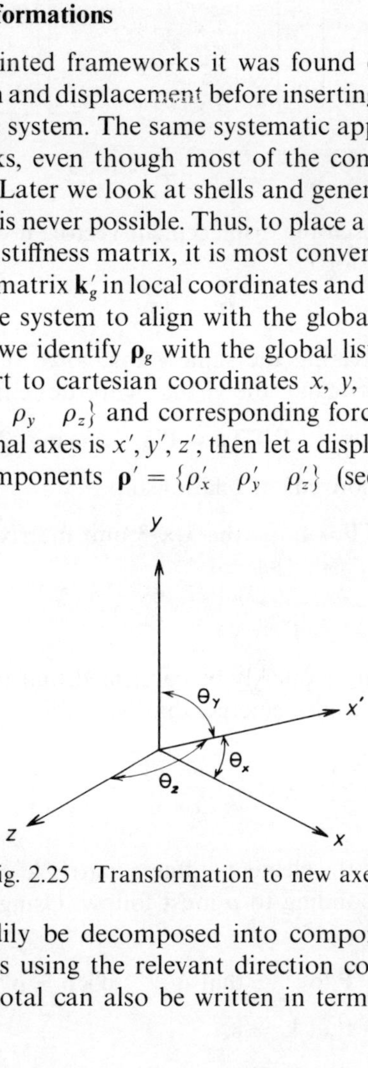

Fig. 2.25 Transformation to new axes

displacement can readily be decomposed into components in any system of orthogonal coordinates using the relevant direction cosines. So the particular component ρ'_x of the total can also be written in terms of the components of $\{\rho_x \quad \rho_y \quad \rho_z\}$.

Thus,

$$\rho'_x = \rho_x \cos\theta_x + \rho_y \cos\theta_y + \rho_z \cos\theta_z$$

To avoid introducing further symbols, we resort to writing the direction cosines as

$$\cos\theta_x = \frac{\partial x'}{\partial x}; \qquad \cos\theta_y = \frac{\partial x'}{\partial y}; \qquad \cos\theta_z = \frac{\partial x'}{\partial z} \qquad \left(\text{cf. } \frac{\partial n}{\partial x} \text{ in equation (1.57}\right)$$

Similarly, we can obtain the components ρ'_y and ρ'_z, so that

$$\boldsymbol{\rho}' = \begin{bmatrix} \rho'_x \\ \rho'_y \\ \rho'_z \end{bmatrix} = \begin{bmatrix} \dfrac{\partial x'}{\partial x} & \dfrac{\partial x'}{\partial y} & \dfrac{\partial x'}{\partial z} \\ \dfrac{\partial y'}{\partial x} & \dfrac{\partial y'}{\partial y} & \dfrac{\partial y'}{\partial z} \\ \dfrac{\partial z'}{\partial x} & \dfrac{\partial z'}{\partial y} & \dfrac{\partial z'}{\partial z} \end{bmatrix} \begin{bmatrix} \rho_x \\ \rho_y \\ \rho_z \end{bmatrix}$$

or

$$\boldsymbol{\rho}' = \mathbf{T}\boldsymbol{\rho} \tag{2.31}$$

We happen to have chosen the displacement vector; it could equally well have been the corresponding force

$$\mathbf{P}' = \mathbf{TP} \tag{2.32}$$

but we are simply describing the same thing, force or displacement, in two different ways; so in particular, the virtual work done must be the same:

$$\mathbf{P}'^{t}\bar{\boldsymbol{\rho}}' = \mathbf{P}^{t}\bar{\boldsymbol{\rho}} \qquad \text{or} \qquad \mathbf{P}^{t}\mathbf{T}^{t}\mathbf{T}\bar{\boldsymbol{\rho}} = \mathbf{P}^{t}\bar{\boldsymbol{\rho}} \qquad \text{using (2.31) and (2.32)}$$

We therefore have the convenient relationship

$$\mathbf{T}^{t}\mathbf{T} = \mathbf{I}, \qquad \text{the } 3 \times 3 \text{ unit matrix.}$$

or

$$\mathbf{T}^{-1} = \mathbf{T}^{t} \tag{2.33}$$

If (2.33) is expanded it may quickly be confirmed that the usual orthogonality properties of direction cosines emerge, that is:

$$\left(\frac{\partial x'}{\partial x}\right)^2 + \left(\frac{\partial x'}{\partial y}\right)^2 + \left(\frac{\partial x'}{\partial z}\right)^2 = 1; \qquad \frac{\partial x'}{\partial x}\frac{\partial y'}{\partial x} + \frac{\partial x'}{\partial y}\frac{\partial y'}{\partial y} + \frac{\partial x'}{\partial z}\frac{\partial y'}{\partial z} = 0, \qquad \text{etc.}$$

Once we have evaluated the element stiffness matrix $\mathbf{k}'$ corresponding to $\boldsymbol{\rho}'$, the conversion to $\mathbf{k}$ corresponding to $\boldsymbol{\rho}$ must follow. Using equivalence of virtual work:

$$\mathbf{P}^{t}\bar{\boldsymbol{\rho}} = \mathbf{P}'^{t}\bar{\boldsymbol{\rho}}', \qquad \text{that is,} \qquad \boldsymbol{\rho}^{t}\mathbf{k}^{t}\bar{\boldsymbol{\rho}} = \boldsymbol{\rho}'^{t}\mathbf{k}'^{t}\bar{\boldsymbol{\rho}}'$$

Using (2.31) and noting that $\mathbf{k}^{t} = \mathbf{k}$,

$$\boldsymbol{\rho}^{t}\mathbf{k}\bar{\boldsymbol{\rho}} = \boldsymbol{\rho}^{t}\mathbf{T}^{t}\mathbf{k}'\mathbf{T}\bar{\boldsymbol{\rho}}$$

or

$$\mathbf{k} = \mathbf{T}^{t}\mathbf{k}'\mathbf{T} \tag{2.34}$$

This transformation is reversible, for suppose we premultiply by $\mathbf{T}$ and postmultiply by $\mathbf{T}^{t}$, then $\mathbf{TkT}^{t} = \mathbf{TT}^{t}\mathbf{k}'\mathbf{TT}^{t}$

$$\text{or} \qquad \mathbf{T}^{t}\mathbf{kT} = \mathbf{k}' \qquad \text{noting (2.33)}$$

The congruent transformation we have met before (2.15) when we converted the natural stiffness k_N of a bar to the stiffness $\mathbf{k} = \mathbf{a}_N^t \mathbf{k}_N \mathbf{a}_N$. The same thing can be done again, but this time using coordinate transformation as the argument. For example, consider the case of the bar in Fig. 2.26. For *small* displacements the bar has no resistance to displacements ρ_2' and ρ_4', so

$$\mathbf{k}' = \frac{AE}{l}\begin{bmatrix} 1 & 0 & -1 & 0 \\ 0 & 0 & 0 & 0 \\ -1 & 0 & 1 & 0 \\ 0 & 0 & 0 & 0 \end{bmatrix}$$

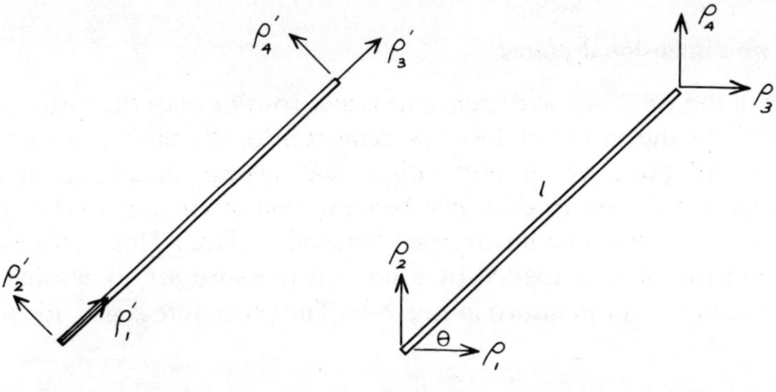

Fig. 2.26

In this case the transformation matrix for the two sets of displacement components at both ends is

$$\mathbf{T} = \begin{bmatrix} \cos\theta & \sin\theta & 0 & 0 \\ -\sin\theta & \cos\theta & 0 & 0 \\ 0 & 0 & \cos\theta & \sin\theta \\ 0 & 0 & -\sin\theta & \cos\theta \end{bmatrix}$$

Using (2.34),

$$\mathbf{k} = \frac{AE}{l}\begin{bmatrix} \cos^2\theta & \cos\theta\sin\theta & -\cos^2\theta & -\cos\theta\sin\theta \\ & \sin^2\theta & -\cos\theta\sin\theta & -\sin^2\theta \\ & & \cos^2\theta & \cos\theta\sin\theta \\ \text{symmetric} & & & \sin^2\theta \end{bmatrix}$$

which agrees with (2.16).

Rotation and moment vectors can be handled in the same way but it would be inappropriate to transform a beam stiffness matrix before we have established the stiffness under an axial-moment vector (torque). We have, however, solved one bending problem with an inclined member (Fig. 2.21), and here a word of caution is necessary. In a simple example it is perfectly easy to enforce the constraint that axial deformations are negligible – we merely equate axial displacements at both ends – but when assembling a large framework stiffness it is inconvenient to reduce the degrees of freedom in this way. Instead, it is usual to allow *all* beam displacements so that a single global displacement at the end of an inclined beam will consist partly of an axial and partly a bending deformation. An element stiffness will then contain a mixture of terms like AE/l which are very large compared with terms like EI/l^3. This can lead to loss of accuracy when using computers of small word-length, but for most present-day computers of even moderate size, the third to fifth significant figure presents no problems.

2.9 Two-dimensional plates

The use of the PVD in two dimensions is also routine once the assumptions have been made on the nature of the displacement field. We choose the bending of an arbitrary flat plate as an illustration, and taking advantage of our beam knowledge we choose to decouple bending and stretching so that the middle plane of the plate is our unstretched 'neutral surface'. This surface is the x–y plane, and the plate is loaded by a normal pressure $p(x, y)$ causing a normal displacement $w(x, y)$ as shown in Fig. 2.27. The procedure is now identical to the

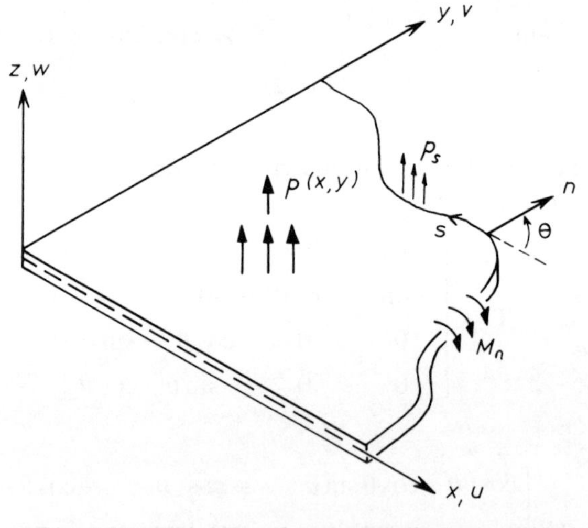

Fig. 2.27 Thin plate forces and displacements

previous beam theory, equation (2.17), assuming normals to the middle surface remain normal, after rotation of the surface by $\partial w/dx$ and $\partial w/\partial y$. So this time we have (cf. (2.17)):

$$u = -z\frac{\partial w}{\partial x}; \qquad v = -z\frac{\partial w}{\partial y}; \qquad w = w(x, y) \quad \text{only} \tag{2.35}$$

We will also take the material to be isotropic and assume no initial strains. These last two restrictions are not necessary but they do avoid excessive algebra. We can justifiably ignore the normal stresses σ_{zz} which are of order p, since they will be much smaller than the stresses σ_{xx}, σ_{yy}.

At this stage the physics of the problem is completely defined and the rest can be left to the PVD plus some algebraic acrobatics. In some ways this attractive aspect of virtual work can be dangerous in that its very simplicity may cloud the physics of the problem. A glance at the standard derivation of the plate equations[21] should convince the reader of this. Indeed, the plate-bending problem remained an elusive enigma for so many years that ultimately a cash prize was offered for its solution,[2] possibly a unique event in structural analysis. However, it does seem logical to accept the benefits of work arguments and be prepared to look at the delivered answers and check their physical meaning if necessary.

Returning now to (2.35), which we substitute into $\boldsymbol{\varepsilon} = \mathbf{D}^t\mathbf{u}$, we obtain just

$$\boldsymbol{\varepsilon} = \{\varepsilon_{xx} \quad \varepsilon_{yy} \quad \varepsilon_{xy}\} = \left\{ -z\frac{\partial^2 w}{\partial x^2} \quad -z\frac{\partial^2 w}{\partial y^2} \quad -2z\frac{\partial^2 w}{\partial x\,\partial y} \right\} \tag{2.36}$$

Putting $\sigma_{zz} = 0$, and inverting the strain–stress relationship (2.1),

$$\sigma_{xx} = \frac{\mathrm{E}}{1-\nu^2}(\varepsilon_{xx} + \nu\varepsilon_{yy}); \qquad \sigma_{yy} = \frac{\mathrm{E}}{1-\nu^2}(\varepsilon_{yy} + \nu\varepsilon_{xx});$$

$$\sigma_{xy} = \frac{\mathrm{E}}{2(1+\nu)}\varepsilon_{xy} \tag{2.37}$$

The PVD (1.69) is

$$\int_V \boldsymbol{\sigma}^t\bar{\boldsymbol{\varepsilon}}\,\mathrm{d}v = \int_A p(x,y)\bar{w}\,\mathrm{d}A + \int_{S_p} \left(m_s\frac{\partial \bar{w}}{\partial n} + p_s\bar{w} \right) \mathrm{d}s \tag{2.38}$$

where m_s and p_s are the known moments and edge shears per unit length, applied over some section S_p of the plate boundary. Substituting (2.36) and (2.37) into (2.38) and integrating over $-t/2 < z < t/2$, we obtain

$$D\int_A \left[\left(\frac{\partial^2 w}{\partial x^2} + \nu\frac{\partial^2 w}{\partial y^2} \right)\frac{\partial^2 \bar{w}}{\partial x^2} + \left(\frac{\partial^2 w}{\partial y^2} + \nu\frac{\partial^2 w}{\partial x^2} \right)\frac{\partial^2 \bar{w}}{\partial y^2} \right.$$
$$\left. + 2(1-\nu)\frac{\partial^2 w}{\partial x\,\partial y}\frac{\partial^2 \bar{w}}{\partial x\,\partial y} - p\bar{w} \right] \mathrm{d}A = \int_{S_p} \left(m_s\frac{\partial \bar{w}}{\partial n} + p_s\bar{w} \right) \mathrm{d}s \tag{2.39}$$

where $D = Et^3/12(1-\nu^2)$ is known as the flexural rigidity of the plate and is analogous to EI in beams. As equation (2.39) is a work statement we should expect the coefficients of virtual curvature to be moments per unit length; that is, about the y and x axes,

$$M_{yy} = D\left(\frac{\partial^2 w}{\partial x^2} + \nu\frac{\partial^2 w}{\partial y^2} \right); \qquad M_{xx} = D\left(\frac{\partial^2 w}{\partial y^2} + \nu\frac{\partial^2 w}{\partial x^2} \right) \tag{2.40}$$

Therein lies the danger of misinterpreting the physics, since the similar coefficient of the virtual rate of twist, $\partial^2\bar{w}/\partial x\,\partial y$, is not the twisting moment, which is in fact only half of this amount,

$$M_{xy} = D(1-v)\frac{\partial^2 w}{\partial x\,\partial y} \tag{2.41}$$

In fact, like the complementary shear stresses whose resultant they are, there are *two* 'complementary' twisting moments contributing to (2.39) on faces normal to both x and y axes. This fact might have been deduced when we realize that there are also two identical rates of twist,

$$\frac{\partial}{\partial x}\left(\frac{\partial\bar{w}}{\partial y}\right) = \frac{\partial^2\bar{w}}{\partial x\,\partial y} = \frac{\partial}{\partial y}\left(\frac{\partial\bar{w}}{\partial x}\right)$$

Equation (2.39) is the simplest form of the PVD which we later use in solving plate problems by approximate methods, since the exact solution involves a fourth-order partial differential equation, as we now demonstrate. As usual we must convert all virtual terms to $\bar{w}$ in (2.39) using the Gauss equation (1.71) again. The details are tedious, so we summarize for the impatient. Taking the first term in (2.39), we apply (1.71), replacing the 'stress term' by the moment $D(\partial^2 w/\partial x^2 + v\partial w^2/\partial y^2)$ and the term $\mathbf{D}^{\mathrm{t}}\bar{\mathbf{u}}$ by $\partial/\partial x(\partial\bar{w}/\partial x)$, so that Gauss converts it to $\partial\bar{w}/\partial x$. The process is repeated to convert $\partial\bar{w}/\partial x$ to $\bar{w}$ as required. The other second derivations of $\bar{w}$ are treated in the same way. The consequent boundary terms $[\bar{\mathbf{u}}^{\mathrm{t}}(\mathbf{D}n)\boldsymbol{\sigma}]$ contain expressions like $(\partial\bar{w}/\partial x)(\partial n/\partial x)$ which have to be further processed since the degrees of freedom on an edge are $\bar{w}$ and $\partial\bar{w}/\partial n$. We therefore use the transformation (2.31),

$$\begin{bmatrix} \dfrac{\partial w}{\partial x} \\[2mm] \dfrac{\partial w}{\partial y} \end{bmatrix} = \begin{bmatrix} \dfrac{\partial x}{\partial n} & \dfrac{\partial x}{\partial s} \\[2mm] \dfrac{\partial y}{\partial n} & \dfrac{\partial y}{\partial s} \end{bmatrix} \begin{bmatrix} \dfrac{\partial w}{\partial n} \\[2mm] \dfrac{\partial w}{\partial s} \end{bmatrix}$$

Finally, $\partial\bar{w}/\partial s$ has to be integrated yet again to $\bar{w}$. Collecting all terms and putting the direction cosines

$$\frac{\partial x}{\partial n} = \cos\theta; \quad \frac{\partial y}{\partial s} = \cos\theta; \quad \frac{\partial x}{\partial s} = -\sin\theta; \quad \frac{\partial y}{\partial n} = \sin\theta \qquad \text{(see Fig. 2.27)}$$

the PVD becomes:

$$\begin{aligned} &\int_A [D\nabla^2(\nabla^2 w) - p]\bar{w}\,\mathrm{d}A \\ &\quad + \int_{S_p} D\left[\left(\frac{\partial^2 w}{\partial x^2} + v\frac{\partial^2 w}{\partial y^2}\right)\cos^2\theta + \left(\frac{\partial^2 w}{\partial y^2} + v\frac{\partial^2 w}{\partial x^2}\right)\sin^2\theta \right. \\ &\quad \left. + 2(1-v)\frac{\partial^2 w}{\partial x\,\partial y}\sin\theta\cos\theta - m_s\right]\frac{\partial\bar{w}}{\partial n}\mathrm{d}s \end{aligned}$$

$$- \int_{S_p} D\left[\left(\cos\theta\frac{\partial}{\partial x} + \sin\theta\frac{\partial}{\partial y}\right)\nabla^2 w + (1-\nu)\frac{\partial}{\partial s}\left\{\left(\frac{\partial^2 w}{\partial y^2} - \frac{\partial^2 w}{\partial x^2}\right)\sin\theta\cos\theta + \frac{\partial^2 w}{\partial x\,\partial y}\cos 2\theta\right\} + p_s\right]\bar{w}\,\mathrm{d}s = 0 \tag{2.42}$$

where $\nabla^2 = \partial^2/\partial x^2 + \partial^2/\partial y^2$ is the Laplace or harmonic operator.

To exclude work done by support reactions as usual, we have to put $\bar{w} = \partial\bar{w}/\partial n = 0$ on $S_u = S - S_p$.

The first integral in (2.42) contains the differential equation of equilibrium, commonly called the biharmonic equation:

$$\nabla^2(\nabla^2 w) = \nabla^4 w = p/D \tag{2.43}$$

The boundary integrals vanish on a clamped edge, as we said above, but if the edge is free to rotate and displace, then $\partial\bar{w}/\partial n$ and $\bar{w}$ are nonzero and their coefficients must vanish. These are the expected equilibrium boundary conditions, so we deduce that the terms in parentheses are the edge moments and shears in equilibrium with m_s and p_s. Neither of these lengthy expressions are easy to establish directly; the shear force in particular is so subtle that it is only true as a form consistent with the kinematic assumptions of (2.35). For a full treatment of the so-called 'Kirchhoff shear' boundary condition, see the *Theory of Plates and Shells.*[21] Solving (2.43) subject to the above boundary conditions is a formidable problem and, in the quoted reference, solutions are restricted to simple geometrical cases like rectangular plates, axisymmetrical plates, and so on. Plate bending is obviously a candidate for our later approximate methods.

2.10 The twisting of open tubes

Returning thankfully to one variable, we will now consider the behaviour of a uniform open-section thin-walled tube subjected to a torque. Open sections will be seen to be pretty inefficient in resisting torsion (closed sections are examined in the next chapter) but are frequently called upon to do so. (Most beams are an open-section simply because that is easier to fabricate, extrude, roll, etc.) The tube is twisted but it is assumed that the section profile is undistorted, so that displacements are functions of one variable only. This is a reasonable assumption for tubes of beam-like proportions such as channels and 'I' sections. Tubes whose cross-sectional dimensions are very much greater than their thickness (aircraft wing boxes and box-girder bridges) may be protected from inevitable section distortions by inserting stiffening ribs or diaphragms, but it is rare to have such sizes as open-sections. Figure 2.28 shows a general open tube subjected to a distributed torque $\partial T/\partial z$ per unit length, and a concentrated torque T_l at the end $z = l$. The axis Oz is the axis about which the tube twists, and according to Maxwell's reciprocal theorem (see Section 3.3) the axis which does not translate when a torque is applied, is also the axis through which a (shear) force will

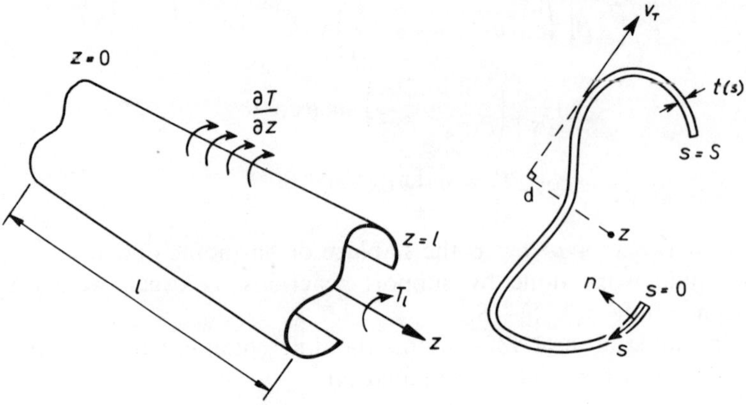

Fig. 2.28 An open tube

produce no twist. This point in the cross-section is known as the *shear centre* and the latter definition is the most convenient one for actually finding it. We simply apply a shear, solve for the shear stresses around the tube using simple beam theory if appropriate, and then find where their resultant acts (see the two examples following).

Since no cross-sectional distortion is admitted, the tangential displacement v_T at any point around the section is given by

$$v_T = d.\theta \tag{2.44}$$

where d is the normal distance of the local tangent from the z-axis (Fig. 2.28). This, of course, is a kinematic assumption to which we are restricted when using the PVD. The question now arises: how is the torque resisted? Well, we have shown (2.36) using only kinematic reasoning, that a thin flat plate may have a shear-strain distribution through its thickness given by $\varepsilon_{xy} = -2z\,\partial^2 w/\partial x\,\partial y$, as seen in Fig. 2.29. Here, then, is a mechanism for resisting torque about the x-axis in Fig. 2.29. We may identify, $\partial w/\partial y$ as the 'rotation' or 'twist' and the shear strains are therefore proportional to the rate of change of twist $\partial/\partial x(\partial w/\partial y)$. The above (Saint-Venant) shears are obviously parallel to the two surfaces $z = \pm t/2$, as they must be since there can be no component σ_{xz} $(= \sigma_{zx})$ on the free surface itself. In

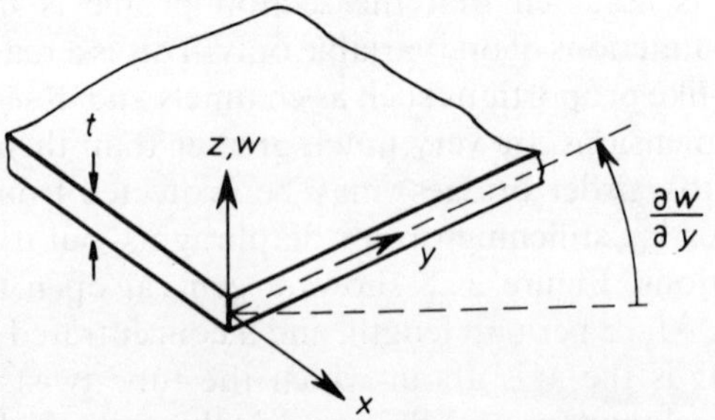

Fig. 2.29 Twist and shear strains

view of this, suppose we now consider the *curved* section of Fig. 2.28, and assume that the local shear strains are similarly distributed through the thickness. If we write

$$\varepsilon_{sz} = -2n\frac{\partial \theta}{\partial z} \tag{2.45}$$

and provided the surfaces $n = \pm t/2$ are virtually parallel to $n = 0$, then these shears will also be parallel to the surface except at the edges $s = 0$ and $s = S$. Near these edges, therefore, there will be an error in assumption (2.45), but if $S \gg t$ the error in the total virtual work will be insignificant since these edge regions are tiny ($t \ll S$) and there is no reason to suppose that the stresses in these regions are larger than anywhere else. Notice this is not saying that these edge shears do not contribute to the equations of equilibrium – in fact they may resist one half of the applied torque. But as we are not applying equilibrium arguments directly, we are concerned only in being able to ignore the virtual work in a tiny region. Having established our strain-displacement field in (2.45), we could go no further but to turn the PVD handle. If we did this we would obtain the classical Saint-Venant solution[16] for thin-walled tubes; however, let us look further at the possibility of further strains, like ε_{zz}. We examine first, as did Wagner[18] in 1929, a kinematic consequence of (2.45), namely that there is no shear strain ε_{sz} along the median line $n = 0$. From the compatibility equations $\boldsymbol{\varepsilon} = \mathbf{D}^{\mathrm{t}}\mathbf{u}$, we have

$$\varepsilon_{sz} = \frac{\partial w}{\partial s} + \frac{\partial v_T}{\partial z} = 0 \tag{2.46}$$

where $w(s, z)$ is the axial displacement. This is sometimes referred to as the 'warping' displacement since it varies with s and so represents a possible departure from plane deformation. It is because $w(s, z)$ may also be a function of z, that axial strains $\varepsilon_{zz} = \partial w/\partial z$ may also arise. On using (2.44), the above equation becomes

$$\frac{\partial w}{\partial s} = -d\theta'$$

or

$$w(s, z) - w_0(0, z) = -\int_0^S d\,\mathrm{d}s\theta'(z)$$

where $w_0(z)$ is the axial displacement at the edge $s = 0$. The integral $\int d\,\mathrm{d}s$ is a known function of s which we cannot evaluate since it depends on the shape of the cross-section. We just abbreviate it as

$$\hat{w}(s) = \int_0^S d\,\mathrm{d}s, \qquad \text{or} \qquad w(s, z) = w_0(z) - \hat{w}(s)\theta'(z) \tag{2.47}$$

$\hat{w}$ clearly represents a warping per unit rate of twist. Both $\theta'(z)$ and $w_0(z)$ are as yet unknown functions of z, in which case we do get axial strains

$$\varepsilon_{zz} = \frac{\partial w}{\partial z}(s, z) = w_0' - \hat{w}(s)\theta'' \tag{2.48}$$

Since the tube is thin, the variation in w_0 and $\hat{w}$ across the thickness can be ignored. However, it is fairly obvious (from equilibrium equation (1.48)) that if σ_{zz} exists then so must σ_{sz} and consequently ε_{sz}, and this appears to violate our premise that $\varepsilon_{sz} = 0$ on the median line. However, it can be argued that these direct and shear stresses are of the type produced in beam bending – as opposed to Saint-Venant behaviour – and we should therefore be able to ignore these shear strains for *slender tubes*, that is, provided $l \gg S$. We can of course always check the validity of this conjecture afterwards when we have a solution. Our PVD, assuming Hooke's law as usual, then becomes

$$\int_V [G\varepsilon_{sz}\bar{\varepsilon}_{sz} + E\varepsilon_{zz}\bar{\varepsilon}_{zz}]\,\mathrm{d}v = \int_0^l \left(\frac{\partial T}{\partial z}\mathrm{d}z\right)\bar{\theta} + T_l\bar{\theta}(l)$$

Substituting (2.45) and (2.48), we get

$$\int_{z=0}^{l}\int_{s=0}^{S}\int_{-t/2}^{t/2} [G4n^2\theta'\bar{\theta} + E(w_0' - \hat{w}\theta'')(\bar{w}_0' - \hat{w}\bar{\theta}'')]\,\mathrm{d}n\,\mathrm{d}s\,\mathrm{d}z$$

$$= \int_0^l \frac{\partial T}{\partial z}\bar{\theta}\,\mathrm{d}z + T_l\bar{\theta}(l)$$

or on integrating with respect to n and s,

$$\int_0^l \left[GJ\theta'\bar{\theta}' - E(Hw_0' - Q\theta'')\bar{\theta}'' + E(Aw_0' - H\theta'')\bar{w}_0' - \frac{\partial T}{\partial z}\bar{\theta}\right]\mathrm{d}z - T_l\bar{\theta}(l) = 0 \tag{2.49}$$

where the various section properties come out naturally as

$$J = \frac{1}{3}\int_0^S t^3\,\mathrm{d}s; \qquad A = \int_0^S t\,\mathrm{d}s; \qquad H = \int_0^S \hat{w}t\,\mathrm{d}s; \qquad Q = \int_0^S \hat{w}^2 t\,\mathrm{d}s$$

Since $\bar{w}_0'$ is arbitrary, we have from (2.49),

$$w_0' = \frac{H}{A}\theta'' \qquad \text{or} \qquad w_0 = \frac{H}{A}\theta' \tag{2.50}$$

so (2.47) now becomes

$$w(s, z) = \left[\frac{H}{A} - \hat{w}(s)\right]\theta' \tag{2.51}$$

This result is capable of a simple interpretation when we put

$$\sigma_{zz} = E\partial w/\partial z = E[H/A - \hat{w}]\theta''$$

Therefore the *total axial force* is given by

$$\int_0^S \sigma_{zz}t\,\mathrm{d}s = E\theta''\left[\frac{H}{A}A - \int_0^S \hat{w}t\,\mathrm{d}s\right] = 0 \qquad \text{as it should be}$$

Now substituting (2.50) back in to the PVD (2.49) and converting $\bar{\theta}''$ to $\bar{\theta}$ in the usual way, we finish with

$$\int_0^l \left(E\Gamma\theta'''' - GJ\theta'' - \frac{\partial T}{\partial z} \right) \bar{\theta}\, dz + [E\Gamma\theta''\bar{\theta}' - E\Gamma\theta'''\bar{\theta} + GJ\theta'\bar{\theta}]_0^l - T_l\bar{\theta}(l) = 0 \quad (2.52)$$

where $\Gamma = Q - H^2/A$ is another section constant known as the 'torsion-bending' constant,[19] and values of which are tabulated for various sections.[22]

We extract from (2.52) the governing differential equation

$$E\Gamma\theta'''' - GJ\theta'' = \frac{\partial T}{\partial z} \quad (2.53)$$

for which we have to provide four boundary conditions at $z = 0$ and l. As usual, these may be of two types:

(a) *Entirely kinematic.* $\theta = 0$ (no twist) and $\theta' = 0$ (fully clamped, $w(s, z) = 0$, noting (2.51)). In this case $\bar{\theta}$ and $\bar{\theta}'$ are also zero and we have satisfied (2.52) completely.

(b) *Indirect equilibrium conditions.* Thus if θ is free at, say, $z = l$, $\bar{\theta}(l) \neq 0$, and from (2.52)

$$T_l = GJ\theta' - E\Gamma\theta''' \quad (2.54)$$

We deduce this must be an equation of torque equilibrium. If θ' is also free, then from the coefficient of $\bar{\theta}'(l)$ we have $E\Gamma\theta''(l) = 0$. This latter condition is not quite so obvious until we note (2.51) and (2.48), and then it becomes clear that if $w(s, z)$ (proportional to θ') is unrestricted there can be no axial stresses σ_{zz} (proportional to θ'').

It is now possible to see under what circumstances the Saint-Venant theory, $\theta' = T/GJ$, is adequate and direct stresses may be ignored. The rate of twist θ' must be a constant and so therefore must be the torque. Also the warping must be unrestricted, for if $\theta' = 0$ at a clamped end then θ' will be a function of z even with constant torque, and then the term $E\Gamma\theta''''$ becomes necessary in (2.53). The direct stresses σ_{zz} are often known as *axial constraint* stresses for this reason. It will be helpful to look at a simple example to demonstrate the behaviour of these constraint stresses.

Example: Simple 'I' beam

Firstly we solve the problem by the crude intuitive approach. We recognize that, when twisted through $\theta(z)$ about the centre axis, the two flanges will move a small horizontal distance θb in opposite directions – Fig. 2.30 greatly exaggerates this deformation. If the end $z = 0$ is firmly clamped, then the two flanges can behave like cantilevers and develop shear forces F given by

$$F = EI\frac{\partial^3}{\partial z^3}(\theta b)$$

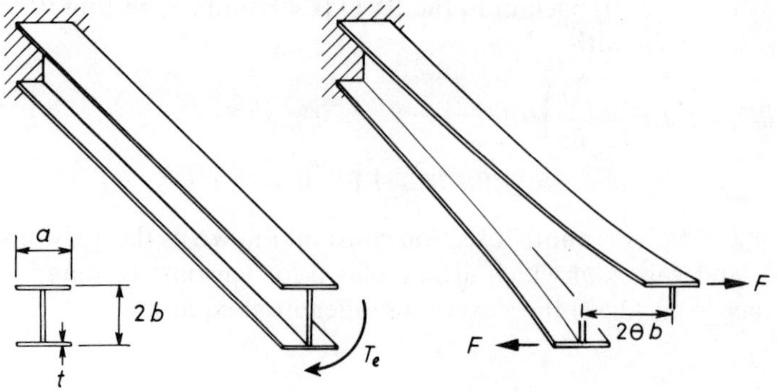

Fig. 2.30 Differential bending

where EI is the flexural rigidity of each flange about its centre; thus

$$F = E\frac{ta^3}{12}b\theta'''$$

The torque resisted is therefore $F2b$ and we have

$$T = E\Gamma\theta''' \qquad \text{where } \Gamma = tb^2a^3/6$$

This mechanism of torsion-bending is sometimes known as differential bending for obvious reasons in this example (see also Problem 2.8), but we draw attention to some shortcomings of this simple approach. The two-flange model is rather simpler than most open-section beams, and even here we have been able to ignore the web which passes through the shear centre and is not going to support the torque. But suppose we move the web to one side of the section: the above differential-bending model would predict the same answer. Clearly a channel has to be solved by equation (2.53), etc.

Example (*Fig. 2.31*)

The first step is to find the shear centre, so we apply a force F through it (no twist), solve the beam shears, and find their resultant. The force F produces bending stresses $\sigma_{zz} = My/I$ which are related to the shear stresses in, say, the upper flange by equation (1.48),

$$\frac{\partial\sigma_{zz}}{\partial z} + \frac{\partial\sigma_{zx}}{\partial x} = 0; \qquad \text{or} \qquad \frac{\partial\sigma_{zx}}{\partial x} = -\frac{\partial M}{\partial z}\frac{y}{I} = -\frac{Fa}{I}$$

Thus $\sigma_{zx} = Fa(a - x)/I$, noting that σ_{zx} must be zero on the free edge $x = a$. Integrating over the whole flange, and putting $I = t8a^3/12 + 2ata^2$, the total force in the flange turns out to be $3F/16$. The same force occurs in the lower flange in the opposite sense, and the force in the vertical web must be F. Thus, taking moments about the shear centre,

$$Fh = \frac{3F}{16}2a, \qquad \text{or} \qquad \frac{h}{a} = \frac{3}{8}$$

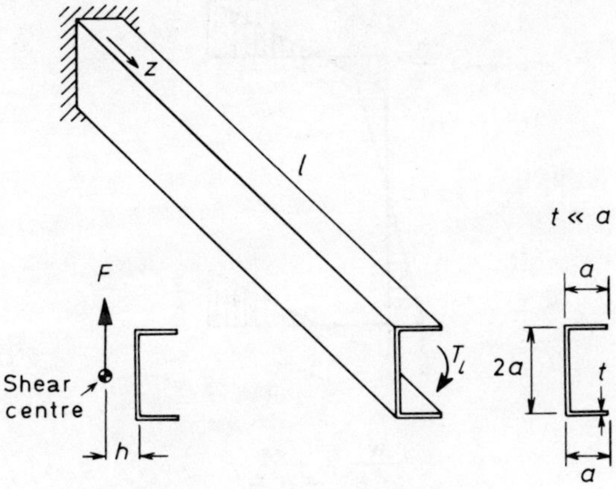

Fig. 2.31 Example

The next step is the normalized warping distribution $\hat{w}(s) = \int d \, \mathrm{d}s$

From A to B: $\hat{w} = as$

From B to C: (restarting s at B) $\hat{w} = a^2 - 3as/8$

From C to D: $\hat{w} = a^2 - 6a^2/8 + as = a^2/4 + as$

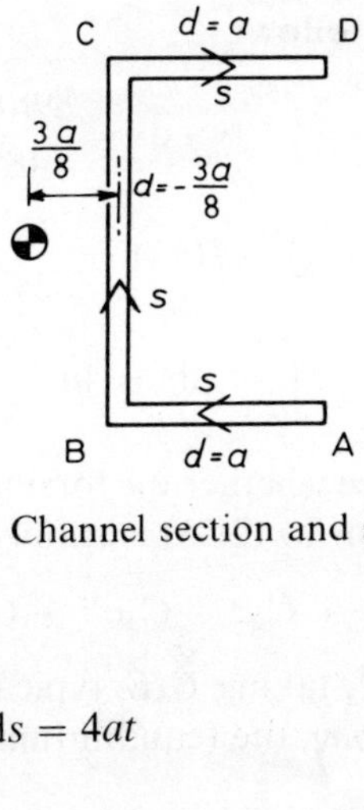

Fig. 2.32 Channel section and shear centre

Following (2.49), $A = \int_0^{4a} t \, \mathrm{d}s = 4at$

$$H = \int_0^{4a} \hat{w}t \, \mathrm{d}s = t[as^2/2]_0^a + t[a^2 s - 3as^2/16]_0^{2a} + t[a^2 s/4 + as^2/2]_0^a = 5ta^3/2$$

Therefore, from (2.51), $w/\theta' = H/A - \hat{w} = \dfrac{5a^2}{8} - \hat{w}$

This variation is sketched in Fig. 2.33.

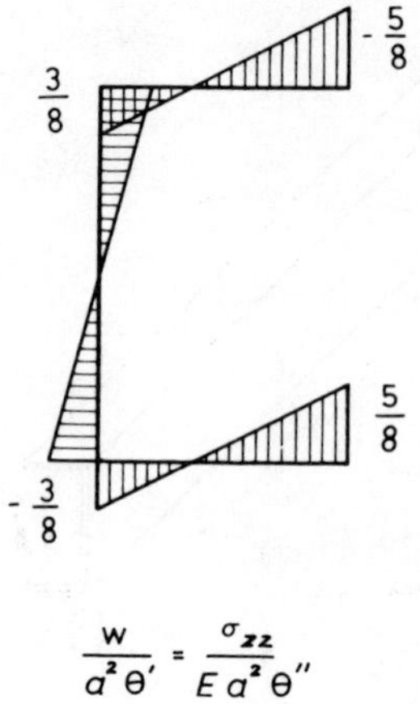

Fig. 2.33

We recall that $\sigma_{zz} = Ew'$, so Fig. 2.33 also shows the variation in axial constraint stress around the section. There is clearly no resultant axial force. It is also apparent that both flanges are undergoing differential bending like the I-beam, but this time not bending about their centres; the web is also bending about its centre. It is a curious fact that once we drop simple beam theory and allow $w(s, z)$ to depart from plane deformation, the component parts of the section still choose to deform in a beam-like fashion.

The section constants now follow:

We find
$$Q = \int_0^{4a} \hat{w}^2 t \, \mathrm{d}s = \frac{89ta^5}{48}$$

$$\Gamma = Q - H^2/A = \frac{7ta^5}{24}$$

$$J = \int_0^{4a} t^3 \, \mathrm{d}s = 4at^3$$

Now we can solve (2.53) and see whether the torsion-bending behaviour is of any significance at all. The solution to (2.53) is simply

$$\theta = C_1 + C_2 z + C_3 \mathrm{e}^{\alpha z} + C_4 \mathrm{e}^{-\alpha z}$$

where $\alpha^2 = GJ/E\Gamma = 36t^2/7a^4$, taking G/E, typically, as 3/8.

Assuming the tube is *very long*, the (equilibrium) boundary conditions at the tip, from (2.52) are:

$$E\Gamma\theta'' = 0; \qquad \therefore \quad C_3 = 0$$

$$GJ\theta' - E\Gamma\theta''' = T_l; \qquad \therefore \quad C_2 = T_l/GJ$$

The kinematic boundary conditions at $z = 0$ are:

$$\theta = 0; \qquad \therefore \quad C_1 = 0$$

$$w = \theta' = 0; \qquad \therefore \quad C_2 - \alpha C_4 = 0, \qquad \therefore \quad C_4 = T_l/GJ\alpha$$

Thus $$\theta(z) = \frac{T_l}{GJ}\left(z + \frac{e^{-\alpha z}}{\alpha}\right)$$

In (2.54) the Saint-Venant contribution to torque is $GJ\theta' = T_l(1 - e^{-\alpha z})$. The torsion-bending contribution is $-E\Gamma\theta''' = T_l e^{-\alpha z}$ (see Figure 2.34). The torsion-bending resistance clearly dominates near the constrained end $z = 0$. Moreover, if

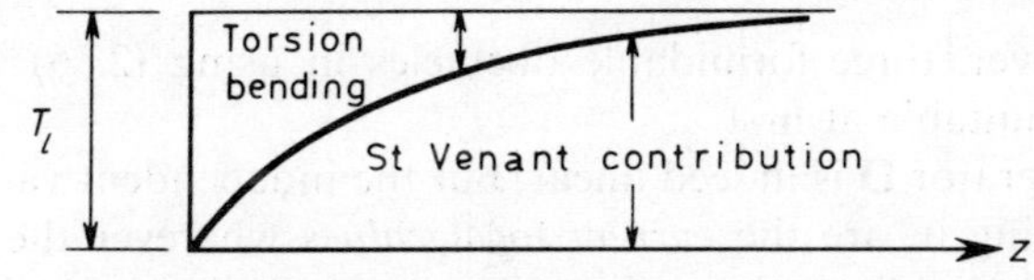

Fig. 2.34 Saint Venant/torsion-bending for long tube

we examine the point at which $e^{-\alpha z} \ll 1$, say $\alpha z = 4$, then the torsion-bending is only negligible at a distance from the constraint of order $4/\alpha$, and

$$\frac{4}{\alpha} = 4/[36t^2/7a^4]^{1/2} = 1.76a(a/t)$$

Thus in a thin-walled tube, where a/t may be large, the torsion-bending effects do persist for a considerable distance before simple Saint-Venant behaviour takes over.

2.11 Gross deformations and buckling

In deriving the equations of compatibility, $\boldsymbol{\varepsilon} = \mathbf{D}^t\mathbf{u}$, we pointed out that strains and displacements must be small, otherwise the elements of the operator **D** would not be linear. Unfortunately, many structures are excessively flexible and displacements may occur which are sufficiently large to cause significant changes in the original geometry.

If we secure a horizontal cable firmly at both ends and then hang a weight on it, a large displacement will *have* to occur before the cable tension can develop any useful components at the 'kink' induced by the applied load.

As another example, if we subject a slender rod to an axial compressive load, we ignore the changes in geometry at our peril. The rod will not just compress, but will bend, and may then develop bending deformations so large that the enhanced bending moments, then produced by the end load, lead to further catastrophic bending deformations. This form of instability, or buckling, has been understood since Euler's time, and we even call such rods 'struts' to emphasize that we know what is going on. Similar buckling can, of course, occur also in plates and shells, as well shall see later.

To apply the PVD, when gross deformations have taken place, is quite permissible provided that we consider *small incremental virtual displacements* $\delta\bar{\mathbf{u}}$ from the grossly deformed position. Then we are entitled to apply linear compatibility equations to the small virtual strains

$$\delta\bar{\boldsymbol{\varepsilon}} = \mathbf{D}^t\,\delta\bar{\mathbf{u}} \tag{2.55}$$

From now on in dealing with large displacements we will drop the bar from the virtual displacements $\delta\bar{\mathbf{u}}$, since they will clearly be distinguishable from the real (large) displacements $\mathbf{u}$. The PVD (1.69) now becomes

$$\int_V \boldsymbol{\sigma}^{\mathrm{t}}\, \delta\boldsymbol{\varepsilon}\, \mathrm{d}v = \int_V \mathbf{p}_v^{\mathrm{t}}\, \delta\mathbf{u}\, \mathrm{d}v + \int_S \mathbf{p}_s^{\mathrm{t}}\, \delta\mathbf{u}\, \mathrm{d}s \tag{2.56}$$

There are, however, three formidable obstacles in using (2.56) which appear almost unsurmountable at first.

Firstly, the operator $\mathbf{D}$ is indeed linear, but the independent variables (x, y, z) used when defining it, are the *current local values* wherever the material has moved to; and the additional small strains $\delta\boldsymbol{\varepsilon}$ should be defined as fractional changes with respect to the current state and not the initial state. Since the current state is as yet unknown, this is embarrassing.

Secondly, if we wish to use (2.56) to solve for the true displacements, we will have to convert stress to strain first, and the stress–strain relationship may have to be re-examined if the strains are large. Fortunately, we are usually spared this complication since, as we have said before, most engineering materials have outstretched their usefulness before their strain is even 1 per cent, so it is permissible to overcome this obstacle by *assuming small strains* while admitting that large displacements may have occurred in the flexible structure.

Finally, even if we can assume the strain to be small, we will still have to be able to find the *true strain* $\boldsymbol{\varepsilon}$ in terms of the true displacements $\mathbf{u}$. This nonlinear relationship is usually too difficult to derive in most structures. We have no alternative but to construct the nonlinear history numerically by applying small successive increments of load $\delta\mathbf{p}_v$, $\delta\mathbf{p}_s$, and solve for the consequent $\delta\mathbf{u}$ at each stage. Thus the increments $\delta\boldsymbol{\varepsilon} = \mathbf{D}^{\mathrm{t}}\,\delta\mathbf{u}$ may be linearized, but the total nonlinear $\mathbf{u} = \sum \delta\mathbf{u}$ is eventually accumulated. This is the technique used in Chapter 5.

There are a few simple but important classes of problems in which we can find the nonlinear compatibility relationship, say symbolically,

$$\boldsymbol{\varepsilon} = f(\mathbf{u}) \tag{2.57}$$

If this can be established, then it is often easier not to find the current value of $\mathbf{D}$ in (2.55), but to form directly

$$\delta\boldsymbol{\varepsilon} = \frac{\partial f}{\partial u}\,\delta u \qquad \text{etc.}$$

Consider, for example, the two-bar frame of Fig. 2.35, where the rise h_0 of the pinned joint is small compared to L. This shallowness ensures that even a small movement from h_0 to h means a significant change in geometry. Normally in this sort of problem we would apply $\delta\boldsymbol{\varepsilon} = \mathbf{D}^{\mathrm{t}}\,\delta\mathbf{u}$ in the form, for each bar, of

$$\delta\varepsilon = \frac{\delta h \sin\theta}{l} \tag{2.58}$$

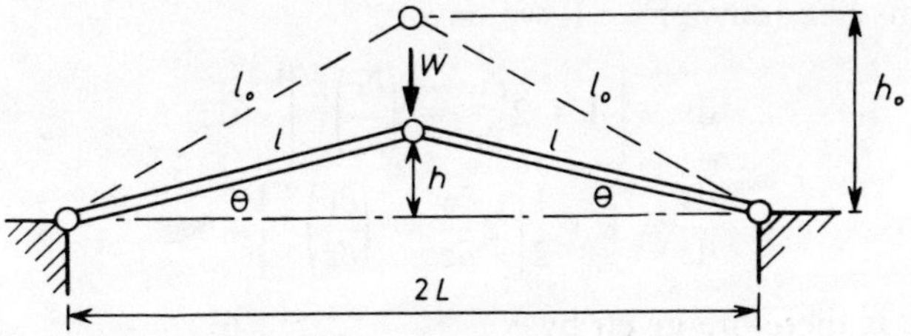

Fig. 2.35 A shallow frame

But θ is not constant, which is what we meant by the *current* value of **D**. However, in this example it is quite simple to write the complete unabridged compatibility relationship:

$$\varepsilon = \frac{l - l_0}{l_0} = \frac{1}{l_0}[(L^2 + h^2)^{1/2} - (L^2 + h_0^2)^{1/2}]$$

If we expand $(L^2 + h^2)^{1/2} = L(1 + h^2/L^2)^{1/2} = L(1 + \frac{1}{2}h^2/L^2 + \ldots)$ and take advantage of $h/L \ll 1$, we may put

$$\varepsilon = \frac{(h^2 - h_0^2)}{2L^2} \tag{2.59}$$

Then immediately,

$$\delta\varepsilon = \frac{h}{L^2}\delta h \tag{2.60}$$

This is the same form as (2.58) once we recognize that $\sin\theta = h/l \simeq h/L$ for $h/L \ll 1$.

It merely remains to substitute into the PVD (2.56) the value of $\delta\varepsilon$ from (2.60), $\sigma = E\varepsilon$, and finally ε from (2.59) to obtain, as a coefficient of δh, the equation giving h in terms of W. We leave this exercise to the reader (see Problem 2.11*) and proceed instead to a more interesting version in Fig. 2.36, the shallow arch. The shallow pin-ended arch, of intial shape $v_0(z)$, has bending as well as extensional deformation. To evaluate the extensional strains, consider firstly an element $\mathrm{d}z$ on the z-axis, which is displaced by $w(z)$ and $v(z)$. We can allow for an initial displacement $v_0(z)$ later. The new length is given by

$$\mathrm{d}s = [(\mathrm{d}z + \mathrm{d}w)^2 + (\mathrm{d}v)^2]^{1/2}$$

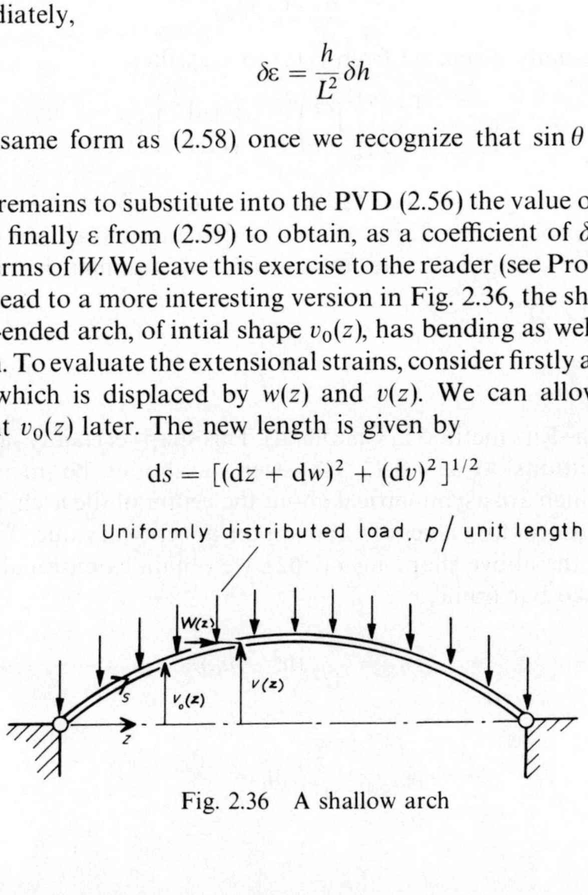

Fig. 2.36 A shallow arch

Ignoring terms like $(\partial w/\partial z)^2 \ll 1$, we have

$$\mathrm{d}s = \left[1 + 2\frac{\partial w}{\partial z} + \left(\frac{\partial v}{\partial z}\right)^2\right]^{1/2} \mathrm{d}z$$

$$\therefore \quad \frac{\mathrm{d}s}{\mathrm{d}z} = 1 + \frac{1}{2}\left[2\frac{\partial w}{\partial z} + \left(\frac{\partial v}{\partial z}\right)^2\right] + \cdots$$

The strain ε_{ss} is therefore given by

$$\varepsilon_{ss} = \frac{\mathrm{d}s - \mathrm{d}z}{\mathrm{d}z} = \frac{\mathrm{d}s}{\mathrm{d}z} - 1 \simeq \frac{\partial w}{\partial z} + \frac{1}{2}\left(\frac{\partial v}{\partial z}\right)^2 \tag{2.61}$$

We now capitalize on the shallowness of the arch, and surmise that the induced axial force produces a constant stress, and therefore a constant strain if the material is linear. Thus ε_{ss} ($\simeq \varepsilon_{zz}$) is a constant in equation (2.61), and if we write

$$\frac{\partial w}{\partial z} = \varepsilon_{zz} - \frac{1}{2}\left(\frac{\partial v}{\partial z}\right)^2, \qquad \text{then integrating,}$$

$$\int_0^l \frac{\partial w}{\partial z}\mathrm{d}z = w(l) - w(0) = 0 = \int_0^L \left[\varepsilon_{zz} - \frac{1}{2}\left(\frac{\partial v}{\partial z}\right)^2\right]\mathrm{d}z$$

$$\therefore \quad \varepsilon_{zz} = \frac{1}{2L}\int_0^L \left(\frac{\partial v}{\partial z}\right)^2 \mathrm{d}z$$

If the arch is actually displaced from $v_0(z)$ to $v(z)$, then

$$\varepsilon_{zz} = \frac{1}{2L}\int_0^l \left[\left(\frac{\partial v}{\partial z}\right)^2 - \left(\frac{\partial v_0}{\partial z}\right)^2\right]\mathrm{d}z \tag{2.62}$$

Equation (2.62) clearly reduces to (2.59) in the case of the two-bar framework.

To keep this problem as simple as the two-bar framework, we will evaluate (2.62) in terms of a single discrete unknown h, by assuming that the deflected shape is

$$v = h \sin\frac{\pi z}{L}$$

(see the Rayleigh–Ritz method in Chapter 4). This shape certainly satisfies all the boundary conditions at $z = 0, L$. However, it should be mentioned that deformations which are asymmetrical about the centre of the arch, are possible; and are found to occur for values of h above a certain critical value,[20] (see Chapter 4). Substituting the above shape into (2.62), we obtain extensional strains very similar to the two bar frame,

$$\varepsilon_{zz} = \frac{\pi^2}{4L^2}(h^2 - h_0^2) \tag{2.63}$$

and therefore

$$\delta\varepsilon_{zz} = \frac{\pi^2}{2L^2} h\,\delta h \tag{2.64}$$

The beam bending strains can be written in terms of the change in curvature, which *if the arch is very shallow*, we may write simply as

$$\frac{\partial^2 v}{\partial z^2} - \frac{\partial^2 v_0}{\partial z^2} = -\frac{\pi^2}{L^2}(h - h_0)\sin\frac{\pi z}{L}$$

Substituting this, together with (2.63) and (2.64), into the PVD, we have

$$\int_0^L \left[AE\frac{\pi^2}{4L^2}(h^2 - h_0^2)\frac{\pi^2}{2L^2}h\,\delta h + \left(EI\frac{\pi^2}{L^2}(h - h_0)\sin\frac{\pi z}{L}\right)\left(\frac{\pi^2}{L^2}\delta h \sin\frac{\pi z}{L}\right)\right]\mathrm{d}z = \int_0^l (-p)\,\delta h \sin\frac{\pi z}{L}\,\mathrm{d}z$$

which, after a little manipulation, can be written as

$$(\tilde{h}^2 - 1)\tilde{h} + \frac{4k^2}{h_0^2}(\tilde{h} - 1) = -\tilde{p} \tag{2.65}$$

in terms of the non-dimensional deflection $\tilde{h} = h/h_0$, and the non-dimensional force $\tilde{p} = 16pL^4/\pi^5 h_0^3 AE$. k is the radius of gyration $(I/A)^{1/2}$ and so the term $(2k/h_0)^2$ may be thought of as a non-dimensional measure of the arch bending stiffness. For values of $2k/h_0$ in excess of unity – that is, for flexurally stiff arches – it is found that the deflection increases monotonically with load, while for values less than unity it is possible for the arch to lose its stiffness completely and 'snap through' under constant load as shown in Fig. 2.37, which is a sketch of equation (2.65) with $(2k/h_0) = 1$ and 0.

This shrewdly chosen example shows dramatically one difference between linear and nonlinear behaviour. In the former, failure can occur only when the

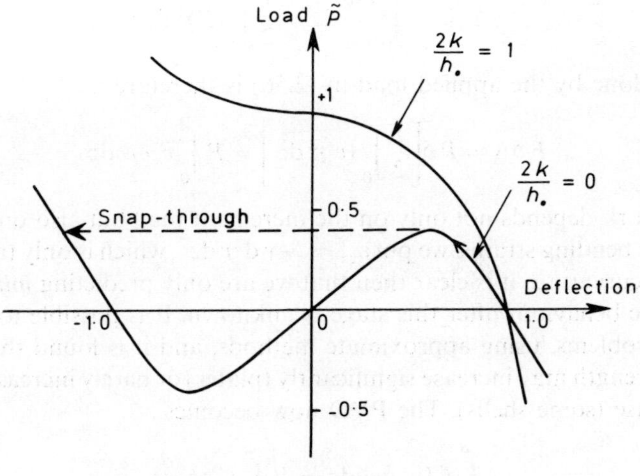

Fig. 2.37 Snap-through

material itself reaches a failing stress. In nonlinear problems, however, it is possible for a structure to lose its resistance (stiffness) to the applied load and become unstable. The resulting large deflections may then, of course, cause ultimate material failure; even so, the load at which stiffness is first lost may never be exceeded significantly, or even at all. If some sort of instability is in the offing, it is therefore natural to enquire whether the PVD can be applied to classical buckling problems. It has already been mentioned that Euler used an energy argument to predict the buckling of a simple strut, so let us humbly follow his example. We take the fictional, but mathematically convenient, perfectly straight strut which suffers no lateral displacement until the critical buckling load is reached, at which stage the strut loses its stiffness and is in a deformed state of *neutral equilibrium*–which can therefore be examined by the PVD. Figure 2.38

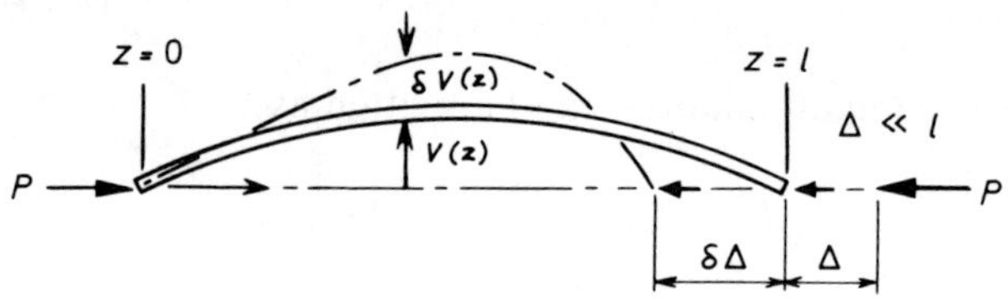

Fig. 2.38 Work done by buckling load

shows such a strut which has deflected to a position $v(z)$ at which stage we impose further virtual displacements $\delta v(z)$. As the lateral displacements $v(z)$ have taken place under constant load, and again assuming a very shallow geometry, we can say that the axial strain ε_{zz} is zero. Therefore, from (2.61),

$$\frac{\partial w}{\partial z} + \frac{1}{2}\left(\frac{\partial v}{\partial z}\right)^2 = 0$$

whence the distance moved by the applied load

$$\Delta = -\int_0^l \frac{\partial w}{\partial z}\,\mathrm{d}z = \frac{1}{2}\int_0^l (v')^2\,\mathrm{d}z$$

The work done by the applied load in (2.56) is therefore

$$P\,\delta\Delta = P\,\delta\left[\frac{1}{2}\int_0^l (v')^2\,\mathrm{d}z\right] = P\int_0^l v'\,\delta v'\,\mathrm{d}z \qquad (2.66)$$

Thus the work depends not only on the increment ($\delta v'$) but also on the history (v'). For the bending strains, we put $\varepsilon_{zz} = -y\,\mathrm{d}^2v/\mathrm{d}z^2$, which is only true for small changes in curvature: it is clear then that we are only predicting *initial buckling* and that the behaviour after this stage is unknown. It is possible to solve post-buckling problems, using approximate methods, and it is found that the post-buckling strength may increase significantly (plates) or barely increase (struts) or even decrease (some shells). The PVD now becomes

$$\int_0^l EIv''\,\delta v''\,\mathrm{d}z = P\int_0^l v'\,\delta v'\,\mathrm{d}z$$

or

$$\int_0^l (EIv'''' + Pv'')\bar{v}\,dz + [EIv''\,\delta v']_0^l - [(EIv''' + Pv')\,\delta v]_0^l = 0 \qquad (2.67)$$

The fourth-order differential equation is therefore

$$EIv'''' + Pv'' = 0$$

for which we provide four boundary conditions. As usual we must supply kinematic boundary conditions directly, otherwise (2.67) provides them. For example, if one end is free ($\delta v \neq 0$) then we have

$$EIv''' + Pv' = 0$$

This is only recognizable as the expected shear-force equilibrium statement provided we account for the geometrical change in end slope, v'. That is the shear force, EIv''', and the component of the applied load *at right angles to the end of the strut*, must together be in equilibrium. Here then is another opportunity to miss a term – or get its sign wrong – when using direct equilibrium arguments.

Had there been a lateral applied load $p(z)$ per unit length, we would have included $\int_0^l p\bar{v}\,dz$ in the PVD so that the differential equation becomes

$$EIv'''' + Pv'' - p(z) = 0 \qquad (2.68)$$

Some interesting facts can be deduced from this last equation. For example, we know that EIv'''' represents the beam's resistance (stiffness × deformation) to the applied load $p(z)$. The second term must therefore represent some sort of similar resistance. Physically it is the existing internal stress (P) being re-aligned to have a lateral component due to the curvature v''. Thus,

$$\text{force} = \text{stiffness} \times \text{deformation}$$
$$\left(P\frac{\partial^2}{\partial z^2}\right) \qquad (v)$$

This type of stiffness, due to re-alignment of existing stresses, is known as *geometric stiffness* and we shall meet it again in Chapter 5. It is often of opposite sign to the elastic stiffness, and when it cancels it precisely ($EIv'''' + Pv'' = 0$) the structure is in trouble and may buckle.

Suppose we solve (2.68),

$$v = C_0 + C_1 z + C_2 \sin\lambda z + C_3 \cos\lambda z + \text{particular integral} \qquad (2.69)$$

where $\lambda^2 = P/EI$, and the particular integral depends on the nature of $p(z)$. Applying the four boundary conditions produces a set of equations for the constants $\mathbf{C} = \{C_0 \quad C_1 \quad C_2 \quad C_3\}$. For example, if a strut of length l is built in at both ends, taking the origin in the middle for algebraic reasons, we have

$$v(-l/2) = v'(-l/2) = v(l/2) = v'(l/2) = 0$$

Substituting into (2.69),

$$\begin{bmatrix} 1 & -\dfrac{l}{2} & -\sin\dfrac{\lambda l}{2} & \cos\dfrac{\lambda l}{2} \\ 1 & \dfrac{l}{2} & \sin\dfrac{\lambda l}{2} & \cos\dfrac{\lambda l}{2} \\ 0 & 1 & \lambda\cos\dfrac{\lambda l}{2} & \lambda\sin\dfrac{\lambda l}{2} \\ 0 & 1 & \lambda\cos\dfrac{\lambda l}{2} & -\lambda\sin\dfrac{\lambda l}{2} \end{bmatrix} \begin{bmatrix} C_0 \\ C_1 \\ C_2 \\ C_3 \end{bmatrix} = \text{function of } p(z) \qquad (2.70)$$

or

$$\mathbf{AC} = \mathbf{p}$$

Solving,

$$\mathbf{C} = \mathbf{A}^{-1}\mathbf{p} = \frac{\text{adj}\,\mathbf{A}}{|\mathbf{A}|}\mathbf{p}$$

Now we can see initial buckling in two ways:

(a) For a strut initially bent by $p(z)$, the deflection ($\mathbf{C}$) becomes very large as $|\mathbf{A}| \to 0$.
(b) For an unbent strut ($\mathbf{p} = \mathbf{0}$), the deflection is zero since the solution of $\mathbf{AC} = \mathbf{0}$ is zero *unless* $|\mathbf{A}| = 0$.

So buckling is again predicted by the same determinant except that in the case of the straight strut the deflections are indeterimate ($\mathbf{C} = \mathbf{0} \div \mathbf{0}$). This confirms that initial buckling is indeed a state of neutral equilibrium. On putting $|\mathbf{A}| = 0$ in the above example, it reduces to

$$2\lambda\sin\frac{\lambda l}{2}\left(\frac{\lambda l}{2}\cos\frac{\lambda l}{2} - \sin\frac{\lambda l}{2}\right) = 0$$

There are two nontrivial possibilities:

(a) $\sin\lambda l/2 = 0$, $\lambda l/2 = \pi$; $P = EI\lambda^2 = 4\pi^2 EI/l^2 = 39.48EI/l^2$
(b) $\tan\lambda l/2 = \lambda l/2$. The solution to this is approximately $\lambda l/2 = 4.49$, or $P = 80.8EI/l^2$.

The strut must therefore buckle at the lowest value of $39.48EI/l^2$. In case (a), for $\lambda l/2 = \pi$, we find on substituting back into (2.70) that

$$C_1 = C_2 = 0, \qquad C_0 = C_3, \qquad \text{so} \qquad v = C_0\left(1 + \cos\frac{2\pi z}{l}\right)$$

This mode shape is of indeterminate amplitude, as we forecast, and is sketched in Fig. 2.39(a).

In case (b), on using $\tan\lambda l/2 = \lambda l/2$, we are left with

$$C_0 = C_3 = 0, \qquad C_1 = -C_2\frac{\lambda l}{2}\cos\frac{\lambda l}{2}$$

With $C_0 = C_3 = 0$, the shape is an odd function of z, and is sketched in Fig. 2.39(b). It is common for buckling and vibration problems to split into symmetric and antisymmetric solutions in this way.

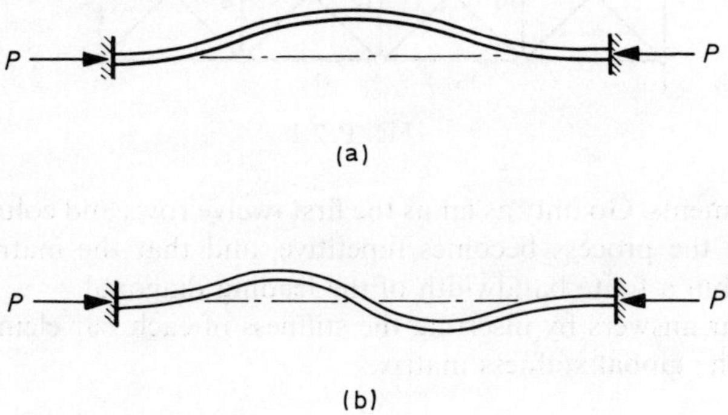

Fig. 2.39 (a) Symmetric buckling. (b) Asymmetric buckling

2.12 Summary

1. Unit displacement theorem: $\int_V \boldsymbol{\sigma}^t\bar{\boldsymbol{\varepsilon}}\,dv = R_i$, where $\bar{\boldsymbol{\varepsilon}}$ are virtual strains $\bar{\boldsymbol{\varepsilon}} = \mathbf{D}^t\bar{\mathbf{u}}$ compatible with unit displacement $r_i = 1$.
2. For a linear elastic structure: $\mathbf{R} = \mathbf{Kr}$ where $k_{ij} = \int_V \boldsymbol{\sigma}_j^t\boldsymbol{\varepsilon}_i\,dv$ and the strains due to r_i or $r_j = 1$ are $\boldsymbol{\varepsilon}_i$ or $\boldsymbol{\varepsilon}_j$.
3. For the gth bar: $\mathbf{P}_g = \mathbf{k}_g\boldsymbol{\rho}_g$; $\boldsymbol{\rho}_g = \mathbf{a}_g\mathbf{r}$; then $\mathbf{K} = \sum_g \mathbf{a}_g^t\mathbf{k}_g\mathbf{a}_g$.
4. Coordinate transformation: If $\boldsymbol{\rho}' = \mathbf{T}\boldsymbol{\rho}$, $\mathbf{k} = \mathbf{T}^t\mathbf{k}'\mathbf{T}$.
5. For gross deformations, but small strains, the PVD may be used as

$$\int_V \boldsymbol{\sigma}^t\,\delta\boldsymbol{\varepsilon}\,dv = \int_V \mathbf{p}_v^t\,\delta\mathbf{u}\,dv + \int_S \mathbf{p}_s^t\,\delta\mathbf{u}\,ds$$

Problems

(Asterisks indicate problems with worked solutions.)

P 2.1

Apply unit displacements $r_1 = 1, r_2 = 1, r_3 = 1$, successively to the framework of Fig. 2.7 in section 2.3, and from the necessary values of the load vector **R** deduce the stiffness matrix. Compare your answer with that given after Fig. 2.7.

*P 2.2**

In this pinjointed framework, all bars are of length l or $\sqrt{2}l$ and have the same area A and modulus E. The nodal displacements are numbered as shown. Establish the global stiffness matrix of the framework directly by using successive

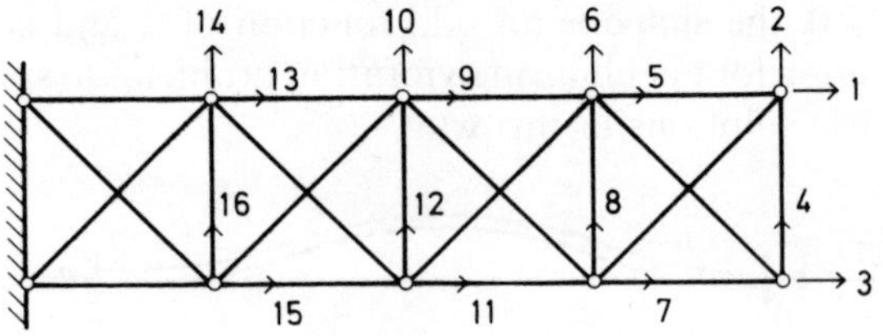

Fig. P. 2.2

unit displacements. Go only as far as the first twelve rows and columns to satisfy yourself that the process becomes repetitive, and that the matrix consists of elements within a finite bandwidth of the leading diagonal.

Check your answers by inserting the stiffness of each bar element (equation (2.16)) into the global stiffness matrix.

P 2.3

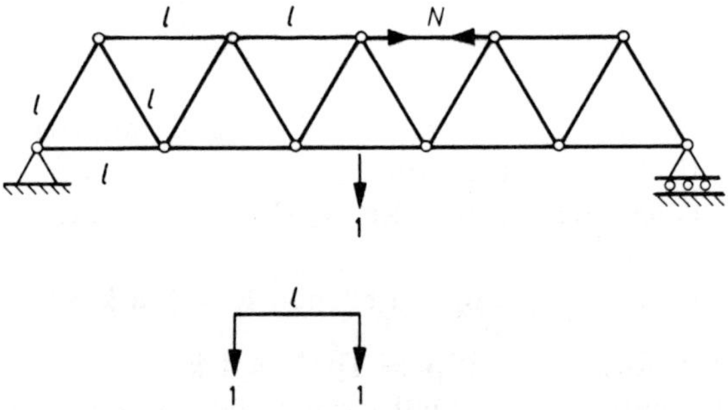

Fig. P. 2.3

All bars in the statically determinate bridge framework shown have the same length l. Using the PVD, sketch the influence line for the depicted bar force N, as a unit load traverses the bottom deck of the bridge, and show that the maximum value of N is $-\frac{12}{5}\sqrt{3}$.

Sketch a similar influence line as two unit loads, spaced l apart, traverse the bridge.

P 2.4

A beam has a quite arbitrary cross-sectional shape; the figure shows the x–y axes which pass through the centroid ($\int x \, dA = \int y \, dA = 0$). The beam is loaded by a distributed load having components p_x and p_y per unit length of the z-axis, and the consequent deflections of the centroid are $u(z)$ and $v(z)$.

Assuming cross-sections remain plane, $w = -yv' - xu'$, where the primes denote differentiation with respect to z. Using the PVD obtain two coupled differential equations for $u(z)$ and $v(z)$ in terms of the second moments of area

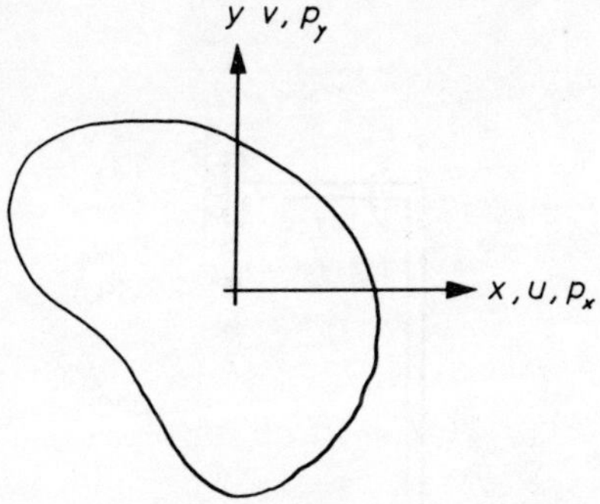

Fig. P. 2.4

$$I_{xx} = \int y^2\,dA; \qquad I_{yy} = \int x^2\,dA; \qquad I_{xy} = \int xy\,dA$$

Show that if $p_x = 0$, the differential equation can be written in the familiar form

$$E\bar{I}_{xx}v'''' - p_y = 0$$

where the 'effective' second moment of area is

$$\bar{I}_{xx} = I_{xx}\left[1 - \frac{I_{xy^2}}{I_{xx}I_{yy}}\right]$$

*P 2.5**

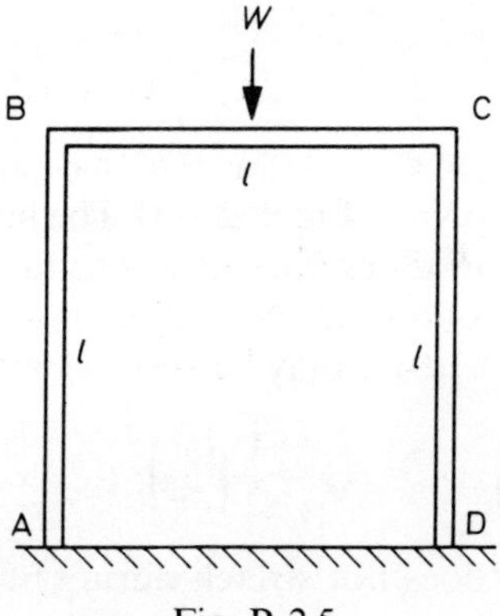

Fig. P. 2.5

All beams in the portal frame have the same flexural rigidity EI and length l. Use the global stiffness matrix derived in the text (equation (2.28)) to solve for the joint displacements. Sketch the bending-moment distribution and show that there is a discontinuity in the bending moment at the joints B and C corresponding to the applied kinematically equivalent couples at these joints. Show that the addition of the self-equilibrating fixed-end solution for BC restores continuity of moments. Sketch the bending moment and the deflected shape.

P 2.6

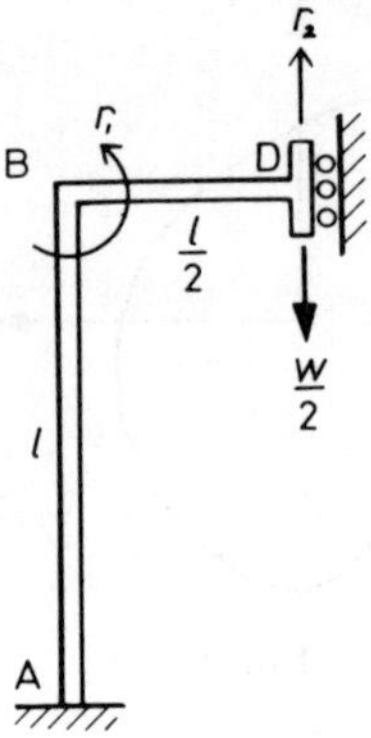

Fig. P. 2.6

Problem 2.5 can be solved, without resorting to artificial kinematically equivalent forces, by splitting the beam BC into two elements so that W is applied at a 'joint' D with a corresponding displacement (r_2). By taking advantage of symmetry, one half of the structure can be solved with just two degrees of freedom, the joint D being constrained not to rotate. Show that the stiffness matrix is

$$\mathbf{K} = \frac{12EI}{l^3}\begin{bmatrix} l^2 & -2l \\ -2l & 8 \end{bmatrix}$$

Show that $r_1 = -Wl^2/48EI$; $r_2 = -Wl^3/96EI$; and confirm that in BD the moments are $M_B = Wl/12$, $M_D = Wl/6$.

P 2.7

The rectangular plate $0 \leqslant x \leqslant a$, $0 \leqslant y \leqslant b$, has applied compressive edge stresses f_{xx}, f_{yy}, and f_{xy} as shown in Fig. P 2.7(a). The initial buckling due to these stresses causes a deflection $w(x, y)$ sufficient to produce edge displacements in a fashion similar to the ends of a strut. Now by considering the geometry of the various triangles in Fig. P 2.7(b) it may be shown that due to w, the strains are

$$\varepsilon_{xx} = \frac{1}{2}\left(\frac{\partial w}{\partial x}\right)^2; \qquad \varepsilon_{yy} = \frac{1}{2}\left(\frac{\partial w}{\partial y}\right)^2; \qquad \varepsilon_{xy} = \frac{\partial w}{\partial x}\frac{\partial w}{\partial y}$$

If the plate middle surface does not stretch during buckling then, for example,

$$\varepsilon_{xx} = 0 = \frac{\partial u}{\partial x} + \frac{1}{2}\left(\frac{\partial w}{\partial x}\right)^2$$

Thus the relative displacements of edges $x = 0$ and $x = a$ are known, and we can write the work done by f_{xx} as:

$$\int_0^b tf_{xx}[-u]_0^a\,dy = \int_0^b\int_0^a tf_{xx}(-\partial u/\partial x)\,dx\,dy = \int_0^b\int_0^a (t/2)f_{xx}(\partial w/\partial x)^2\,dx\,dy$$

and similarly for f_{yy} and f_{xy}.

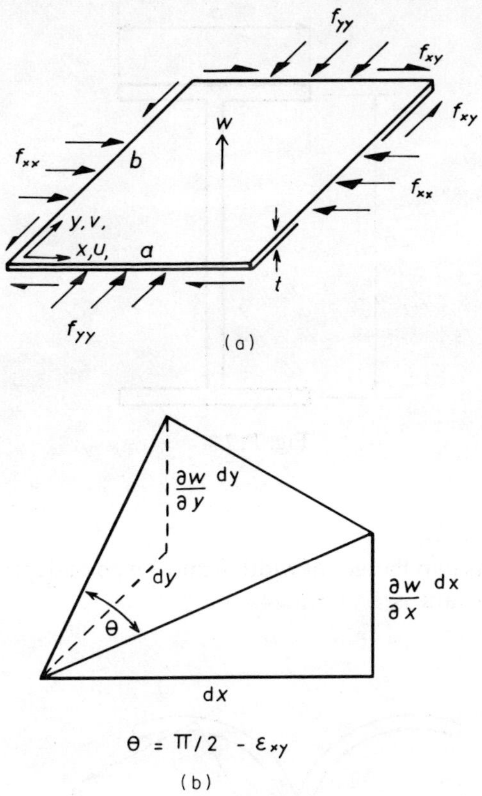

Fig. P. 2.7

By considering virtual displacements δw, show that the governing differential equation for initial buckling is

$$D\nabla^2(\nabla^2 w) + tf_{xx}\frac{\partial^2 w}{\partial x^2} + tf_{yy}\frac{\partial^2 w}{\partial y^2} + 2tf_{xy}\frac{\partial^2 w}{\partial x\,\partial y} = 0$$

Show further that the shear-force boundary condition for a free edge, say $x = a$, is

$$D\left[\frac{\partial^3 w}{\partial x^3} + (2 - \nu)\frac{\partial^3 w}{\partial x\,\partial y^2}\right] + tf_{xx}\frac{\partial w}{\partial x} + tf_{xy}\frac{\partial w}{\partial y} = 0$$

*P 2.8**

A beam has the cross-section shown, where the flanges and web have a thickness t much smaller than a or b. Sketch the distribution of w/θ', the warping per rate of twist, around the tube, and show that the torsion-bending constant $\Gamma = tb^2a^3/6$.

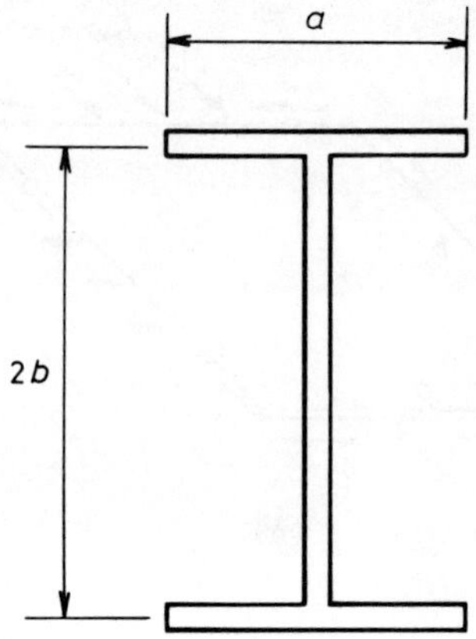

Fig. P. 2.8

P 2.9

A channel section has two flanges of width b and a web of depth $3b$. Show that its torsion bending constant is $\Gamma = 3tb^5/4$.

P 2.10

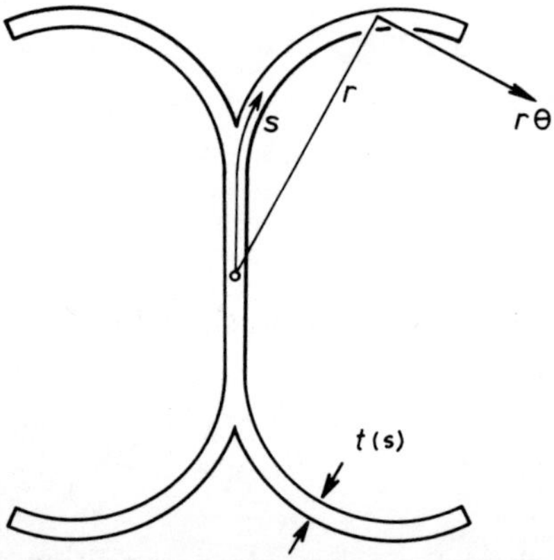

Fig. P. 2.10

A thin-walled open tube of doubly symmetrical cross-section (to reduce algebra) is subjected to a compressive stress σ which is not large enough to cause flexural buckling, but it is possible that the torsionally weak tube may twist. If the twist is $\theta(z)$, then any point on the section will displace an amount $r\theta$ as shown in the figure, and we may repeat the arguments of the previous two problems and write the zero axial strain (during twisting) as

$$\varepsilon_{zz} = \frac{\partial w}{\partial z} + \tfrac{1}{2}(r\theta')^2 = 0$$

Hence the relative movements of similar positions at the ends $z = 0$ and $z = l$ will be

$$w(0) - w(l) = -[w]_0^l = -\int_0^l \frac{\partial w}{\partial x}\,\mathrm{d}z = \frac{1}{2}\int_0^l r^2\theta'^2\,\mathrm{d}z$$

Deduce that the work done by the applied stress, σ, during a virtual displacement $\delta\theta(z)$ will be

$$\sigma I_p \int_0^l \theta'\,\delta\theta'\,\mathrm{d}z$$

where $I_p = \int r^2 t\,\mathrm{d}s$ is the polar second moment of area. Show that the differential equation of neutral equilibrium is

$$E\Gamma\theta'''' + (\sigma I_p - GJ)\theta'' = 0$$

and find the various possible boundary conditions.

Show that a tube whose ends are prevented from rotating, but are completely free to warp, will buckle in a twisting mode at a stress

$$\sigma = [GJ + \pi^2 E\Gamma/l^2]/I_p$$

*P 2.11**

For the shallow two-bar framework of Fig. 2.35, obtain the relationship between $(W/AE)(L/h_0)^3$ and (h/h_0), and sketch it. Show that the framework becomes unstable as W approaches $0.38AE(h_0/L)^3$ and suggest what happens when W is a much larger positive (or negative) number than this.

3

THE PRINCIPLE OF VIRTUAL FORCES

3.1 Introduction

The PVF may be developed in an analogous fashion, and with similar benefits, to that of the PVD in Chapter 2. The parallels in the two sorts of argument are instructive and illuminating, but just in case the reader is totally convinced by the duality of the two approaches, a word of caution and realism is appropriate at this stage. It is tempting to write this chapter on virtual forces in an identical manner to Chapter 2, and simply exchange the words forces for displacements, stresses for strains, equilibrium for compatibility, and statics for kinematics. But there are differences in the two approaches; and the main difference is a consequence of the structure itself – whether it is statically determinate or statically indeterminate.

We recall that the PVD only worked when we enforced compatibility requirements relating strains to displacements. In a general structure this takes the form $\boldsymbol{\varepsilon} = \mathbf{D}^t\mathbf{u}$, or for frameworks – and other structures with finite degrees of freedom – it might take the form $\boldsymbol{\Delta} = \mathbf{ar}$ (equation (1.27)). We have said that if this can be done, then the structure is *kinematically determinate* in that strains can be found in terms of displacements. This can be done for *any* structure and if we cannot succeed in so doing then our labelling is insufficiently precise or the 'structure' is a mechanism. In applying the PVF we will have to enforce equilibrium requirements, that is, relate stresses to applied forces; but the stresses cannot be found directly, using equilibrium arguments only, if the structure is *statically indeterminate*. In our discrete frameworks, for example, the equations of equilibrium $\mathbf{a}^t\mathbf{N} = \mathbf{R}$ (1.38) were not sufficient for solving $\mathbf{N}$ ($\mathbf{a}$ was not square) in such problems. There will therefore be clear differences in applying the PVF to statically determinate or statically indeterminate structures.

3.2 The unit-load method

The unit-load theorem is a special case of the PVF and is a powerful technique for finding a solitary displacement in a structure when the stresses and strains are already known. It is therefore particularly suitable for simple statically determinate structures where the stresses can be readily found without ever introducing the displacements as unknowns. In such structures, in fact, there is little reason for finding the displacements everywhere if they do not enter into the calculations. Displacements as such are not a local failure criteria although it is common to impose some sort of restriction on the deformations. For example, aerodynamic surfaces should not have their character changed adversely by deforming, aerials should not become significantly misaligned, doors should not spontaneously open or jam, and heavy traffic should not find its exit from a floppy bridge becoming progressively steeper. To check this sort of limit – often called serviceability – it is usually sufficient to evaluate just a few displacements, and for this purpose the unit-load theorem is tailor-made.

Suppose then a structure has in it strains $\boldsymbol{\varepsilon}$ which are due to any cause whatsoever – loads or temperature – and we require a solitary displacement r (Fig. 3.1(a)) due to these effects. Corresponding to the desired displacement, in both direction and nature, we apply a unit load to the structure *and only a unit load* (Fig. 3.1(b)). Let the virtual stress system in equilibrium with this unit load be $\bar{\boldsymbol{\sigma}}$. Then the PVF in its general form (1.73) becomes simply

$$\mathbf{r} = \int_V \bar{\boldsymbol{\sigma}}^{\mathrm{t}} \boldsymbol{\varepsilon} \, \mathrm{d}v \tag{3.1}$$

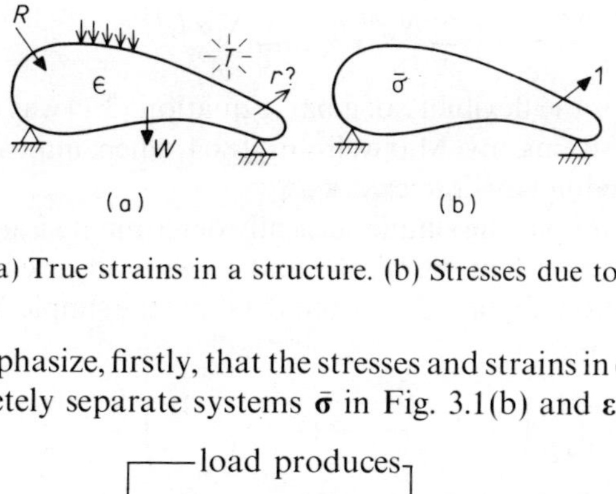

Fig. 3.1 (a) True strains in a structure. (b) Stresses due to unit load

Again we must emphasize, firstly, that the stresses and strains in equation (3.1) are due to the completely separate systems $\bar{\boldsymbol{\sigma}}$ in Fig. 3.1(b) and $\boldsymbol{\varepsilon}$ in Fig. 3.1(a).

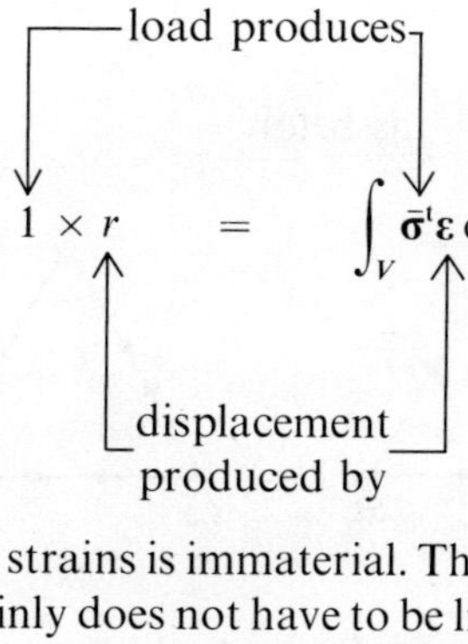

Secondly, the nature of the strains is immaterial. The stress–strain law is not even mentioned in (3.1). It certainly does not have to be linear, as has been stated more than once in the literature. The unit-load theorem is demonstrably a very convenient formula for finding the displacement of a structure, once the true strains are known. It avoids using direct geometrical arguments or having to integrate equations like $\mathbf{D}^{\mathrm{t}}\mathbf{u} = \boldsymbol{\varepsilon}$; it substitutes instead the exercise in equilibrium reasoning of finding another stress system $\bar{\boldsymbol{\sigma}}$. This stress system is virtual, but if the structure is statically determinate then there is a unique solution $\bar{\boldsymbol{\sigma}}$ to the stresses caused by a unit load – in which case $\bar{\boldsymbol{\sigma}}$ must be *true* stresses and the 'bar' is unnecessary. If the structure is statically indeterminate, and ignoring for the moment the question of how we determine the true strains $\boldsymbol{\varepsilon}$, then there is no longer a unique stress field $\bar{\boldsymbol{\sigma}}$ in equilibrium with the unit load; in which case it is our privilege to find the simplest stress system possible. Consider, for example, the ubiquitous pinjointed framework with uniform members (N, A, l, E, α, T, as usual). The true strain in the gth bar is

$$\varepsilon = \frac{N_g}{A_g E_g} + \alpha T_g$$

We apply a unit load and somehow solve for bar forces $\bar{N}_g$. The unit-load theorem (3.1) becomes

$$r = \sum_g \left(\frac{\bar{N}_g}{A_g}\right)\left(\frac{N_g}{A_g E} + \alpha T_g\right) A_g l_g$$

or

$$r = \sum_g \bar{N}_g N_g f_g + \sum_g \bar{N}_g \alpha T_g l_g \tag{3.2}$$

where $f_g = l_g / A_g E$ is the flexibility of a bar. Equation (3.2) was actually first used, without thermal strains, by Maxwell[3] in 1864 when imposing compatibility requirements in redundant frameworks.

Consider, for example, the simple statically determinate framework of Fig. 2.7 again, and suppose we had wished to find r_1 without attempting a complete displacement solution. Figures 3.2(a) and (b) show the simple N and $\bar{N}$ solutions. Using (3.2),

$$r_1 = \frac{l}{AE}\left[(1)\left(\frac{-W}{\sqrt{3}}\right) + (-1)\left(\frac{-W}{\sqrt{3}}\right) + \frac{1}{2}\left(\frac{W}{2\sqrt{3}}\right)\right]$$

$$= \frac{Wl}{4\sqrt{3}AE} \quad \text{as before}$$

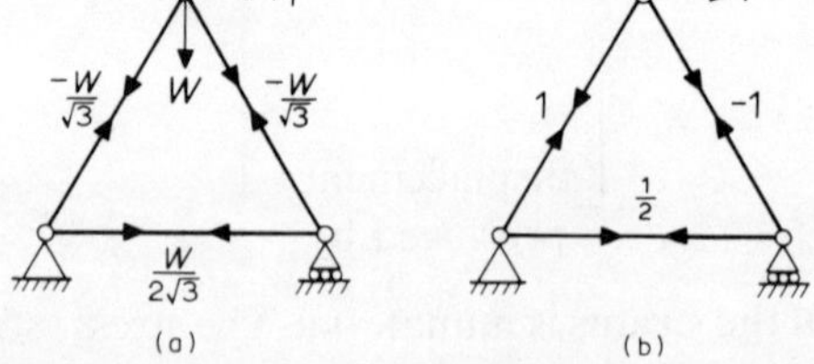

Fig. 3.2 (a) Forces N. (b) Forces $\bar{N}$

As mentioned, if we require the deflection of a joint in a highly redundant structure, much effort can be saved by choosing the simplest possible solution for $\bar{N}$ which satisfies equilibrium. Figure 3.3 might be such a structure. There are a

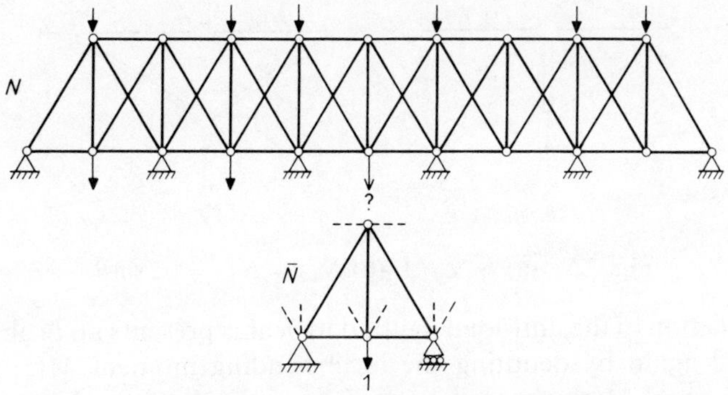

Fig. 3.3 Displacement in a highly redundant framework

large number of possible stress systems $\bar{N}$ and only one is shown in the figure, where we have selected a simple five-bar substructure to equilibrate the unit load. All the other bars and supports have been discarded in the figure but they are to be imagined as simply having zero stresses in them, and thus excluded from the summation in (3.2) which is consequently confined to the five bars shown. The fact that some bars and supports have had their stresses put to zero does not violate the necessary condition that the $\bar{\boldsymbol{\sigma}}$ stresses must satisfy equilibrium everywhere in the structure. The bar forces $\bar{N}$ in Fig. 3.3 satisfy *all* joint equilibrium requirements both in the four joints shown and in all the others discarded. To illustrate how any of the alternative *statically equivalent* stress systems $\bar{N}$ will still deliver the same displacement, let us resurrect the problem of Fig. 1.10, simplifying it to a single applied load W as shown in Fig. 3.4. The bar forces were previously found to be $N_{12} = N_{13} = W \sin^2\theta/(1 + 2\sin^3\theta)$ and $N_{14} = W/(1 + 2\sin^3\theta)$. If we wish to find the vertical deflection of joint (1) then we could use either of the virtual stress systems shown in Figs 3.5(a) or (b). Then using (3.2):

in (a),
$$r = (1)\left[\frac{W}{1 + 2\sin^3\theta}\right]\frac{l\sin\theta}{AE} = \frac{Wl\sin\theta}{AE(1 + 2\sin^3\theta)}$$
as before (equation (1.16))

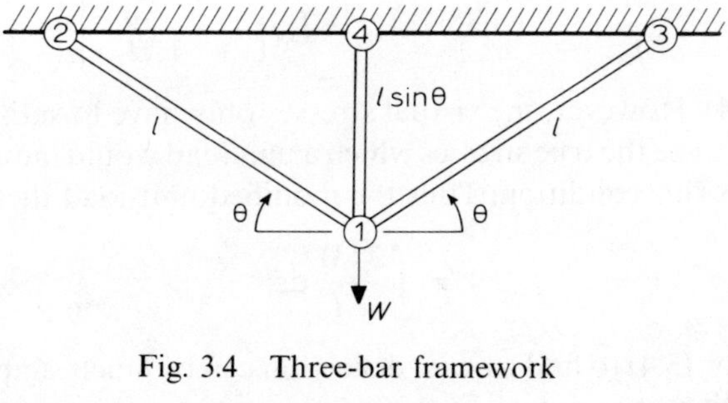

Fig. 3.4 Three-bar framework

or

in (b),
$$r = 2\left(\frac{1}{2\sin\theta}\right)\left[\frac{W\sin^2\theta}{1 + 2\sin^3\theta}\right]\frac{l}{AE} = \frac{Wl\sin\theta}{AE(1 + 2\sin^3\theta)} \qquad \text{again}$$

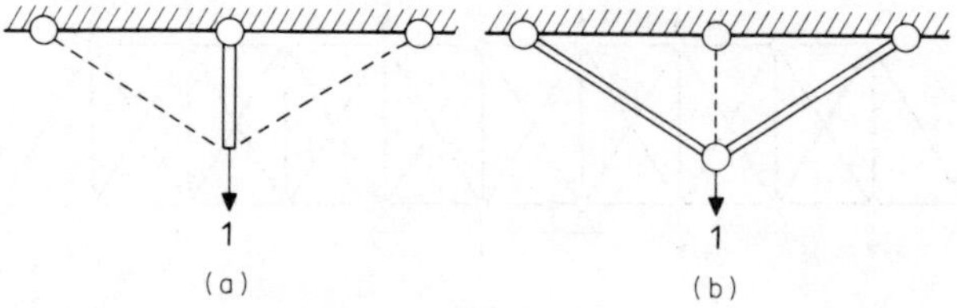

Fig. 3.5 (a) $\bar{N}_{14} = 1$. (b) $\bar{N}_{12} = \bar{N}_{13} = 1/2\sin\theta$

The application of the unit-load method to beams presents no problem; it may be simplified again by denoting the local bending moment $M(z)$ due to the applied forces and $\bar{M}(z)$ due to the relevant unit load acting alone. Suppose an axial force N is also present, then we may write the stress σ_{zz} as

$$\sigma_{zz} = \frac{N}{A} + \frac{M}{I}y \qquad \text{in the usual notation,}$$

where it is assumed that the z-axis passes through the centroid, and the x and y axes are principal. Then equation (3.2) predicts

$$r = \int_V \bar{\boldsymbol{\sigma}}^{\mathrm{t}}\boldsymbol{\varepsilon}\,\mathrm{dv} = \int_A \int_l \left(\frac{\bar{N}}{A} + \frac{\bar{M}y}{I}\right)\left(\frac{N}{AE} + \frac{My}{EI}\right)\mathrm{d}z\,\mathrm{d}A$$

or

$$r = \int_l \left(\frac{\bar{N}N}{AE} + \frac{\bar{M}M}{EI}\right)\mathrm{d}z \tag{3.3}$$

The integral $\int_A y\,\mathrm{d}A$ vanishes because of the choice of centroidal axes, so that the deformations due to axial forces and moments do not couple in (3.3). Notice that both terms are still clearly 'work': the first term is virtual force $\bar{N}$ times actual displacement Nl/AE while the second is virtual moment times actual curvature M/EI. This concise form is not true if bending is not about principal axes, of course; in that case the true curvature can be written in terms of an *effective* second moment of area $\bar{I}_{xx}$, as

$$\frac{M}{E\bar{I}_{xx}}, \qquad \text{where} \quad \bar{I}_{xx} = I_{xx}\left[1 - \frac{I_{xy}^2}{I_{xx}I_{yy}}\right]$$

(see Problem 2.4). However, the virtual stresses only have to satisfy equilibrium and do not have to be the true stresses which a unit load would induce. The virtual moment satisfies this condition. Thus the modified unit-load theorem is simply

$$r = \int_l \frac{\bar{M}M}{E\bar{I}}\mathrm{d}z \tag{3.4}$$

The use of (3.3) or (3.4) to find a beam deflection can be much simpler than direct integration of $\mathbf{D}^{\mathrm{t}}\mathbf{u} = \boldsymbol{\varepsilon} = \boldsymbol{\phi\sigma}$, or in this case

$$\frac{\partial^2 v}{\partial z^2} = \frac{-M(z)}{EI}$$

and therefore

$$v(z) = -\int\int \frac{M(z)}{EI} \mathrm{d}z\,\mathrm{d}z + C_1 z + C_0 \tag{3.5}$$

This is particularly so if either $M(z)$ or $I(z)$ are discontinuous. The above expression then consists of several parts, all having different constants of integration which have to be chosen to ensure continuity. (Macaulay's method[23] is a technique for coping with discontinuous solutions, but it does very rapidly become cumbersome.) In contrast, equation (3.2), being an integral, simply has to be divided into its component parts if any terms in the integrand are discontinuous. The beam shown in Fig. 3.6, for example, has a discontinuous rigidity. If we use the conventional method, applying (3.5) twice with $M = Wz$ leads to

$$o < z < l: \quad v = \frac{-Wz^3}{6EI} + C_1 z + C_0$$

$$l < z < 2l: \quad v = \frac{-Wz^3}{12EI} + C_2 z + C_3$$

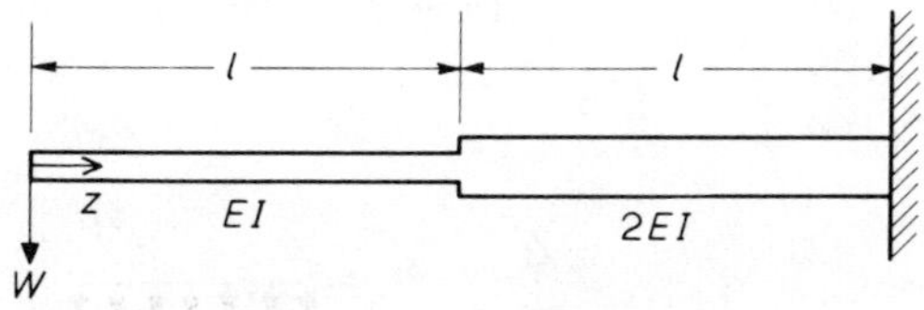

Fig. 3.6 Displacement of a discontinuous structure

Imposing the conditions that $v(z)$ and $v'(z)$ are continuous at $z = l$, and $v(2l) = v'(2l) = 0$, we find the four constants. In particular, the constant $C_0 = -3Wl^3/2EI$ is the value of the tip deflection $v(0)$. Contrast this approach with the unit-load method, using $\bar{M}(z) = z$ obviously in this case, where we have

$$r = \int_0^{2l} \frac{\bar{M}M}{EI} \mathrm{d}z = \frac{1}{EI}\int_0^l Wz^2\,\mathrm{d}z + \frac{1}{2EI}\int_l^{2l} Wz^2\,\mathrm{d}z = \frac{3Wl^3}{2EI} \qquad \text{again}$$

Discontinuous loading also makes the conventional method become unwieldy, while the unit-load integrals may remain fairly straightforward. It is not always convenient to have to evaluate the bending moment as explicit functions, indeed the theorem may have to be applied when the moment $M(z)$ is a diagram, graph, or table. The most common variations in $M(z)$ are a constant, linear, or parabolic, and for these shapes it is useful to have a handy table of the integral of such products, as in Table 3.1.

Example

Figure 3.7(a) shows a uniform beam subjected to two concentrated loads W and a uniformly distributed load W/l per unit length. Suppose we require the clockwise

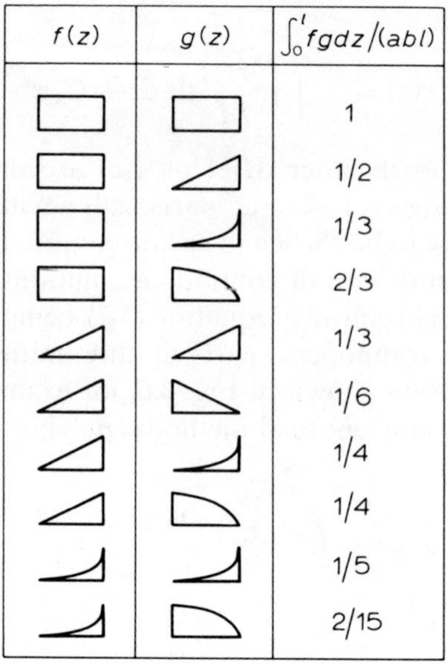

Table 3.1

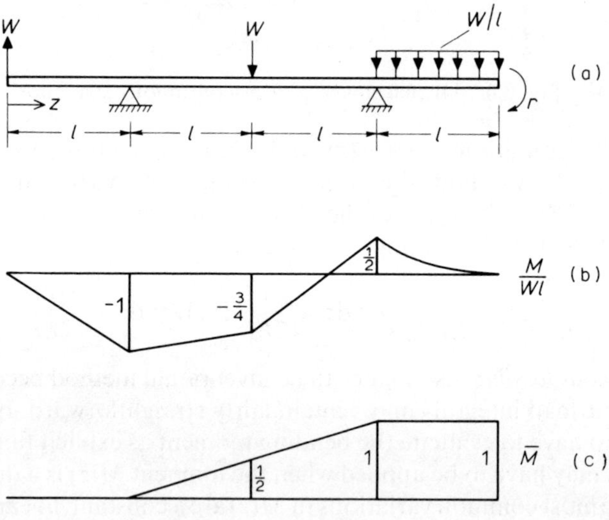

Fig. 3.7

rotation of the end $z = 4l$. We must then construct $\bar{M}(z)$ due to a unit moment placed here – a moment being that 'load' which does virtual work over the rotation r. The two bending-moment diagrams are readily found as shown in Figs 3.7(b) and (c).

Evaluating $r = \int_0^{4l} \bar{M}M\,\mathrm{d}z/EI$ using the table:

$$\begin{aligned}
\frac{rEI}{Wl^2} &= (-\tfrac{3}{4}\tfrac{1}{2})\tfrac{1}{2} + (-\tfrac{1}{4}\tfrac{1}{2})\tfrac{1}{6} && l < z < 2l \\
&\quad + (\tfrac{1}{2}(-\tfrac{3}{4}))1 + (\tfrac{1}{2}\tfrac{5}{4})\tfrac{1}{2} + (\tfrac{1}{2}(-\tfrac{3}{4}))\tfrac{1}{2} + (\tfrac{1}{2}\tfrac{5}{4})\tfrac{1}{3} && 2l < z < 3l \\
&\quad + (1.\tfrac{1}{2})\tfrac{1}{3} && 3l < z < 4l \\
&= -\tfrac{1}{12}
\end{aligned}$$

that is, an anticlockwise rotation of $Wl^2/12EI$

3.3 The flexibility matrix of a framework

In an analogous fashion to the stiffness matrix $\mathbf{K}$ first introduced in equation 1.41, we may define a *flexibility* matrix $\mathbf{F}$ relating a set of displacements $\mathbf{r}$ to the corresponding forces $\mathbf{R}$ by

$$\mathbf{r} = \mathbf{F}\mathbf{R} \tag{3.6}$$

assuming of course that the behaviour of the structure is linear.

The flexibility matrix is not, in truth, as useful a concept as the stiffness matrix. We have already said that *all* displacements, due to all likely loads, are rarely required. Furthermore knowledge of $\mathbf{F}$ does not help us find internal stresses. However, we shall see later that $\mathbf{F}$ can be as useful a concept as $\mathbf{K}$ in solving certain dynamic problems, so it is of more than passing interest.

Expanding (3.6) as

$$\begin{bmatrix} r_1 \\ r_2 \\ \vdots \\ r_i \\ \vdots \\ r_l \end{bmatrix} = \begin{bmatrix} f_{11} & \cdots & f_{1j} & \cdots & f_{1l} \\ f_{21} & & & & \vdots \\ \vdots & & & & \vdots \\ f_{i1} & \cdots & f_{ij} & \cdots & f_{il} \\ \vdots & & & & \vdots \\ f_{l1} & \cdots & f_{lj} & \cdots & f_{ll} \end{bmatrix} \begin{bmatrix} R_1 \\ R_2 \\ \vdots \\ R_j \\ \vdots \\ R_i \\ \vdots \\ R_l \end{bmatrix}$$

we see that simply from this definition we can generate the ith column of $\mathbf{F}$ by imposing a single force $R_i = 1$ and finding the resulting displacements

$$\mathbf{r} = \{f_{1i} \;\; f_{2i} \;\; f_{3i} \;\; \ldots \;\; f_{ji} \;\; \ldots \;\; f_{li}\}$$

Now the single jth displacement of this set can be found by the unit-load theorem, applying a unit load $R_j = 1$ and causing stresses $\boldsymbol{\sigma}_j$ throughout. Thus, if the load $R_i = 1$ produces stresses and strains $\boldsymbol{\sigma}_i$ and $\boldsymbol{\varepsilon}_i$ we must have

$$f_{ji} = \int_V \boldsymbol{\sigma}_j^{\mathrm{t}}\,\boldsymbol{\varepsilon}_i\,\mathrm{d}v \tag{3.7}$$

This is the most general form for an element in the flexibility matrix. Because the problem is linear, from (2.2) $\boldsymbol{\varepsilon} = \boldsymbol{\phi\sigma}$, so

$$f_{ji} = \int_V \boldsymbol{\sigma}_j^{\mathrm{t}} \boldsymbol{\phi} \boldsymbol{\sigma}_i \, \mathrm{d}v \tag{3.8}$$

But f_{ji} is a scalar quantity so we may take its transpose,

$$f_{ji} = \int_V \boldsymbol{\sigma}_i^{\mathrm{t}} \boldsymbol{\phi}^{\mathrm{t}} \boldsymbol{\sigma}_j \, \mathrm{d}v$$

Now (3.8) is equally true for $f_{ij} = \int_v \boldsymbol{\sigma}_i^{\mathrm{t}} \boldsymbol{\phi} \boldsymbol{\sigma}_i \, \mathrm{d}v$, so comparing the last two expressions, and noting that $\boldsymbol{\phi}$ is a symmetrical matrix, we conclude that $f_{ij} = f_{ji}$. In other words, **F** is symmetrical like **K**. The statement $f_{ij} = f_{ji}$ is usually known as Maxwell's reciprocal theorem, which can be stated physically as 'The displacement at point i due to a unit load at j is precisely the same as the displacement at j due to a unit load at i.' Having explained the elements of **F** in this direct fashion, we can also deduce that **F** is likely to be fully populated – in contrast to **K** – since any load R_i is likely to cause displacements everywhere, even though some parts of the structure may not actually be strained.

The use of (3.7) to generate elements of **F** for a pinjointed framework is straightforward. If loads R_i and R_j cause bar forces N_{gi} and N_{gj} in the gth bar, then we have

$$f_{ji} = \sum_g \frac{N_{gi} N_{gj} l_g}{A_g E} \tag{3.9}$$

Suppose we develop this idea further. As long as the structure is statically determinate we may solve for the force N_g in any bar in terms of the applied loads, that is (cf. equation (1.24)),

$$N_g = \mathbf{b}_g \mathbf{R} \tag{3.10}$$

The PVF applied to the whole framework will be

$$\bar{\mathbf{R}}^{\mathrm{t}} \mathbf{r} = \sum_g \bar{N}_g \Delta_g$$

where $\Delta_g = N_g l_g / A_g E = f_g N_g$, using the bar flexibility notation again. Substituting (3.10), the work becomes

$$\bar{\mathbf{R}}^{\mathrm{t}} \mathbf{r} = \sum_g \bar{\mathbf{R}}^{\mathrm{t}} \mathbf{b}_g^{\mathrm{t}} f_g \mathbf{b}_g \mathbf{R}$$

or

$$\bar{\mathbf{R}}^{\mathrm{t}}(\mathbf{r} - \mathbf{FR}) = 0 \qquad \text{and therefore} \quad \mathbf{r} = \mathbf{FR},$$

where

$$\mathbf{F} = \sum_g \mathbf{b}_g^{\mathrm{t}} f_g \mathbf{b}_g \tag{3.11}$$

Thus this appears superficially to be analogous to $\mathbf{K} = \sum_g \mathbf{a}_g^t \mathbf{k}_g \mathbf{a}_g$ – a point to which we must return.

This assembly idea may be extended to any sort of statically determinate structure, since the PVF knows no limit in application. Consider then a bar in a stiff-jointed framework, where now bending strains must be considered. Indeed, we first consider *only* bending deformations since these will predominate for slender beams anyway. The stress in a beam element is no longer to be found in terms of a single axial force N_g but is related to the bending moment which varies along the beam. Now we attempt to define the virtual work in a typical beam entirely in terms of the four forces $\mathbf{P}_g$ at the ends of the beam in exactly the same way as was done with the beam displacements $\boldsymbol{\rho}_g$ in section 2.6. But now we have to ensure that equilibrium is satisfied directly and this places a constraint on the pattern of forces $\mathbf{P}_g = \{P_1 \quad P_2 \quad P_3 \quad P_4\}$ in Fig. 3.8. (The notation corresponds to the displacement notation used in Fig. 2.18.) Thus we have two equations of equilibrium like

$$P_1 + P_3 = 0; \qquad P_2 + P_4 + P_3 l = 0$$

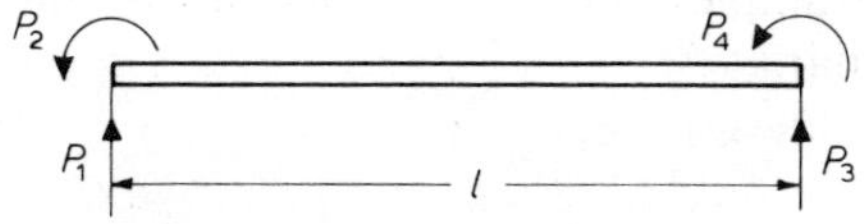

Fig. 3.8 $\mathbf{P}_g$ forces for a beam

There are consequently only two forces needed to specify the stresses within the element, and there are several ways of choosing them. We could choose the two end moments as

$$\mathbf{P}_g = \{P_2 \quad P_4\}, \qquad \text{whence} \quad P_1 = -P_3 = (P_2 + P_4)/l$$

or the 'forces' at one end only, say

$$\mathbf{P}_3 = \{P_4 \quad P_3\}, \qquad \text{whence} \quad P_1 = -P_3, \; P_2 = -P_4 - P_3 l$$

We can now prepare to apply the PVF to a whole framework by writing the integral as a summation over all g beams:

$$\int_V \bar{\boldsymbol{\sigma}}^t \boldsymbol{\varepsilon} \, dv = \sum_g \int_{V_g} \bar{\boldsymbol{\sigma}}^t \boldsymbol{\varepsilon} \, dv$$

Suppose we select the second description in Fig. 3.9, then the bending moment at any point z from the left-hand end is

$$M = -P_3 z - P_4 = [-z \quad -1]\{P_3 \quad P_4\}$$

The stress, assuming principal bending, is therefore

$$\sigma = \frac{M}{I} y = \frac{y}{I} [-z \quad -1] \mathbf{P}_g \tag{3.12}$$

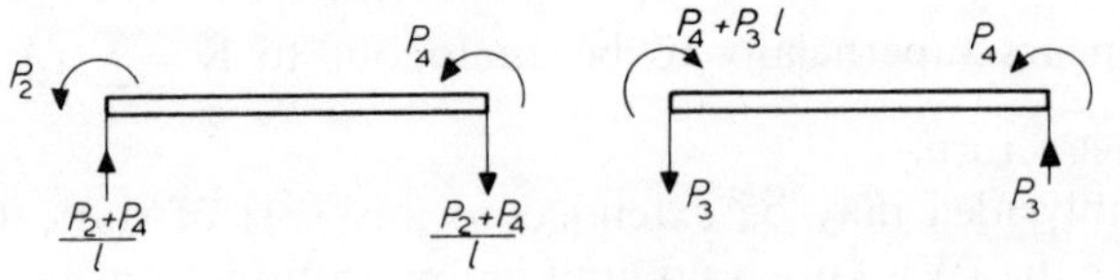

Fig. 3.9 Alternative $\mathbf{P}_g$

This equation is typical of the sort of statement we must always start with in assembling a total flexibility matrix from the set of component elements whether the elements be bars, beams, shear panels, tubes, or general finite elements as in Chapter 6. The stress has been found in terms of the element forces $\mathbf{P}_g$, and in terms of some function of x, y, or z which represents the distribution of stress through the element. It is convenient to summarize this as

$$\boldsymbol{\sigma} = \boldsymbol{\beta}\mathbf{P}_g \tag{3.13}$$

where in this case

$$\boldsymbol{\beta} = \frac{y}{I}[-z \quad -1]$$

In this particular example there is only one stress component σ_{zz} and so $\boldsymbol{\sigma}$ happens to be a scalar.

Putting

$$\bar{\boldsymbol{\sigma}} = \boldsymbol{\beta}\bar{\mathbf{P}}_g \qquad \text{and} \qquad \boldsymbol{\varepsilon} = \frac{1}{E}\boldsymbol{\beta}\mathbf{P}_g$$

the virtual work in the element now becomes

$$\int_{V_g} \bar{\boldsymbol{\sigma}}^t \boldsymbol{\varepsilon}\, dv = \bar{\mathbf{P}}_g^t \int_{V_g} \frac{1}{E}\boldsymbol{\beta}^t\boldsymbol{\beta}\, dv\, \mathbf{P}_g$$

where the integral

$$\int_{V_g} \boldsymbol{\beta}^t\boldsymbol{\beta}\, dv = \int_A \frac{y^2\, dA}{I^2} \int_0^l \{-z \quad -1\}[-z \quad -1]\, dz$$

$$= \frac{1}{I}\begin{bmatrix} \frac{l^3}{3} & \frac{l^2}{2} \\ \frac{l^2}{2} & l \end{bmatrix}$$

The above virtual work must also be equal to $\bar{\mathbf{P}}_g^t \boldsymbol{\rho}_g$, where $\boldsymbol{\rho}_g$ are the displacements corresponding to $\mathbf{P}_g$, and it is logical to write them as

$$\boldsymbol{\rho}_g = \mathbf{f}_g \mathbf{P}_g \tag{3.14}$$

where $\mathbf{f}_g$ is the beam element flexibility. Thus the virtual work is $\bar{\mathbf{P}}_g^t \mathbf{f}_g \mathbf{P}_g$ and comparing this with the above, we deduce that

$$\mathbf{f}_g = \frac{1}{E}\int_{V_g} \boldsymbol{\beta}^t\boldsymbol{\beta}\, dv = \frac{l}{6EI}\begin{bmatrix} 2l^2 & 3l \\ 3l & 6 \end{bmatrix} \tag{3.15}$$

The flexibility for alternative descriptions $\mathbf{P}_g$ is as straightforward (see Problem 3.4).

At this stage we have a check on both our arithmetic and our philosophy. Recalling equation (2.27), we have the value of beam stiffness $\mathbf{k}_g$ for a free beam having four displacement freedoms $\boldsymbol{\rho}_g$. By suitably supporting the beam ($\rho_1 = \rho_2 = 0$) we may deduce the stiffness for the two freedoms $\boldsymbol{\rho}_g = \{\rho_3 \quad \rho_4\}$ present in our model. Thus, deleting row/columns 1 and 2 in (2.27),

$$\mathbf{k}_g = \frac{2EI}{l^3}\begin{bmatrix} 6 & -3l \\ -3l & 2l^3 \end{bmatrix}$$

But $\mathbf{P}_g = \mathbf{k}_g\boldsymbol{\rho}_g = \mathbf{k}_g\mathbf{f}_g\mathbf{P}_g$, so $\mathbf{k}_g\mathbf{f}_g = \mathbf{I}$. On substituting (3.15) this is confirmed: the flexibility is indeed the inverse of the stiffness matrix. (We note that we had to make the stiffness matrix invertible by putting $\rho_1 = \rho_2 = 0$ and so suppressing any possible rigid-body displacements.)

The flexibility of a complete framework can now be assembled since we have the PVF at the stage $\bar{\mathbf{R}}^t\mathbf{r} = \sum \bar{\mathbf{P}}_g^t\mathbf{f}_g\mathbf{P}_g$, and it remains only to find the local beam forces $\mathbf{P}_g$ in terms of $\mathbf{R}$. This is written, as for pinjointed frameworks (3.10), as

$$\mathbf{P}_g = \mathbf{b}_g\mathbf{R} \tag{3.16}$$

except that $\mathbf{P}_g$ is not a scalar. So this time the flexibility

$$\mathbf{F} = \sum_g \mathbf{b}_g^t\mathbf{f}_g\mathbf{b}_g \tag{3.17}$$

contains matrix flexibilities. An example should illustrate the pros and cons of a systematic assembly like (3.17).

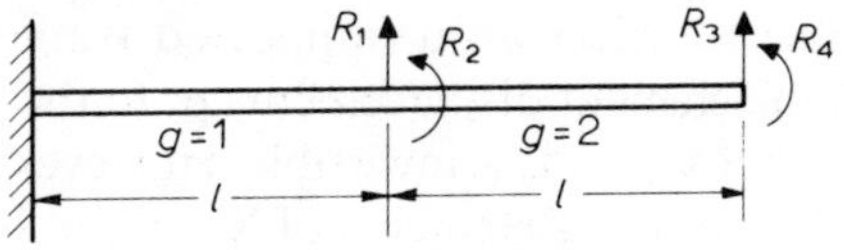

Fig. 3.10 Example

Example

Element $g = 1$:

$$\mathbf{P}_1 = \{R_1 + R_3 \quad R_2 + R_4 + R_3 l\}$$

$$\therefore \quad \mathbf{b}_1 = \begin{bmatrix} 1 & 0 & 1 & 0 \\ 0 & 1 & l & 1 \end{bmatrix}$$

Using (3.15),

$$\mathbf{b}_1^t\mathbf{f}_1\mathbf{b}_1 = \frac{l}{6EI}\begin{bmatrix} 2l^2 & 3l & 5l^2 & 3l \\ & 6 & 9l & 6 \\ \text{symm.} & & 14l^2 & 9l \\ & & & 6 \end{bmatrix}$$

Element $g = 2$: $\mathbf{P}_2 = \{R_3 \quad R_4\}$

$$\therefore \quad \mathbf{b}_2 = \begin{bmatrix} 0 & 0 & 1 & 0 \\ 0 & 0 & 0 & 1 \end{bmatrix}$$

$$\mathbf{b}_2^t \mathbf{f}_2 \mathbf{b}_2 = \frac{l}{6EI} \begin{bmatrix} 0 & 0 & 0 & 0 \\ & 0 & 0 & 0 \\ \text{symm.} & & 2l^2 & 3l \\ & & & 6 \end{bmatrix}$$

$$\therefore \quad \mathbf{F} = \sum_{g=1,2} \mathbf{b}_g^t \mathbf{f}_g \mathbf{b}_g = \frac{l}{6EI} \begin{bmatrix} 2l^2 & 3l & 5l^2 & 3l \\ & 6 & 9l & 6 \\ & & 16l^2 & 12l \\ \text{symm} & & & 12 \end{bmatrix}$$

As predicted, the flexibility matrix is fully populated.

Although the above procedure is systematic and readily programable, the number of arithmetic steps is large, and the generation of the $\mathbf{b}_g$ matrices becomes tedious and not even particularly routine for complex structures with perhaps many sorts of elements. It is therefore tempting to contemplate whether it is possible to construct an unsupported element flexibility like we did for the stiffness $\mathbf{k}_g$ in (2.16). The advantage in doing this was that the set of four element displacements $\boldsymbol{\rho}_g$ were all individually components also of $\mathbf{r}$ so that $\mathbf{a}_g$ in $\boldsymbol{\rho}_g = \mathbf{a}_g\mathbf{r}$ was Boolean. The equivalence of element work work in terms of both $\boldsymbol{\rho}_g$ and $\mathbf{r}$ meant that the element stiffness terms merely had to be inserted into the appropriate positions in the global stiffness $\mathbf{K}$. No congruent transformations $\mathbf{a}^t\mathbf{k}\mathbf{a}$ were necessary.

It is indeed possible to select an unsupported (singular) element flexibility matrix $\mathbf{f}_g$ relating *all* the element displacements $\boldsymbol{\rho}_g$ to the forces $\mathbf{P}_g$. In fact, $\mathbf{k}_g\mathbf{f}_g$ is still $\mathbf{I}$ even though neither $\mathbf{k}_g$ nor $\mathbf{f}_g$ is invertible. However, it is still not possible to identify the components of $\mathbf{P}_g$ as elements of $\mathbf{R}$, so that $\mathbf{b}_g$ would be Boolean. In fact, if we think about it, the equations of equilibrium can be obtained from the PVD as

$$\sum \bar{\boldsymbol{\rho}}_g^t \mathbf{P}_g = \bar{\mathbf{r}}^t \mathbf{R}, \qquad \text{or putting} \quad \bar{\boldsymbol{\rho}}_g = \mathbf{a}_g \bar{\mathbf{r}},$$

$$\sum \mathbf{a}_g^t \mathbf{P}_g = \mathbf{R}$$

These equations are not going to produce a Boolean relationship $\mathbf{P}_g \doteq \mathbf{b}_g\mathbf{R}$

3.4 Statically indeterminate pinjointed frameworks – The force method

The perceptive reader will be aware that so far we have restricted ourselves to statically determinate frameworks in which we can directly write the stress or internal forces $\mathbf{P}_g$ in terms of the applied loads, $\mathbf{P}_g = \mathbf{b}_g\mathbf{R}$. If the framework is statically indeterminate then this is no longer possible. Indeed the problem of solving the stresses then becomes much more important than finding deflections or flexibilities. We have already shown in Chapter 2 how to solve indeterminate frameworks by the displacement method, in which case the displacements are solved everywhere before the stresses. We also mentioned (equation (1.45)) that a displacement solution can be set up using the equilibrium equations to assemble the stiffness matrix. The question is now posed, 'Can we set up a *force solution* of statically indeterminate frameworks using only equilibrium equations and forces as unknowns?' The motive for so doing is not purely academic. It turns out that the number of equations to be solved is usually far less than in the displacement method, although they are not as routine or straightforward to solve. The method will be more suited to moderately sized structures. If a robust routine method is desired, and cheap access to a computer is available, then the displacement method is usually preferred.

We recall (1.74) that the general PVF is

$$\int_V \bar{\boldsymbol{\sigma}}_i^t \boldsymbol{\varepsilon}\, \mathrm{d}v = \int_{S_u} \bar{\mathbf{P}}_s^t \tilde{\mathbf{u}}\, \mathrm{d}s \tag{1.74}$$

where the unknown displacements have been eliminated entirely by allowing virtual forces only at those few places on the surface, Su, where these displacements $\tilde{\mathbf{u}}$ are prescribed. We will look at such a problem but in the majority of cases, wherever displacements are given, these are zero – at the supports, say. In this case (1.74) becomes

$$\int_V \bar{\boldsymbol{\sigma}}_i^t \boldsymbol{\varepsilon}\, \mathrm{d}v = 0 \tag{3.18}$$

The force solution of statically indeterminate frames can be developed quite generally using matrix notation for brevity. However, some of the subtle ideas may be lost in the algebra; and some of them may be lost in the PVF which tends to remove the need for thought completely. We will therefore develop the theme on a specific problem before generalizing.

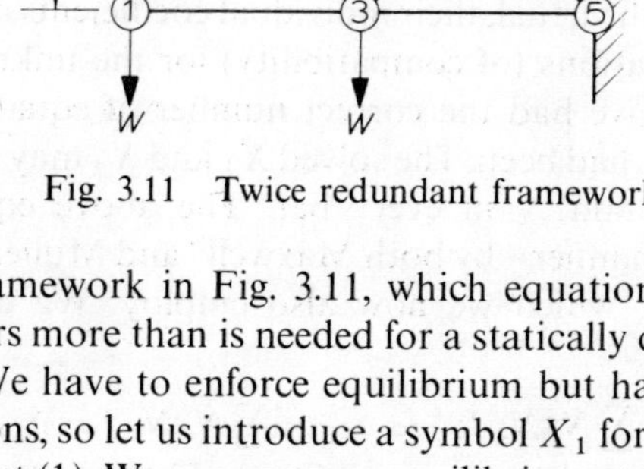

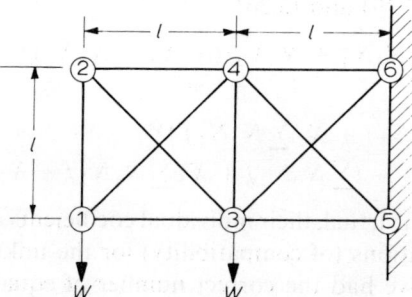

Fig. 3.11 Twice redundant framework

Consider the framework in Fig. 3.11, which equation (1.18) shows has two redundant members more than is needed for a statically determinate connection of the six joints. We have to enforce equilibrium but have more unknown bar forces than equations, so let us introduce a symbol X_1 for (say) the unknown bar force N_{14} meeting at (1). We can now use equilibrium at (1) to solve for N_{12} and

N_{13} in terms of X_1 and W and similarly at (2), but at (3) or (4) we meet another redundant impasse. Denoting again the force N_{34} as an unknown redundancy X_2, we may use equilibrium at (4) and (3) to complete the solution of all bar forces in terms of W, X_1, and X_2. The complete picture is shown in Fig. 3.12. The force N in any typical bar is therefore a linear combination of the applied loads W and of X_1 and X_2. We write this symbolically as

$$N = N_0 + N_1X_1 + N_2X_2 \tag{3.19}$$

where N_0 is the contribution due to the applied loads W.

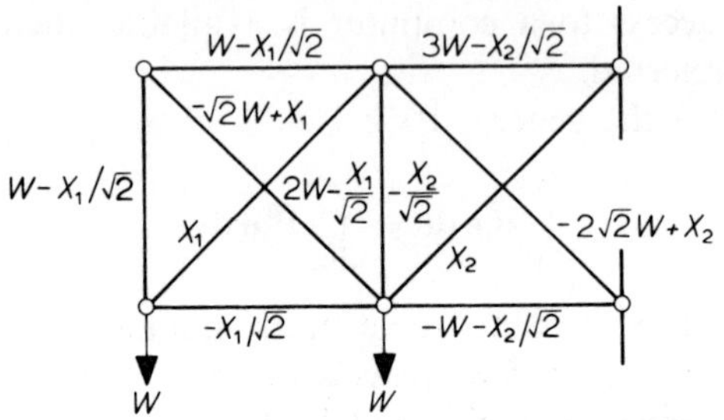

Fig. 3.12 Forces in terms of W, X_1, X_2

We now have to select a virtual stress system $\bar{\boldsymbol{\sigma}}$ which must satisfy equilibrium but require no virtual applied loads. But (3.19) is derived entirely from equilibrium equations and therefore if we put $N_0 = 0$, and hence remove the effect of the applied loads, we have a ready-made self-equilibrating virtual stress system,

$$\bar{N} = N_1\bar{X}_1 + N_2\bar{X}_2 \tag{3.20}$$

The PVF (3.18) then becomes, summing over all bars and assuming Hooke's law,

$$\sum \frac{\bar{N}}{A}\frac{N}{AE}Al = 0 = \sum \bar{N}Nf$$

where $f = l/AE$. Using (3.19) and (3.20),

$$\sum(N_0 + N_1X_1 + N_2X_2)(N_1\bar{X}_1 + N_2\bar{X}_2)f = 0$$

or

$$(\sum N_0N_1f + X_1\sum N_1N_1f + X_2\sum N_2N_1f)\bar{X}_1$$
$$+ (\sum N_0N_2f + X_1\sum N_1N_2f + X_2\sum N_2N_2f)\bar{X}_2 = 0$$

Since $\bar{X}_1$ and $\bar{X}_2$ are both virtual, their individual coefficients must vanish and we therefore have two equations (of compatibility) for the unknowns X_1 and X_2. Obviously we would have had the correct number of equations whatever the number of redundancies had been. The solved X_1 and X_2 may then be substituted back into (3.19) to obtain N in every bar. The above equations were first derived – in a different manner – by both Maxwell[3] and Müller-Breslau[24] using a more concise notation, which we now also employ. We denote the various summations as

$$\delta_{10} = \sum N_0N_1f; \qquad \delta_{12} = \sum N_1N_2f = \delta_{21};$$
$$\delta_{11} = \sum N_1N_1f; \qquad \delta_{22} = \sum N_2N_2f \tag{3.21}$$

Our equations then become

$$\begin{aligned}\delta_{11}X_1 + \delta_{12}X_2 + \delta_{10} &= 0 \\ \delta_{21}X_1 + \delta_{22}X_2 + \delta_{20} &= 0\end{aligned} \tag{3.22}$$

This force solution is a fairly impressive demonstration of the PVF as an alternative strategy. The displacement solution would produce eight equations for the unknown displacements of joints (1), (2), (3), and (4).

Now we explain in physical terms some of the algebraic manipulation above, although we should warn that this physical interpretation is not possible or profitable on more complex structures.

Firstly, there is a more convenient way of generating N_0, N_1, and N_2 solutions which for complex frameworks will rapidly test our ingenuity at introducing further unknowns X. Equation (3.19) may be simply split into its component parts N_0, N_1, and N_2 by putting:

(a) $X_1 = X_2 = 0$ and applying the loads W.
(b) $X_1 = 1$, $X_2 = 0$, and no applied loads.
(c) $X_1 = 0$, $X_2 = 1$, and no applied loads.

These systems may be physically imagined by introducing 'cuts' in the chosen redundant bars and applying successively W, $X_1 = 1$, and $X_2 = 1$, as shown in Fig. 3.13(a), (b), (c). Such a set of solutions, even for a large number of redundancies, is relatively easy to construct. Each one is an equilibrium solution of a statically determinate framework. This statically determinate 'cut' structure is known as the *basic system*.

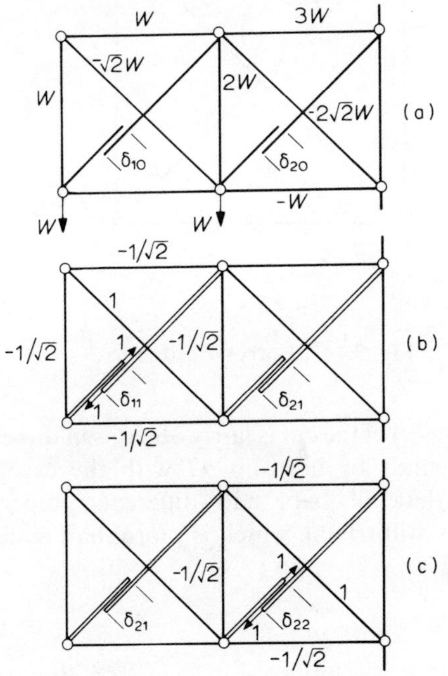

Fig. 3.13 (a) N_0 due to applied loads. (b) N_1 due to $X_1 = 1$. (c) N_2 due to $X_2 = 1$

In each loading case the 'cuts' will obviously undergo relative displacements, and have been shown in Fig. 3.13 to slide over each other. The choice of the depicted labels for these 'cut' displacements is not coincidental. Consider for example δ_{10} in Fig. 3.13(a), due to the bar forces N_0. If we wish to find δ_{10} by the unit-load theorem we would apply successively unit loads to both forces and solve $\bar{N}$. But this is precisely what N_1 is in Fig. 3.13(b). Thus by (3.2) with $\bar{N} = N_1$, we know that

$$\delta_{10} = \sum N_1 N_0 f, \qquad \text{which agrees with (3.21)}$$

Similarly for δ_{20}, δ_{21}, etc. Thus, equations (3.22) which we knew in principle were compatibility statements, are simply restoring the cuts by making the total relative displacement in (a) + X_1(b) + X_2(c) equal to zero.

The force method can easily cope with initial or thermal strains since it is the *total* strain, due to applied forces, temperature, and so on, which is used in the PVF. For example, consider the frame in Fig. 3.14 having all bars heated by T degrees except for the bar (1)–(2). Choosing this bar as the redundancy X_1, there are no N_0 forces due to applied loads, and the forces N_1 due to $X_1 = 1$ are as shown in the figure. Thus we have

$$\delta_{10} = \int_V \bar{\boldsymbol{\sigma}}^t \boldsymbol{\varepsilon}\, dv = \sum N_1 \alpha T l = 2(-\sqrt{2})(\alpha T \sqrt{2} l) + 3(1)(\alpha T l) = -\alpha T l$$

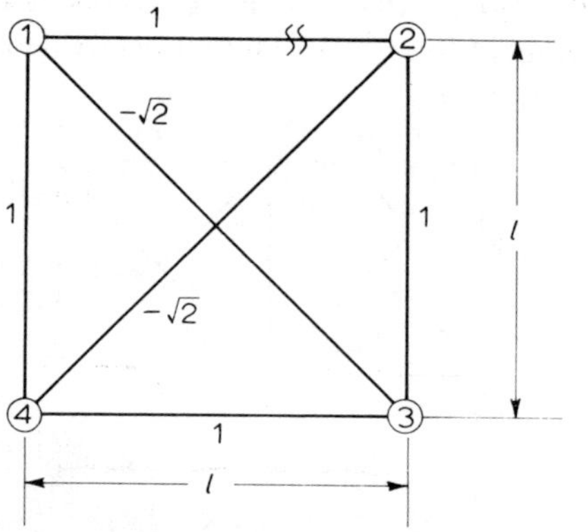

Fig. 3.14 Forces N_1 due to $X_1 = 1$

The 'contraction' $(-\alpha T l)$ of the cut is fairly obvious in this example anyway, since the entire frame expands by a strain αT with the exception of the cut bar. However, imagine that all bars had different temperature increases: the summation for δ_{10} is still trivial, which is more than could now be said of any other way of finding it.

Also,

$$\delta_{11} = \sum N_1 N_1 f = 4(1)^2 \frac{l}{AE} + 2(-\sqrt{2})^2 \frac{\sqrt{2} l}{AE} = 4(1 + \sqrt{2}) \frac{l}{AE}$$

Thus (3.22),

$$\delta_{11}X_1 + \delta_{10} = 0, \qquad \text{delivers} \quad X_1 = AE\alpha T/4(1+\sqrt{2})$$

The force method can be extended to other structures. We have deliberately developed the analysis thus far by introducing unknowns X to satisfy equilibrium requirements, and only later using the concept of 'cuts'. These 'cuts' are either inappropriate or impossible to imagine in many structures; however, they may be imagined in beams and stiff-jointed frames.

3.5 Statically indeterminate beams

The main difference between beams, stiff-jointed frameworks, and the previous pinjointed variety is that the stress is no longer constant in any member. It varies linearly across the section and in almost any way along the length. The problem of satisfying equilibrium requirements then falls into the category mentioned in the introduction to this chapter. The equations for $\boldsymbol{\sigma}$ must be a special case of the general $\mathbf{D}\boldsymbol{\sigma} + \mathbf{p}_v = \mathbf{0}$ which when integrated give $\boldsymbol{\sigma}$ as a function of the applied load $\mathbf{p}_v$ (the particular integral) and as other functions of integration (the complementary function) whose magnitudes have to be determined. These magnitudes will be our redundancies X_k whose selection we discuss shortly. Continuing the same theme as the previous section, we use subscripts 'o' for the stresses in equilibrium with the applied load and k for the redundant self-equilibrating systems. Thus the particular integral will be $\boldsymbol{\sigma}_0(x, y, z)$ and the complementary function $\boldsymbol{\sigma}_k(x, y, z)$. The general solution will look like

$$\boldsymbol{\sigma}(x, y, z) = \boldsymbol{\sigma}_0(x, y, z) + \sum_k X_k \boldsymbol{\sigma}_k(x, y, z) \tag{3.23}$$

The true strains in a linear material, including initial strains $\boldsymbol{\eta}$, from (2.2) are

$$\boldsymbol{\varepsilon} = \boldsymbol{\phi}\boldsymbol{\sigma} + \boldsymbol{\eta} = \boldsymbol{\phi}\boldsymbol{\sigma}_0 + \sum X_k \boldsymbol{\phi}\boldsymbol{\sigma}_k + \boldsymbol{\eta} \tag{3.24}$$

Again, having solved (3.23) the selection of $\bar{\boldsymbol{\sigma}}_i$ is trivial; we merely put all $\boldsymbol{\sigma}_0 = 0$, then

$$\bar{\boldsymbol{\sigma}}_i = \sum \bar{X}_k \boldsymbol{\sigma}_k \tag{3.25}$$

On substituting (3.25) and (3.24) into PVF (3.18), we obtain

$$\sum_k \bar{X}_k \int_V \boldsymbol{\sigma}_k^{\mathrm{t}} (\boldsymbol{\phi}\boldsymbol{\sigma}_0 + \sum_k X_k \boldsymbol{\phi}\boldsymbol{\sigma}_k + \boldsymbol{\eta})\, \mathrm{d}v = 0$$

As this is zero for all possible virtual $\bar{X}_k$, we have 'k' equations for the X_k unknowns; the jth equation being

$$\int_V \boldsymbol{\sigma}_j^{\mathrm{t}} (\boldsymbol{\phi}\boldsymbol{\sigma}_0 + \boldsymbol{\eta})\, \mathrm{d}v + \sum_k X_k \int_V \boldsymbol{\sigma}_j^{\mathrm{t}} \boldsymbol{\phi}\boldsymbol{\sigma}_k\, \mathrm{d}v = 0 \tag{3.26}$$

We have assumed linear elasticity in putting $\boldsymbol{\varepsilon} = \boldsymbol{\phi}\boldsymbol{\sigma}$. If the stress–strain law is some nonlinear relationship $\boldsymbol{\varepsilon} = f(\boldsymbol{\sigma})$ then the PVF still delivers k equations in the unknowns X_k; however, they are nonlinear, and we prefer to leave the solution of nonlinear sets to numerical methods in Chapter 5.

For simple beams and the like, the stress systems in equation (3.23) can be interpreted again in terms of 'cuts' which make the chosen redundancies $X_k = 0$. Thus $\boldsymbol{\sigma}_0$ are the stresses in the statically determinate basic system due to all the applied loads, and $\boldsymbol{\sigma}_k$ due to $X_k = 1$ with all the other $X = 0$ and with the applied loads removed. Clearly $\boldsymbol{\phi}\boldsymbol{\sigma}_0 + \boldsymbol{\eta}$ is the total strain, say $\boldsymbol{\varepsilon}_0$, in the basic system; we may therefore put

$$\delta_{j0} = \int_V \boldsymbol{\sigma}_j^{\mathrm{t}} \boldsymbol{\varepsilon}_0 \, \mathrm{d}v \tag{3.27}$$

and interpret this as before as the relative displacements of the jth cut due to applied loads and initial strains. Similarly,

$$\delta_{jk} = \int_V \boldsymbol{\sigma}_j^{\mathrm{t}} \boldsymbol{\phi} \boldsymbol{\sigma}_k \, \mathrm{d}v = \int_V \boldsymbol{\sigma}_j^{\mathrm{t}} \boldsymbol{\varepsilon}_k \, \mathrm{d}v \tag{3.28}$$

is the displacement at the jth cut due to a unit load $X_k = 1$.

Equations (3.26) then take the same form as (3.22) in the pinjointed framework:

$$\sum_k X_k \delta_{jk} + \delta_{j0} = 0 \tag{3.29}$$

All equations (3.26)–(3.29) are quite general, but if we care to specify that the structure is beam-like, we may now logically denote as:

M_0 – the bending moment at any point due to applied forces.
M_j or M_k – the bending moment due to $X_j = 1$ or $X_k = 1$ alone.

Then

$$\delta_{jk} = \int_V \left(\frac{M_j y}{I}\right)\left(\frac{1}{E}\right)\left(\frac{M_k y}{I}\right) \mathrm{d}v = \int_l \frac{M_j M_k}{EI} \mathrm{d}z \tag{3.30}$$

and

$$\delta_{j0} = \int_l \frac{M_j M_0}{EI} \mathrm{d}z \tag{3.31}$$

The contribution to δ_{j0} of the initial strains in (3.26) cannot be evaluated until we know the nature and magnitudes of $\boldsymbol{\eta}$.

Suppose we attempt to solve a few beam problems using the concept of cuts to generate M_k, and consider the simple redundant beams illustrated in Fig. 3.15. If frame (a) is completely severed through a section it becomes statically determinate as required, and in doing so we release *three* internal forces X_1, X_2, and X_3, as shown in Fig. 3.16. In beam (b) we release only *two* if axial forces are ignored. In beam (c), which is only singly redundant, we could cut the connection between the beam and a support or perhaps insert a 'pin' which puts the moment there to zero.

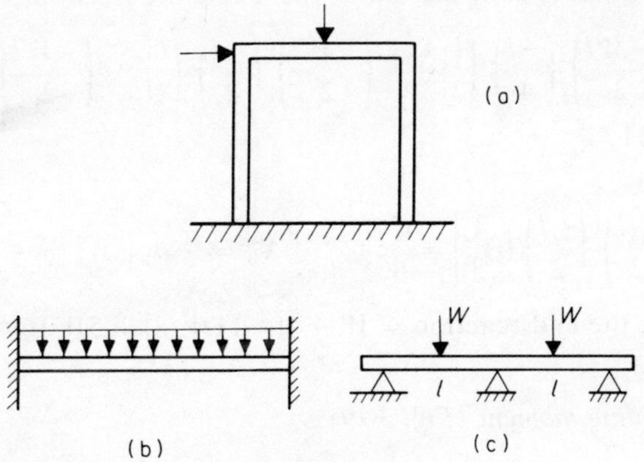

Fig. 3.15 Redundant beams

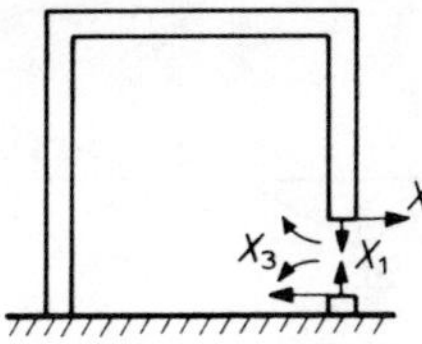

Fig. 3.16 Choice of redundancies

These possible basic systems are summarized in Fig. 3.17. It is apparent that both the type and position of the chosen redundancies are somewhat arbitrary, and yet the final answer should not depend upon this choice. For example, suppose we analyse (c) using the above alternatives.

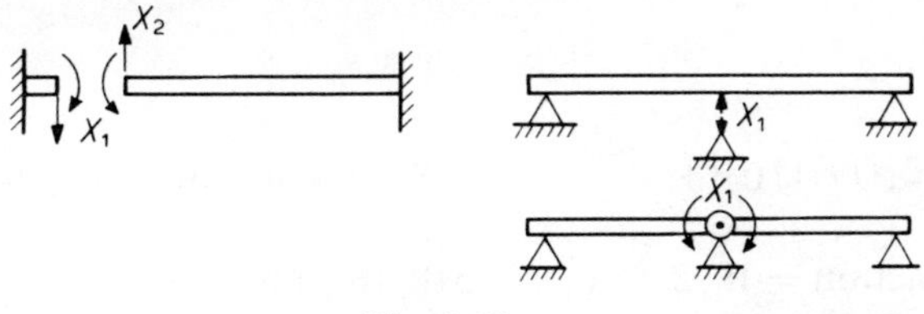

Fig. 3.17

The central support reaction (*Fig.* 3.18)

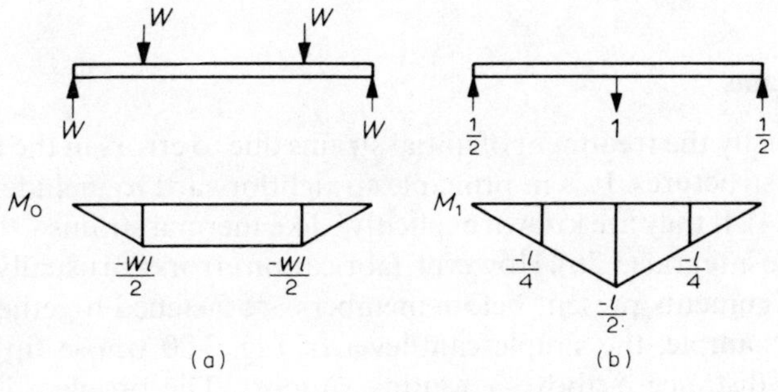

Fig. 3.18

Using (3.30) and (3.31) with the aid of the Table 3.1 (section 3.2):

$$EI\delta_{10} = 2\left[\left(\frac{-Wl}{2}\right)\left(\frac{-l}{4}\right)\left(\frac{l}{2}\right)\frac{1}{3} + \left(\frac{-Wl}{2}\right)\left(\frac{-l}{4}\right)\left(\frac{l}{2}\right) + \left(\frac{-Wl}{2}\right)\left(\frac{-l}{4}\right)\left(\frac{l}{2}\right)\frac{1}{2}\right]$$

$$= Wl^3\frac{11}{48}$$

$$EI\delta_{11} = 2\left[\left(\frac{-l}{2}\right)\left(\frac{-l}{2}\right)(l)\frac{1}{3}\right] = \frac{l^3}{6}; \quad \therefore \quad X_1 = -\delta_{10}/\delta_{11} = -11W/8$$

So for example, the end reaction $= W + \frac{1}{2}(-11W/8) = 5W/16$

The central bending moment (*Fig. 3.19*)

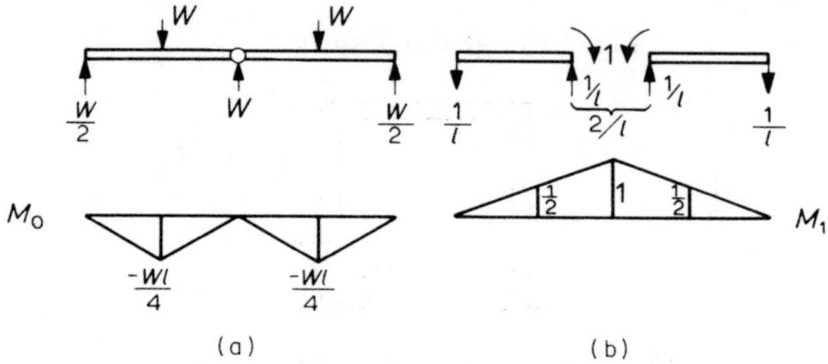

Fig. 3.19

$$EI\delta_{10} = 2\left[\left(\frac{-Wl}{4}\right)\left(\frac{1}{2}\right)\left(\frac{l}{2}\right)\frac{1}{3} + \left(\frac{-Wl}{4}\right)\left(\frac{1}{2}\right)\left(\frac{l}{2}\right)\frac{1}{2} + \left(-\frac{Wl}{4}\right)\left(\frac{1}{2}\right)\left(\frac{l}{2}\right)\frac{1}{6}\right]$$

$$= \frac{-Wl^2}{8}$$

$$EI\delta_{11} = 2[(1)(1)(l)\tfrac{1}{3}] = \frac{2l}{3}; \quad \therefore \quad X_1 = -\delta_{10}/\delta_{11} = 3Wl/16$$

The end reaction $= W/2 - X_1/l = 5W/16$ as before.

The choice of redundancies does therefore seem to be arbitrary. Unfortunately, this is not so when there are a large number of them, and we return to this point in section 3.8.

3.6 Initial strains

We mention briefly the treatment of initial strains due to errors in the fabrication process of real structures. It is in principle straightforward to include any initial strains $\boldsymbol{\eta}$ in (3.24). If they are known explicitly – like thermal strains – they are just inserted into the integral (3.26). However, fabrication errors are usually defined as 'gaps' or misalignments present before members are fastened together at joints. Consider, for example, the simple cantilever of Fig. 3.20 whose tip is initially poised a finite distance δ above a waiting support. The problem is statically

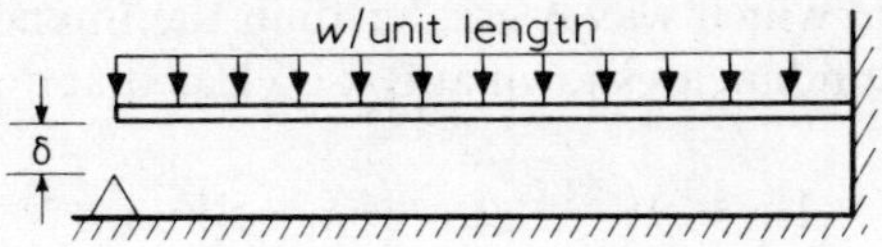

Fig. 3.20 Structure with 'lack of fit'

determinate until the load is sufficient to close the gap and induce a reaction R. One way to solve the problem is to treat it as a case of prescribed displacement δ at the tip. Then using (1.74) we have

$$\int_0^l \frac{\bar{M}M}{EI}\mathrm{d}z = \bar{R}(-\delta)$$

(The right-hand side is the ultimate in *virtual* work in that the *actual* work done by R is zero since it is incapable of moving!) We now introduce the redundant force connection X_1 at, say, the tip; in which case by equilibrium, $\bar{X}_1 = -\bar{R}$. Putting $M = M_0 + X_1 M_1$ as usual, we get

$$\int_0^l \frac{M_0 M_1}{EI}\mathrm{d}z + X_1 \int_0^l \frac{M_1^2}{EI}\mathrm{d}z - \delta - 0$$

However, we could just as easily treat the problem as an initial gap δ which has to be closed by the correct sum of δ_{10} and $X_1\delta_{11}$ in the usual notation. Thus $X_1\delta_{11} + \delta_{10} = \delta$; which is the same as the above.

On evaluating

$$\delta_{10} = wl^4/8EI \qquad \text{and} \qquad \delta_{11} = l^3/3EI$$

we find

$$X_1 = 3EI\delta/l^3 - 3Wl/8$$

As we have posed the problem, the support does not sustain positive values of X_1, that is no contact is made, until $3wl/8$ exceeds $3EI\delta/l^3$.

Initial 'gaps' may of course be more involved than this, and they will always depend on the choice of X_k. For example, in Fig. 3.21 the joint A may have been

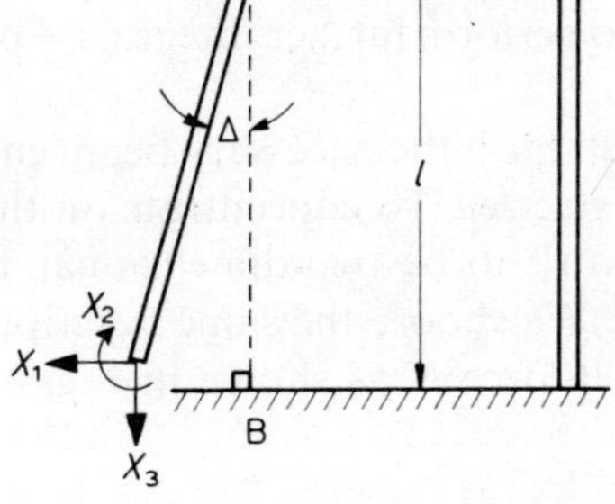

Fig. 3.21 Fabrication error

welded into an angle which was Δ greater than the intended right angle. If we choose the three redundancies shown at B it is clear that due to initial strains we have

$$\delta_{10} = \Delta l; \qquad \delta_{20} = \Delta; \qquad \delta_{30} = 0$$

Now a word of caution when considering initial strains in slender beams and frameworks. Normally it is considered justifiable to neglect axial deformations compared with bending deformations. This seems reasonable when we consider both effects of M and N together; thus

$$\delta_{jk} = \int_V \boldsymbol{\sigma}_j^t \boldsymbol{\varepsilon}_k \, dv = \int_l \left(\frac{M_j M_k}{EI} + \frac{N_j N_k}{AE} \right) dz$$

But if the axial forces and shear forces in the framework are of order N then the moments will be of order Nl, where l is a typical beam length. The ratio of the first term to the second in the integrand is therefore

$$\frac{l^2 A}{I} = \left(\frac{l}{k}\right)^2, \qquad \text{where } k \text{ is the section radius of gyration}$$

This ratio is very large for slender beam, and so we conclude that bending strains predominate, and δ_{jk} can be taken as the first term only. But we must be careful when evluating δ_{j0} due to initial strains $\boldsymbol{\eta}$,

$$\delta_{j0} = \int_V \left(\frac{M_j y}{I} + \frac{N_j}{A} \right) \eta \, dv = \int_l N_j \eta \, dz$$

if the strains η are uniform over the section. Consequently, had we included only bending strains in δ_{j0} there would have been no contribution at all from the initial strains.

3.7 Complex frameworks

Faced with stiff-jointed frameworks typical of modern multi-storey buildings, we have to be a little more systematic than in the previous section. It is necessary to use the ideas developed in the stiffness and flexibility matrix assembly, whereby the stresses in a beam element are summarized in terms of the forces at the end of the element, and so the virtual-work integrals can be evaluated over the element in terms of them. The total virtual work of the complete structure can then be summed without the need to perform further integrals, a procedure ideally suited to a computer.

The first step then is to establish the necessary beam end forces $\mathbf{P}_g$ from which we can obtain the internal stresses. To concentrate on the assembly process we will firstly take the framework to be two-dimensional, and all beams to have principal axes in this plane. We choose the same description of end forces $\mathbf{P}_g$ as (3.12) and now add an axial force N as shown in Fig. 3.22.

Thus

$$\mathbf{P}_g = \{P_3 \quad P_4 \quad N\} \tag{3.32}$$

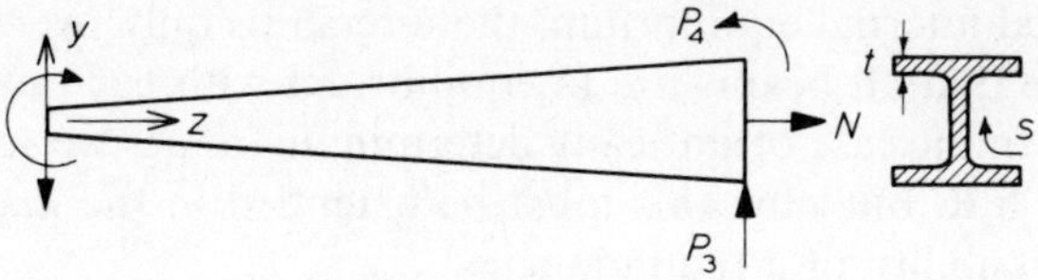

Fig. 3.22 $\mathbf{P}_g$ for a beam with bending, tension, and shear

The extensional deformations due to N are likely to be of secondary importance but we include N for interest, as it would be necessary if the beam were not too slender – in which case the shear strains might be significant so we include them also. This latter is a formidable task in general, but if the beam section is *thin-walled* then the shear stresses σ_{sz} may be taken as uniform across the thickness $t(s)$. The equation of equilibrium (1.48) then may be written as

$$t\frac{\partial \sigma_{zz}}{\partial z} + \frac{\partial}{\partial s}(t\sigma_{sz}) = 0 \tag{3.33}$$

where the thickness t must be incorporated with σ_{sz} as a product if both vary with s. The direct stress $\sigma_{zz} = My/I$; and if the beam is only slightly tapered we may write

$$\frac{\partial \sigma_{zz}}{\partial z} = \frac{\partial}{\partial z}\left(\frac{My}{I}\right) = \frac{y}{I}\frac{\partial M}{\partial z} = \frac{y}{I}P_3$$

Substituting into (3.33),

$$\sigma_{sz} = -\frac{P_3}{It}\int_0^s ty\,\mathrm{d}s$$

where the shear stress must be zero at the free tip of the section $s = 0$. (If the section is a closed tube then the shear stress is zero on an axis of symmetry and the above expression is still valid.) We abbreviate the integral as

$$\int_0^s ty\,\mathrm{d}s = D(s)$$

so

$$\sigma_{sz} = \frac{-D(s)}{It}P_3 \tag{3.34}$$

The stress–force relation, $\boldsymbol{\sigma} = \boldsymbol{\beta}\mathbf{P}_g$, of equation (3.13) has now been extended to

$$\begin{bmatrix}\sigma_{zz}\\ \\ \sigma_{sz}\end{bmatrix} = \begin{bmatrix}\dfrac{-y(l-z)}{I} & \dfrac{-y}{I} & \dfrac{1}{A}\\ \dfrac{-D(s)}{It} & 0 & 0\end{bmatrix}\begin{bmatrix}P_3\\ P_4\\ N\end{bmatrix} \tag{3.55}$$

and the material flexibility, in $\boldsymbol{\varepsilon} = \boldsymbol{\phi}\boldsymbol{\sigma}$, to

$$\boldsymbol{\phi} = \begin{bmatrix}\dfrac{1}{E} & \dfrac{1}{G}\end{bmatrix}$$

Having satisfied internal equilibrium, there remains only overall equilibrium in which we state that all beams are in equilibrium with each other and with the applied loads. In the case of statically determinate frames we summarized this in (3.16) as $\mathbf{P}_g = \mathbf{b}_g\mathbf{R}$, but now this must be extended in the manner of (3.19) to include the possibility of n redundancies,

$$\mathbf{X} = \{X_1 \quad X_2 \quad X_3 \quad \ldots \quad X_n\} \tag{3.36}$$

where these redundancies are necessary when the total number of bar forces $\mathbf{P}_g$ in the entire structure exceeds the number of equations of equilibrium by n. We can always write for the gth beam

$$\mathbf{P}_g = \mathbf{b}_{0g}\mathbf{R} + \mathbf{b}_{xg}\mathbf{X} \tag{3.37}$$

The meanings of the matrices $\mathbf{b}_{og}$ and $\mathbf{b}_{xg}$ should be clear from this expression – even if the choice of $\mathbf{X}$ is not yet so. If $\mathbf{X}$ was a set of discrete internal forces, then by putting them to zero – perhaps by 'cuts' – we infer that $\mathbf{b}_{og}$ is the set of beam forces for the gth element due to unit values of the $\mathbf{R}$ acting alone upon the basic system. Similarly, each column of $\mathbf{b}_{xg}$ is the set of forces $\mathbf{P}_g$ induced by a single redundancy in $\mathbf{X}$, with all applied loads $\mathbf{R} = \mathbf{0}$. For example, in Fig. 3.23 there are three elements, that is, nine unknowns. Ignoring the two clamped ends where the reactions are unknown, there are two sets of three equilibrium equations at the loaded points R_1 and R_2, hence $9 - 6 = 3$ redundancies. Suppose we choose, as these, the beam forces at the position R_1 as shown. The $\mathbf{b}_{0g}$ and $\mathbf{b}_{xg}$ systems are then as follows:

$$g = 1: \qquad \mathbf{b}_{01} = \begin{bmatrix} 1 & 0 \\ 0 & 0 \\ 0 & 0 \end{bmatrix} \qquad \mathbf{b}_{x1} = \begin{bmatrix} 1 & 0 & 0 \\ 0 & 1 & 0 \\ 0 & 0 & 1 \end{bmatrix}$$

$$g = 2: \qquad \mathbf{b}_{02} = \begin{bmatrix} 0 & 0 \\ 0 & 0 \\ 0 & 0 \end{bmatrix} \qquad \mathbf{b}_{x2} = \begin{bmatrix} 1 & 0 & 0 \\ -l & 1 & 0 \\ 0 & 0 & 1 \end{bmatrix}$$

$$g = 3: \qquad \mathbf{b}_{03} = \begin{bmatrix} 0 & -1 \\ 0 & 0 \\ 0 & 0 \end{bmatrix} \qquad \mathbf{b}_{x3} = \begin{bmatrix} 0 & 0 & 1 \\ -l & 1 & -2l \\ -1 & 0 & 0 \end{bmatrix}$$

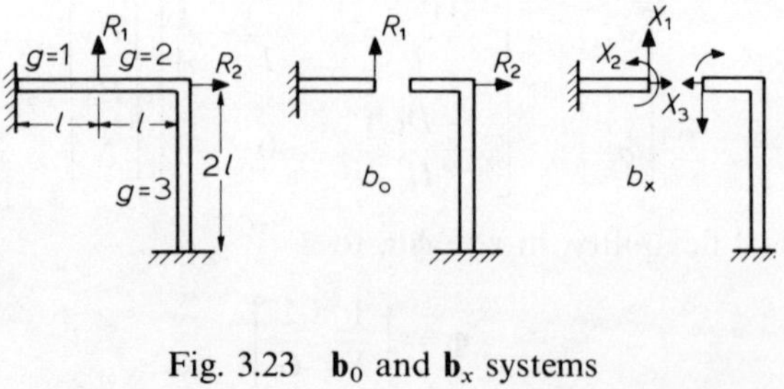

Fig. 3.23 $\mathbf{b}_0$ and $\mathbf{b}_x$ systems

The generation of these matrices is quite straightforward. For example, in the vertical beam, $g = 3$, if we require the last column in $\mathbf{b}_{x3}$ we apply $X_3 = 1$ and find the three forces at the foot of the beam as $P_3 = 1$, $P_2 = -2l$, and $N = 0$.

We now have the true stresses $\boldsymbol{\sigma} = \boldsymbol{\beta}\mathbf{P}_g = \boldsymbol{\beta}\mathbf{b}_{0g}\mathbf{R} + \boldsymbol{\beta}\mathbf{b}_{xg}\mathbf{X}$, and the self-equilibrating stresses $\bar{\boldsymbol{\sigma}}_i = \boldsymbol{\beta}\bar{\mathbf{P}}_g = \boldsymbol{\beta}\mathbf{b}_{xg}\bar{\mathbf{X}}$, ready to insert into the PVF (3.18). To add a little interest we suppose that there are also initial strains in $\boldsymbol{\varepsilon} = \boldsymbol{\phi}\boldsymbol{\sigma} + \boldsymbol{\eta}$. Then the PVF becomes

$$\int_V \bar{\boldsymbol{\sigma}}_i^{\mathrm{t}}\boldsymbol{\varepsilon}\,\mathrm{d}v = 0 = \sum_g \int_{V_g} \bar{\boldsymbol{\sigma}}_i^{\mathrm{t}}(\boldsymbol{\phi}\boldsymbol{\sigma} + \boldsymbol{\eta})\,\mathrm{d}v$$

$$= \sum_g \int_{V_g} \bar{\mathbf{X}}^{\mathrm{t}}\mathbf{b}_{xg}^{\mathrm{t}}\boldsymbol{\beta}^{\mathrm{t}}(\boldsymbol{\phi}\boldsymbol{\beta}\mathbf{b}_{0g}\mathbf{R} + \boldsymbol{\phi}\boldsymbol{\beta}\mathbf{b}_{xg}\mathbf{X} + \boldsymbol{\eta})\,\mathrm{d}v$$

We note that $\bar{\mathbf{X}}$, $\mathbf{X}$, and $\mathbf{R}$ can be taken outside the summation over the beam elements. Also, the beam flexibility (3.15) emerges once again from the integral as

$$\mathbf{f}_g = \int_{V_g} \boldsymbol{\beta}^{\mathrm{t}}\boldsymbol{\phi}\boldsymbol{\beta}\,\mathrm{d}v \tag{3.38}$$

and the initial strains produce a further element matrix in the form of a set of displacements

$$\mathbf{h}_g = \int_{V_g} \boldsymbol{\beta}^{\mathrm{t}}\boldsymbol{\eta}\,\mathrm{d}v \tag{3.39}$$

The PVF has now become

$$\bar{\mathbf{X}}^{\mathrm{t}}\left[\left(\sum_g \mathbf{b}_{xg}^{\mathrm{t}}\mathbf{f}_g\mathbf{b}_{0g}\right)\mathbf{R} + \left(\sum_g \mathbf{b}_{xg}^{\mathrm{t}}\mathbf{f}_g\mathbf{b}_{xg}\right)\mathbf{X} + \left(\sum_g \mathbf{b}_{xg}^{\mathrm{t}}\mathbf{h}_g\right)\right] = 0$$

for all $\bar{\mathbf{X}}$. Hence we have n equations for the solution for $\mathbf{X}$ again. We denote for brevity the summations as

$$\mathbf{D}_{x0} = \sum_g \mathbf{b}_{xg}^{\mathrm{t}}\mathbf{f}_g\mathbf{b}_{0g} \tag{3.40}$$

$$\mathbf{D}_{xx} = \sum_g \mathbf{b}_{xg}^{\mathrm{t}}\mathbf{f}_g\mathbf{b}_{xg} \qquad \text{(a symmetric matrix)} \tag{3.41}$$

$$\mathbf{H}_x = \sum_g \mathbf{b}_{xg}^{\mathrm{t}}\mathbf{h}_g \tag{3.42}$$

Then the equations for $\mathbf{X}$ are: $\mathbf{D}_{x0}\mathbf{R} + \mathbf{D}_{xx}\mathbf{X} + \mathbf{H}_x = \mathbf{0}$

$$\text{or} \qquad \mathbf{X} = -\mathbf{D}_{xx}^{-1}\mathbf{D}_{x0}\mathbf{R} - \mathbf{D}_{xx}^{-1}\mathbf{H}_x \tag{3.43}$$

Having solved for the redundancies we may substitute back into (3.37) for the forces:

$$\mathbf{P}_g = [\mathbf{b}_{0g} - \mathbf{b}_{xg}\mathbf{D}_{xx}^{-1}\mathbf{D}_{x0}]\mathbf{R} - \mathbf{b}_{xg}\mathbf{D}_{xx}^{-1}\mathbf{H}_x \tag{3.44}$$

The first term is the beam element forces induced by the applied loads, and the second the forces induced by the initial strains. The element stresses then follow: $\boldsymbol{\sigma} = \boldsymbol{\beta}\mathbf{P}_g$. The entire analysis has been performed using equilibrium equations and the PVF. Displacements as such have never been mentioned. However, if we wish in addition to find the displacements after solving $\mathbf{P}_g$, we may use the PVF again as we did for statically determinate frames (equation (3.17),

$$\text{i.e.:} \qquad \bar{\mathbf{R}}^t\mathbf{r} = \int_V \bar{\boldsymbol{\sigma}}^t\boldsymbol{\varepsilon}\,dv$$

We recall that $\bar{\boldsymbol{\sigma}}$ has only to be statically equivalent to $\bar{\mathbf{R}}$, so we can place $\bar{\mathbf{X}} = \mathbf{0}$ in (3.37) and put $\bar{\boldsymbol{\sigma}} = \boldsymbol{\beta}\bar{\mathbf{P}}_g = \boldsymbol{\beta}\mathbf{b}_{0g}\bar{\mathbf{R}}$. The true strains are still

$$\boldsymbol{\varepsilon} = \boldsymbol{\phi\sigma} + \boldsymbol{\eta} = \boldsymbol{\phi\beta}\,[\mathbf{b}_{0g}\mathbf{R} + \mathbf{b}_{xg}\mathbf{X}] + \boldsymbol{\eta}$$

then

$$\bar{\mathbf{R}}^t\mathbf{r} = \sum_g \int_{V_g} \bar{\mathbf{R}}^t\mathbf{b}_{0g}^t\boldsymbol{\beta}^t\,[\boldsymbol{\phi\beta}\mathbf{b}_{0g}\mathbf{R} + \boldsymbol{\phi\beta}\mathbf{b}_{xg}\mathbf{X} + \boldsymbol{\eta}]\,dv$$

$$= \bar{\mathbf{R}}^t\left[\left(\sum_g \mathbf{b}_{0g}^t\mathbf{f}_g\mathbf{b}_{0g}\right)\mathbf{R} + \left(\sum_g \mathbf{b}_{0g}^t\mathbf{b}_{xg}\right)\mathbf{X} + \sum_g \mathbf{b}_{0g}^t\mathbf{h}_g\right]$$

or

$$\mathbf{r} = \mathbf{D}_{00}\mathbf{R} + \mathbf{D}_{x0}^t\mathbf{X} + \mathbf{H}_0 \tag{3.45}$$

where

$$\mathbf{D}_{00} = \sum_g \mathbf{b}_{0g}^t\mathbf{f}_g\mathbf{b}_{0g} \qquad \text{(symmetric)} \tag{3.46}$$

and

$$\mathbf{H}_0 = \sum_g \mathbf{b}_{0g}^t\mathbf{h}_g \tag{3.47}$$

On substituting (3.43) for $\mathbf{X}$, (3.45) becomes

$$\mathbf{r} = [\mathbf{D}_{00} - \mathbf{D}_{x0}^t\mathbf{D}_{xx}^{-1}\mathbf{D}_{x0}]\mathbf{R} + [\mathbf{H}_0 - \mathbf{D}_{x0}^t\mathbf{D}_{xx}^{-1}\mathbf{H}_x] \tag{3.48}$$

The first term in brackets is the displacements due to the applied loads, and the second due to the initial strains $\boldsymbol{\eta}$. From the first we see that the flexibility matrix must be

$$\mathbf{F} = \mathbf{D}_{00} - \mathbf{D}_{x0}^t\mathbf{D}_{xx}^{-1}\mathbf{D}_{x0} \tag{3.49}$$

We now consider briefly the various element and structural matrices which emerged from this extensive force solution.

Firstly, the beam element flexibility matrix $\mathbf{f}_g$, equation (3.38). We already have $\boldsymbol{\beta}(y, s, z)$ in (3.35) and so we find, assuming a *uniform* beam,

$$\mathbf{f}_g = \begin{bmatrix} \frac{l^3}{3EI} + \frac{l}{A'G} & \frac{l^2}{2EI} & 0 \\ & \frac{l}{EI} & 0 \\ \text{symm.} & & \frac{l}{AE} \end{bmatrix} \tag{3.50}$$

The new section property $A' = I^2/\int(D^2/t)\,\mathrm{d}s$ has dimensions of area, but it is also physically apparent that it is an effective 'shear area' when we consider it in isolation in the above matrix. It simply relates the shear force P_3 to the corresponding displacement due to shear strains,

$$\rho_3 = \frac{l}{A'G} P_3 \tag{3.51}$$

which is what we would expect if a uniform shear stress P_3/A' existed in a beam of length l. The ratio of the shear term to the bending term $(l/A'G) \div (l^3/3EI) = 3(E/G)(A/A')(k^2/l^2)$, where k is the radius of gyration again. Mostly E/G and A/A' are of order one and $(k/l)^2$ much less, so shear flexibility can be ignored. (This is not so in sandwich beams where the effective shear modulus of a weak core can make E/G large.) We see above that the separate effects of the direct stresses and shear stresses are additive, which is not unexpected since they both relate to the same force component

$$\rho_3 = [f_B\,(\text{bending}) + f_S\,(\text{shear})]\; P_3$$

In fact this 'series assembly' of flexibilities need not merely refer to different strain components in the same structure; it can also be applied to the assembled flexibility of different structures connected in series and loaded by the same force. This is in contrast to stiffnesses which add only when elements in parallel have a common displacement (see Fig. 3.24). The zeros in (3.50) indicate that there is no coupling between the bending and stretching effects which we have previously obtained separately. If the beam is bent about all axes – including torsion – then the force matrix $\mathbf{P}_g$ has to be extended and the flexibility $\mathbf{f}_g$ evaluated in the usual way (see Problem 3.5).

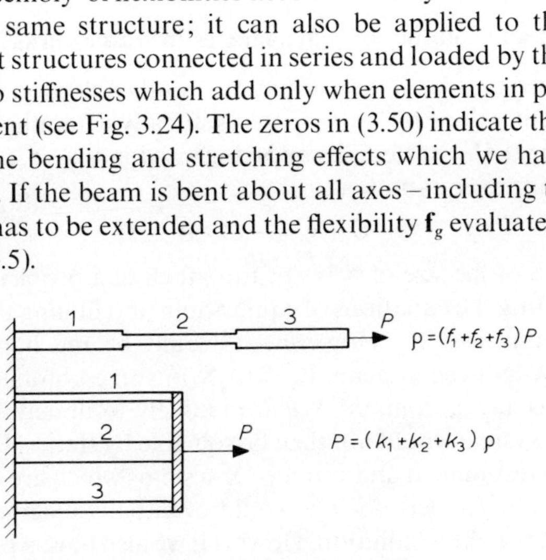

Fig. 3.24 Series and parallel systems

Returning to the initial strains which are accumulated naturally in the matrix $\mathbf{H}_x$, and which then merely supplement the loads $\mathbf{R}$, all that is required is to evaluate the element integral $\mathbf{h}_g$ of equation (3.39), which can only be done when the strains $\boldsymbol{\eta}$ are specified. Suppose, for example, that a beam has imposed a temperature variation which varies linearly through the beam depth like

$$T(y) = T_0 + T_1 \frac{y}{d}$$

where T_0 and T_1 are constants, and d is some typical dimension of the section. For an isotropic material we have

$$\boldsymbol{\eta} = \{\eta_{zz} \quad \eta_{zs}\} = \{\alpha T \quad 0\}$$

Therefore,

$$\boldsymbol{\beta}^t \boldsymbol{\eta} = \begin{bmatrix} \dfrac{-y(l-z)}{I} & \dfrac{-D(s)}{It} \\ \dfrac{-y}{I} & 0 \\ \dfrac{1}{A} & 0 \end{bmatrix} \begin{bmatrix} \alpha T_0 + \alpha T_1 \dfrac{y}{d} \\ 0 \end{bmatrix}$$

And therefore

$$\mathbf{h}_g = \int_0^l \int_A \boldsymbol{\beta}^t \boldsymbol{\eta} \, \mathrm{d}A \, \mathrm{d}z = \left\{ \frac{-\alpha T_1 l^2}{2d} \quad \frac{-\alpha T_1 l}{d} \quad \alpha T_0 l \right\}$$

Due to the 'thermal curvature' $\alpha T_1/d$, the end $z = l$ displaces downwards by $\rho_3 = -\alpha T_1 l^2/2d$, and rotates by $\rho_4 = (-\alpha T_1/d)l$, while the centroidal axis extends simply by $\alpha T_0 l$.

3.8 Selection and solution of redundancies in frameworks

We have implied that both the estimation and the selection of redundancies $\mathbf{X}$ is most conveniently done by 'cutting' to a statically determinate basic system and then generating the $\mathbf{b}_{xg}$ systems by applying unit loads across the cuts. $\mathbf{X}$ may then be solved from the equation

$$\mathbf{D}_{xx}\mathbf{X} = -\mathbf{D}_{x0}\mathbf{R} - \mathbf{H}_x \tag{3.43}$$

The estimation of the size of $\mathbf{X}$ is not too much of a problem. We could always resort to counting the equations of equilibrium and finding the shortfall from the number of unknowns $\mathbf{P}_g$. However, it would be much more convenient to generate the $\mathbf{X}$-induced systems $\mathbf{P}_g = \mathbf{b}_{xg}\mathbf{X}$ in some automatic way so that the sizing of $\mathbf{X}$ was also automatic. We are unlikely to underestimate the size of $\mathbf{X}$ since the basic system would not then be solvable by statics. But it might easily be possible to overestimate it and produce $\mathbf{X}$ systems which are linear combinations of others. This linear dependency would eventually emerge (expensively) as a singular $\mathbf{D}_{xx}$ in the above solution. However, we also have a problem if $\mathbf{D}_{xx}$ is *near singular* or badly conditioned for inversion (see Appendix A).

We will discuss the *inversion* of $\mathbf{D}_{xx}$, although it is uncommon to do more than solve (3.43) for particular $\mathbf{R}$, unless perhaps we required $\mathbf{F}$. In Chapter 2 it was pointed out that the solution of $\mathbf{Kr} = \mathbf{R}$ poses little problems in general, in that $\mathbf{K}$ is usually favourably conditioned for inversion, having a finite diagonal bandwidth if we organize our book-keeping well. The same is not unfortunately true of $\mathbf{D}_{xx}$. We have already shown that the flexibility $\mathbf{F}$ is fully populated and the same may be true of

$$\mathbf{D}_{xx} = \sum_g \mathbf{b}^t_{xg}\mathbf{f}_g\mathbf{b}_{xg}$$

We see that $\mathbf{D}_{xx}$ will only be banded if the self-equilibrating systems due to $X_1 = 1, X_2 = 1, \ldots, X_j = 1$, individually produce forces $\mathbf{P}_g = \mathbf{b}_{xg}X_i$ in a few bars in a small localized region. In other words, *the self-equilibrating systems should overlap as little as possible.* There will always be some overlap, since if there were none it would imply that the redundant systems were entirely surrounded by elements in which the forces were statically determinate, so the redundant substructures could be solved one at a time. (This would mean that $\mathbf{D}_{xx}$ then had *only* diagonal terms.)

The aim should be to construct $\mathbf{X}$ systems with the minimum overlap that is physically possible. For simple pinjointed frameworks the production of $\mathbf{X}$ systems from 'cut' unit loads usually leads to localized solutions. In Fig. 3.25 the systems are all clearly local and the generation of $\mathbf{b}_{xg}$ can be automated. The $\mathbf{b}_{2g}$ system due to $X_2 = 1$, for instance, is confined to the shaded bars. But in Fig. 3.26

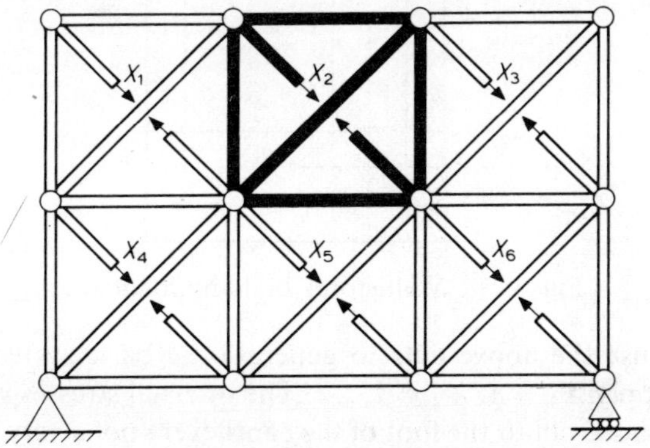

Fig. 3.25 Localized self-equilibrating systems

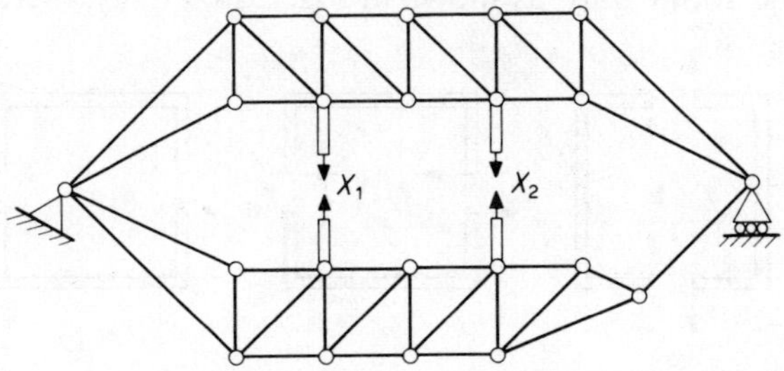

Fig. 3.26 Bad choice of redundancies

the chosen X_1 and X_2 would be bad since $X_1 = 1$ and $X_2 = 1$ excite forces everywhere. The framework in Fig. 3.26 does, however, illustrate again one of the reasons for the early popularity of the force method as opposed to the displacement or stiffness method. There are 22 joints and therefore 44 equations of equilibrium, but 43 unknown bar forces and 3 reactions – hence 2 redundancies. The matrix $\mathbf{D}_{xx}$ is then 2×2 whereas the stiffness formulation involves 41 unknowns at the 22 joints. This example is somewhat extreme and the differences are not usually as pronounced in, for instance, stiff-jointed frameworks. Consider, for example, the common building framework of which one plane is shown in Fig. 3.27, with m floors and n bays. To generate a basic system it is apparent we could 'cut' every floor and leave standing statically determinate column cantilevers – like the last one shown. Three unknown forces would then exist at each cut – and hence $\mathbf{X}$ is of order $3mn$. To check this we note that there are mn floor beam elements and $m(n+1)$ column elements, and since $\boldsymbol{\rho}_g$ for each element has 3 components, there are $3m(2n+1)$ unknown forces. But there are $m(n+1)$ joints at which there are 3 equations of equilibrium, hence the degree of redundancy is $3m(2n+1) - 3m(n+1) = 3mn$. Now if we use the displacement method there will be $3m(n+1)$ unknowns (2 displacements and a rotation at each joint), hence the stiffness matrix $\mathbf{K}$ exceeds $\mathbf{D}_{xx}$ in size by $3m(n+1) - 3mn = 3m$. The force method therefore only merits consideration for very tall thin buildings.

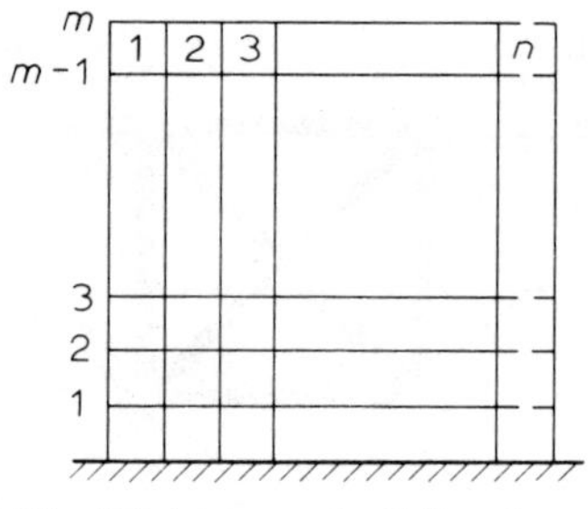

Fig. 3.27 Multistorey building framework

Suppose we use the above cuts to generate the basic system a fairly simple routine – and apply $X_1 = 1, X_2 = 1, \ldots$. The internal stresses will extend all the way from the loaded cut to the foot of the cantilever – not a very localized system. It is, however, fairly easy to generate self-equilibrating $\mathbf{b}_{xg}$ systems which are confined to one 'room' only, as shown in Fig. 3.28. Firstly, there is no doubt that

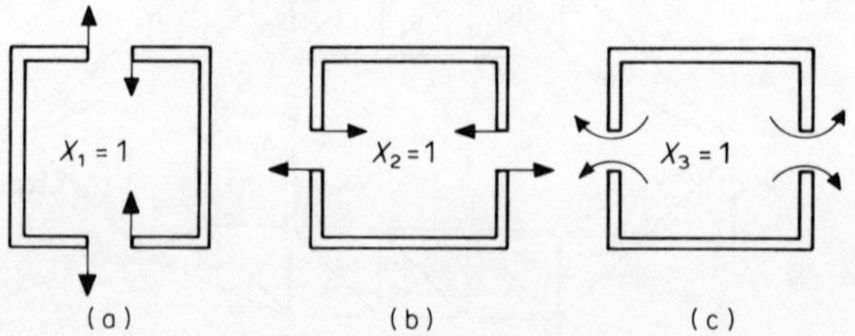

Fig. 3.28 Localized bending systems

(a), (b), and (c) produce internal stresses which are confined to the square region shown. Secondly, we have to be certain that all **X** are independent variables, thus none of the above is capable of being constructed from a combination of the other two. There is also no possibility of constructing a further pattern of stresses in a different **X** system. One might be tempted for example to load the faces in (a) by axial forces, as shown in Fig. 3.29. But this produces stresses identical to Fig. 3.28(b) even though the cuts do not correspond.

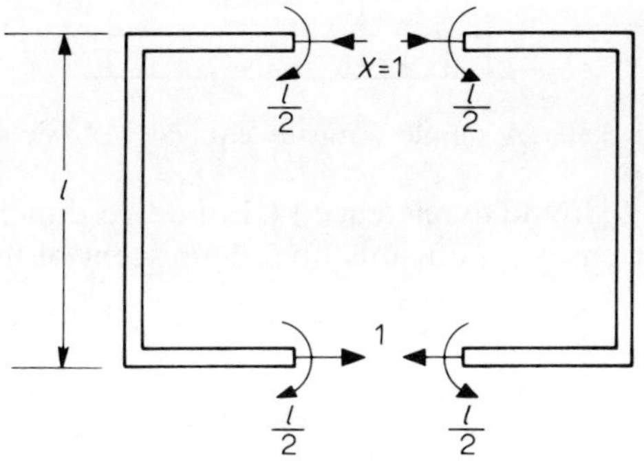

Fig. 3.29

The system shown in Fig. 3.28 can obviously be generated in a repetitive manner throughout the framework, remembering to include the ground-floor system. (The 'floor' itself does not have to be included in the virtual-work summation if it is imagined to have zero flexibility, that is $\boldsymbol{\varepsilon}$ is zero in $\int \bar{\boldsymbol{\sigma}}^t \boldsymbol{\varepsilon}\, dv$.) We see that simple cuts are no longer attractive and we should feel free to construct any convenient self-equilibrating systems and not just those produced by forces at fixed cuts.

We are also completely free to choose unrelated $\mathbf{b}_{0g}$ systems in equilibrium with **R** and here it is of some convenience to generate solutions which are fairly close to the expected true solution; in other words, make the magnitude of **X** as small as possible and render the solution less sensitive to its accurate determination. The $\mathbf{b}_{0g}$ systems may then be quite different from the $\mathbf{b}_{xg}$: for example, the cuts in Fig. 3.27 would undoubtedly result in horizontal forces producing moments much higher than would occur in the uncut framework, and large values of **X** would be required to restore the balance. We could therefore insert a number ($3mn$) of pinjoints, as shown in Fig. 3.30, which would produce a more realistic set $\mathbf{b}_{0g}$ closer to the likely true internal forces even though devoid of bending moments in most places.

The concept of generating local and systematic self-equilibrating systems was first applied by Argyris and Kelsey[14] to stressed-skin box-type structures typical of aircraft wings and also of box girder bridges. These boxes were idealized as assemblies of rods or 'booms' carrying the direct stresses and connected by pure shear panels – a fairly drastic idealization (see also diffusion panels, etc., in the section 3.10). Self-equilibrating systems for three-dimensional boom-panel

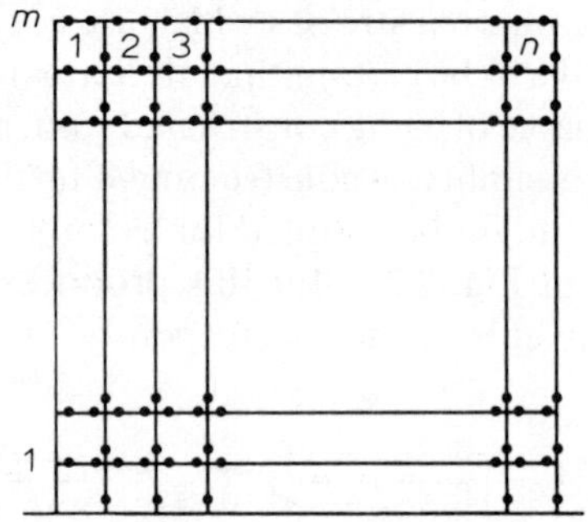

Fig. 3.30 A simple statically equivalent basic system

configurations can be found in reference 14. For a two-dimensional flat plate the nature of the **X** system is an obvious one reminiscent of frameworks (see Fig. 3.31).

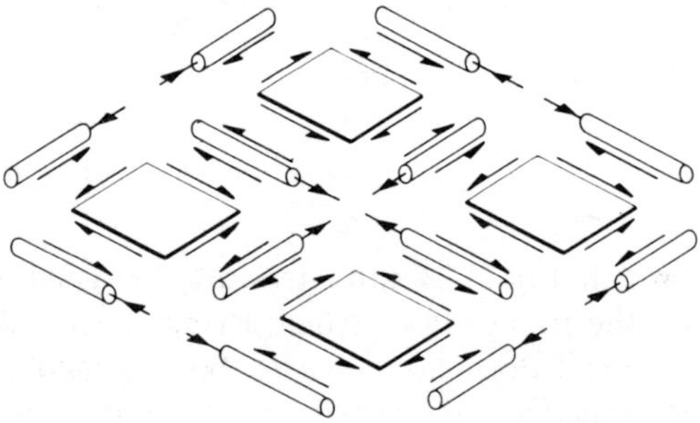

Fig. 3.31 A localized boom-panel system

It is not always straightforward to generate in a systematic way, a series of local self-equilibrating **X** systems. Firstly, even a regular periodic structure may be punctured by a few irregular cut-outs. These are frequent in aircraft and box girders, and are necessary to provide access or visibility. Secondly, the structure may have to be modelled using many different sorts of elements with perhaps little regularity of form and geometry.

The first (cut-out) problem is solved simply and elegantly[14] by 'filling in' the cut-outs and thus restoring the regularity for generating the $\mathbf{b}_{0g}$ and $\mathbf{b}_{xg}$ solutions. The stresses in the fictitious members are then removed as a 'postmortem' exercise. This can be done by selecting a set of initial strains in the elements of the structure which cause stresses that precisely cancel those in the fictitious components.

Briefly then, we collect separately all the h nodal forces associated with the fictitious members as $\mathbf{P}_h$ and the associated $\mathbf{b}_{0g}$ and $\mathbf{b}_{xg}$ are likewise collected together as $\mathbf{b}_{0h}$ and $\mathbf{b}_{xh}$. Now we confine our initial strains $\mathbf{h}_g$ entirely to those unwanted members so that in equation (3.42),

$$\mathbf{H}_x = \sum_h \mathbf{b}_{xg}^t \mathbf{h}_g = \mathbf{b}_{xh}^t \mathbf{H}_h, \qquad \text{say}$$

Then the loads $\mathbf{P}_h$ can be written from (3.44) as

$$\mathbf{P}_h = [\mathbf{b}_{0h} - \mathbf{b}_{xh}\mathbf{D}_{xx}^{-1}\mathbf{D}_{x0}]\mathbf{R} - \mathbf{b}_{xh}\mathbf{D}_x^{-1}\mathbf{b}_{xh}^t\mathbf{H}_h$$

and these must be zero.

Therefore,

$$\mathbf{H}_h = [\mathbf{b}_{xh}^t\mathbf{D}_{xx}^{-1}\mathbf{b}_{xh}]^{-1}\,[\mathbf{b}_{0h} - \mathbf{b}_{xh}\mathbf{D}_{xx}^{-1}\mathbf{D}_{x0}]\mathbf{R} \tag{3.52}$$

The order of the new matrix, in brackets, to be inverted in (3.52) is only equal to the number of nodal forces $\mathbf{P}_h$ to be removed and is presumably small compared with the total number of $\mathbf{P}_g$. The values of $\mathbf{H}_h$ in (3.52) are then simply inserted – where they exist – in the right-hand side of (3.44) when evaluating all the element nodal forces and then stresses.

The second problem, of generating **X** systems in complex irregular structures, is far more difficult and the search for physically derived self-equilibrating systems can be counter-productive. We must then return to our first arguments (Fig. 3.12), where we simply introduced unknowns **X** into all the equations of equilibrium, as and when the need arose. This particular approach was first formalized by Denke,[25, 26] in which he used the Jordan elimination procedure on the augmented equations of equilibrium to separate, by a series of transformations, the assembled $\mathbf{b}_{xg}$ and $\mathbf{b}_{0g}$ matrices for the complete structure. The **X** systems are generated automatically, and by systematic row and column manipulation it is possible to produce a well-conditioned $\mathbf{D}_{xx}$. The reader can find worked examples in reference (28). It is, in fact, possible to go one stage further and demand an optimum of both $\mathbf{b}_{xg}$ *and* $\mathbf{b}_{0g}$ systems. As mentioned before, we could select a best basic system as that which delivered $\mathbf{P}_g = \mathbf{b}_{0g}\mathbf{R}$ differing as little as possible from the true loads. The consequent $\mathbf{P}_g = \mathbf{b}_{xg}\mathbf{X}$ then have only little destroyed compatibility to restore – or the system is *the stiffest load path* for R. Now of course we have introduced *stiffness* and not just equations of equilibrium into our definition of 'best'. Denke[26] and Robinson and Haggenmacher[27] both weight the equilibrium equations using the stiffness $\mathbf{f}_g^{-1}$ in slightly different ways to achieve this objective. However, any automatic selection of **X** systems which is purely algebraic, and based on the equations of equilibrium, needs considerable manipulation, and the popularity of the force method has steadily declined for this reason except in those areas where **X** systems can be chosen by physical insight.

There is another alternative, treated in Chapter 6, which is useful in approximate analysis using finite elements. The flexibility of an element is inverted and then this stiffness incorporated into the global stiffness for solving displacements in the usual manner. In the case of frameworks with real exact elements there is no difference in the stiffness obtained this way, and the method has no advantages.

3.9 Continuous systems

Pursuing the analogy between forces in the PVF and displacements in the PVD, it is natural to emulate section 2.5 and turn to continuous force systems, firstly in one dimension and then in two. The one-dimensional displacement problems were slender beams and tubes, but when we turn to a force description of these problems we have already found they do not generally involve continuous unknown functions. By necessarily satisfying equilibrium conditions, the stresses everywhere are found in terms of a discrete number of unknown internal forces and moments. We have to try very hard to devise a beam-type problem with a continuous force unknown, but the following example will suffice. Figure 3.32 shows a uniform beam supported on a continuous elastic foundation. The support is not a known applied force since the stress in it depends upon the deflection $v(z)$ of the beam: it may therefore be treated as a continuous unknown force variable. The applied loading and the spring stiffness are denoted as $w(z)$ and k per unit length respectively, so the force per unit length induced in the supporting spring is $p = kv$. By considering a small element, the equilibrium conditions are:

$$\frac{-\mathrm{d}F}{\mathrm{d}z} + p + w = 0 \qquad \text{and} \qquad \frac{\mathrm{d}M}{\mathrm{d}z} = F$$

Therefore

$$-M'' + p + w = 0$$

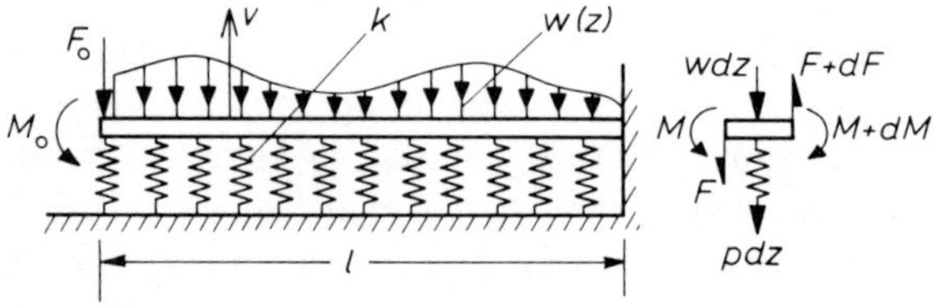

Fig. 3.32 Beam on continuous spring

The virtual work now consists of two parts:

The beam: Assuming the usual linear stress distribution through the section, then purely from equilibrium reasoning, $\sigma_{zz} = My/I$.

Thus $\displaystyle\int_V \bar{\boldsymbol{\sigma}}^{\mathrm{t}}\boldsymbol{\varepsilon}\,\mathrm{d}v = \int_0^l \frac{\bar{M}M}{EI}\mathrm{d}z$ as usual

Notice that we have deliberately avoided using the compatibility condition that the strain $\boldsymbol{\varepsilon} = -yv''$.

The spring:

$$\int_V \bar{\boldsymbol{\sigma}}^{\mathrm{t}}\boldsymbol{\varepsilon}\,\mathrm{d}v = \int_0^l \bar{p}\frac{p}{k}\mathrm{d}z$$

We now use our equilibrium equations to express stress in terms of applied forces and unknown redundancies. Thus we can eliminate the spring 'stress' in terms of unknown bending moments

$$p = -w + M'' \qquad \text{and} \qquad \bar{p} = \bar{M}''$$

The PVF is now

$$\int_V \bar{\boldsymbol{\sigma}}^t \boldsymbol{\varepsilon} \, dv = \int_0^l \left[\frac{\bar{M}M}{EI} + \bar{M}'' \left(\frac{-w + M''}{k} \right) \right] dz = 0$$

Integrating by parts as usual,

$$\int_0^l \left[\frac{M}{EI} + \left(\frac{-w'' + M''''}{k} \right) \right] \bar{M} \, dz + \left[\left(\frac{-w + M''}{k} \right) \bar{M}' \right]_0^l - \left[\left(\frac{-w' + M'''}{k} \right) \bar{M} \right]_0^l = 0 \quad (3.53)$$

From the integral we have the fourth-order differential equation

$$M'''' + \frac{k}{EI} M = w''$$

There are therefore four necessary boundary conditions which in this example are of two types: equilibrium and kinematic. At $z = 0$ all the applied forces are prescribed; $M(0) = M_0$ and $M'(0) = F_0$. We simply note in passing that no virtual applied forces are allowed so $\bar{M}(0)$ and $\bar{M}'(0)$ are zero in (3.53). The remaining terms $\bar{M}(l)$ and $\bar{M}'(l)$ are not zero since this end of the beam is connected to a rigid support, so their coefficients must be zero. Therefore, $M''(l) - w(l) = 0$ and $M'''(l) - w'(l) = 0$, and the problem is solved completely.

The last two boundary conditions must be kinematic even though the PVF has relieved us of the responsibility for deriving them. Nevertheless, it is interesting to confirm this by noting that the beam deflection may be written as

$$v = p/k = (M'' - w)/k$$

The boundary conditions therefore imply that $v(l) = 0$ and $v'(l) = 0$ as we might have expected.

We now turn to another fairly useful class of problems which involve continuous force functions of one dependent variable. We refer to the idealization of 'stressed-skin structures' or 'thin-walled panels' which for many years have been used in the analysis of aircraft structures. It is common practice to stiffen these panels by adding to them slender stiffeners and so significantly raise the buckling stress of the thin panel. These stiffeners are essentially stiff only along their own axis, as opposed to the two-dimensional panel. The stiffened panel of Fig. 3.33 depicts the various possible stress components, the stiffeners having only the longitudinal component σ_{zz}. The panel stresses are of course functions of s and z, but we will now make two fairly reasonable assumptions to simplify the analysis. Firstly, most aircraft wing-boxes and box-girders are internally stiffened by ribs or diaphragms in the transverse (s) direction, so that strains ε_{ss} are small – we will (initially) ignore them. Secondly, we will simplify the equilibrium equation

$$\frac{\partial \sigma_{zz}}{\partial z} + \frac{\partial \sigma_{sz}}{\partial s} = 0 \qquad (1.48)$$

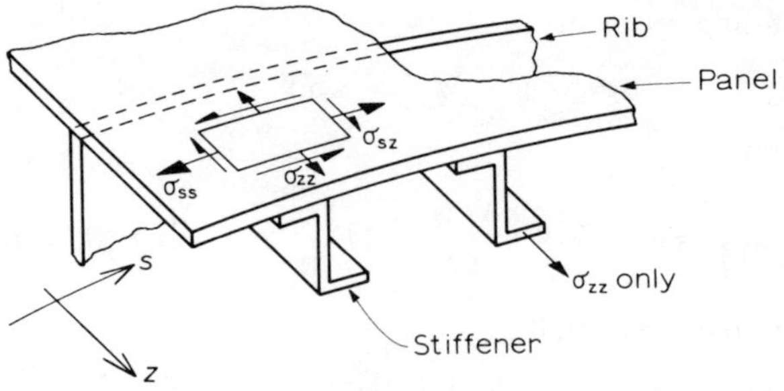

Fig. 3.33 Stiffened panel

by letting the discrete stiffeners carry *all* the longitudinal stresses σ_{zz}. The panel's contribution to these stresses is allowed for by 'adding' their area to the stiffeners to form an 'effective boom'. Thus we do include the panel's contribution to the virtual-work integral, but we move the points of action of the σ_{zz} stresses somewhat – this will not be a disastrous approximation if our booms are fairly closely spaced. If the actual stiffeners are not closely spaced then we merely place completely fictional booms at more frequent intervals. Since σ_{zz} is now zero in the panels, we see from (1.48) that σ_{sz} is constant between booms – hence the expression 'pure shear panels'. It is actually a little more convenient to use the variable $t\sigma_{sz} = q$, the so-called 'shear flow'. The equivalent of (1.48) is now a discrete 'jump' across a boom, giving us an equilibrium relationship between the gradient of the boom load (P) and the adjacent shear flows (see Fig. 3.34).

Thus

$$q_1 - q_2 = \frac{\mathrm{d}P}{\mathrm{d}z} \tag{3.54}$$

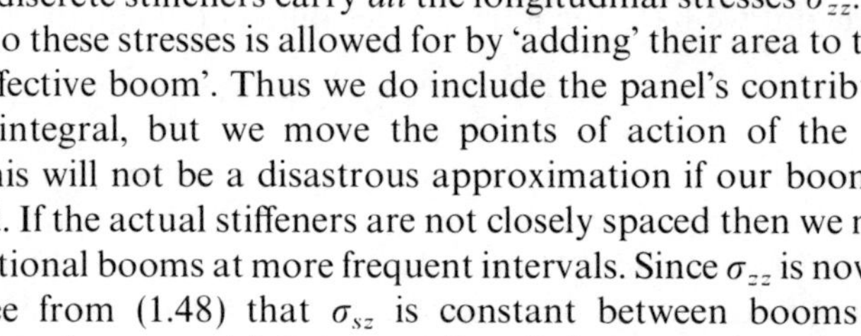

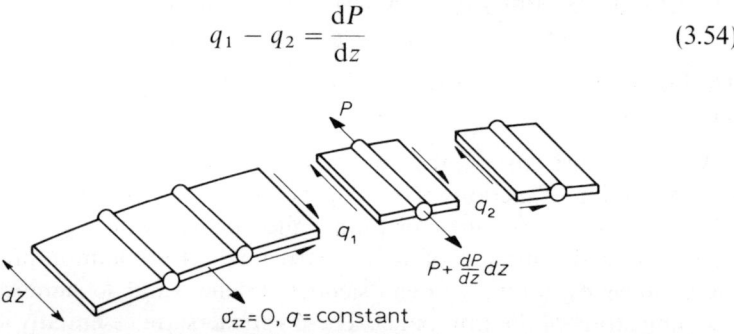

Fig. 3.34 Idealized stiffened shell

This idealization is tailor-made for the PVF since all stresses are now functions of a single variable, z. We now look at some typical boom-panel examples, and consider firstly a rectangular section, closed thin-walled tube subjected to a pure torque. This example is chosen to illustrate the sort of behaviour that can be expected in wings or box-girder bridges. The simple Bredt–Batho theory predicts that a closed tube resists a torque T by a shear flow $q = T/2A$ everywhere, where A is the cross-sectional area enclosed by the tube. Unfortunately, this very simple shear-stress distribution produces axial longitudinal displacements around the tube perimeter, and these, unless completely free to develop, will result in further 'axial constraint stresses' as we have seen in the open tube (section 2.10). The axial, or warping, displacement $w(z)$ will produce the most significant stress and strain when the gradient $\varepsilon_{zz} = \partial w/\partial z$ is large, and hence this will be felt when either the loading or the structure is discontinuous. This can happen at the supports of a bridge, or at the centre section of a complete wing where axial displacements will be zero if both wings are identically loaded.

Let us drastically idealize our uniform tube as four pure-shear panels with four identical corner booms. We could improve upon this idealization by introducing more booms, but this simple model will exhibit the salient features of torsional constraint stresses. We must now apply directly the equations of equilibrium to a cross-section, Fig. 3.35. By resolving in three directions and taking moments about three axes, the only possible symmetrical pattern of shear flows and boom loads is that shown in the figure, together with the remaining equation

$$(q_1 + q_2)ab = T$$

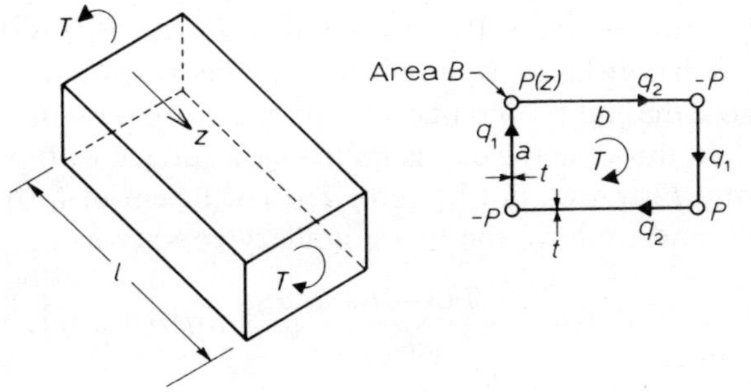

Fig. 3.35 A four-boom tube

The boom equilibrium equation is simply $q_1 - q_2 = P'(z)$, so we have

$$q_1 = \frac{T}{2ab} + \frac{P'(z)}{2}; \quad q_2 = \frac{T}{2ab} - \frac{P'(z)}{2} \tag{3.55}$$

Notice the form of the equations: they are analogous to the earlier discrete equations in frameworks except that the unknown (redundancy) $P(z)$ is now a continuous function. The statically equivalent stresses–with $P(z) = O$–are simply the Bredt–Batho solution $q = T/2ab$ which therefore has the merit of being a fairly simple basic solution at least in equilibrium with the applied load. The old concept of 'cuts', to generate basic systems and self-equilibrating stresses is clearly no longer helpful.

Equilibrium arguments have now been exhausted, and the definition of the problem is complete; the rest is left to the PVF. Assuming Hookes' law to find ε, and generating $\bar{\boldsymbol{\sigma}}$ by putting $T = 0$, we find:

in the vertical panels: $\varepsilon = \dfrac{T}{2abtG} + \dfrac{P'}{2Gt}$; $\bar{\sigma}_i = \dfrac{\bar{P}'}{2t}$

in the horizontal panels: $\varepsilon = \dfrac{T}{2abtG} - \dfrac{P'}{2Gt}$; $\bar{\sigma}_i = \dfrac{-\bar{P}'}{2t}$

in the four booms: $\varepsilon = \pm \dfrac{P}{BE}$; $\bar{\sigma}_i = \pm \dfrac{\bar{P}}{B}$

Substituting into the PVF and converting $\bar{P}'$ to $\bar{P}$ in the usual way, we obtain after some manipulation,

$$\int_0^l \left[\frac{8GtP}{BE} - (a+b)P''\right]\bar{P}\,\mathrm{d}z + \left[\left(\frac{T(a-b)}{ab} + (a+b)P'\right)\bar{P}\right]_0^l = 0 \quad (3.56)$$

The differential equation of compatibility is therefore

$$P'' - \alpha^2 P = 0, \qquad \text{where} \qquad \alpha^2 = 8Gt/BE(a+b),$$

$$\text{or} \qquad P(z) = C_1 \cosh \alpha z + C_2 \sinh \alpha z \quad (3.57)$$

Now the boundary conditions. Suppose for instance that the tube is axially unconstrained at both ends $z = 0$ and $z = l$; then $P(0) = P(l) = 0$, $C_1 = C_2 = 0$, and so $P(z) = 0$ everywhere, as simple Batho theory would have us believe. However, suppose the end $z = l$ is free, but the end $z = 0$ is *rigidly* clamped ($\varepsilon = 0$ in the support). In this case the end is no longer a surface with prescribed loads and consequently $\bar{P}(0)$ need not be zero. The coefficient of $\bar{P}(0)$ in (3.56) must therefore vanish and we have the two boundary conditions:

$$P(l) = 0; \qquad \frac{T(a-b)}{ab} + (a+b)P'(0) = 0$$

The complete solution is then

$$P(z) = \frac{-T(a-b)}{ab(a+b)\alpha}(\sinh \alpha z - \tanh \alpha l \cosh \alpha z) \quad (3.58)$$

If we take typical values $G/E = \frac{3}{8}$, B of order at, and b of order a, then α is of order $1/a$. (If b is actually equal to a then $P(z) = 0$, and the tube becomes a special case of a 'Neuber' non-warping tube.[30]) Consequently, if the tube is very long ($l \gg a$) then $\alpha l \gg 1$ and $\tanh \alpha l \simeq 1$; whence

$$P = \frac{T(a-b)}{ab(a+b)\alpha}\mathrm{e}^{-\alpha z} \qquad \text{and} \qquad q_1 = \frac{T}{ab}\left(1 - \frac{a-b}{a+b}\mathrm{e}^{-\alpha z}\right)$$

The axial stresses induced by the constraint at $z = 0$ decay rapidly (at $z = 2a$, $\mathrm{e}^{-\alpha z}$ is of order $\mathrm{e}^{-2} = 14$ per cent). This is not surprising: the self-equilibrating system $P(z)$ shown in Fig. 3.36, having no resultant, should have a limited zone of influence according to Saint Venant. Having used the PVF to solve the stresses, the displacements can be found using the unit-load theorem (3.1). Suppose for example that we require the twist $\theta(z)$. We apply a unit torque at section z and

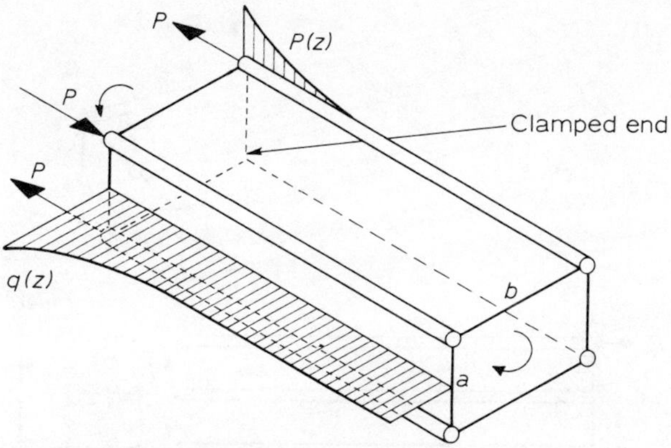

Fig. 3.36 Axial constraint stresses

react it at the clamped end. But remember that the stresses $\bar{\sigma}$ are virtual and therefore only have to be statically equivalent. The simplest possible solution is of course obtained from (3.55) putting $\bar{P} = 0$, $T = 1$; that is, $q_1 = q_2 = 1/2ab$. The unit-load theorem then produces

$$\theta(z) = \int_V \bar{\boldsymbol{\sigma}}^t \boldsymbol{\varepsilon} \, dv = \int_0^z \frac{1}{2abt}\left(\frac{2aq_1 + 2bq_2}{Gt}\right) t \, dz$$

$$= \frac{T}{2a^2b^2tG}\left[(a+b)z - \frac{(a-b)^2}{\alpha(a+b)} e^{-\alpha z}\right]$$

This fairly trivial example demonstrates that the simple Batho theory of torsion is only valid if axial warping displacements are free to depart from a plane. Constraint stresses can also occur in bending problems where, ironically, simple theory assumes that displacements do remain plane! Wide-flanged beams or boxes in particular suffer this constraint effect (see Problem 3.13).

3.10 Diffusion problems

A problem common in thin-walled structures is how to introduce gently a concentrated load into a thin plate ('skin' in aircraft) which is ill-equipped to receive it. The answer is to provide a heavier diffusing member which is stiff in the direction of the applied load and which is long enough for the induced load to diffuse out gradually into the thin skin to which the member is attached. Also a diffusion can occur when the structure rather than the load is discontinuous: obvious instances are mechanical joints or cut-outs in plates and shells to provide access or visibility. At such discontinuities stress concentrations are invariably produced unless the diffusion process is adequately understood and then designed for. The simple boom-panel idealization allows us to obtain a quick understanding since the diffusion process is (almost) one-dimensional. Consider for example the rectangular box in Fig. 3.37, in which the tensile loads in the upper surface are rudely interrupted by a necessary cut-out. The local diffusion

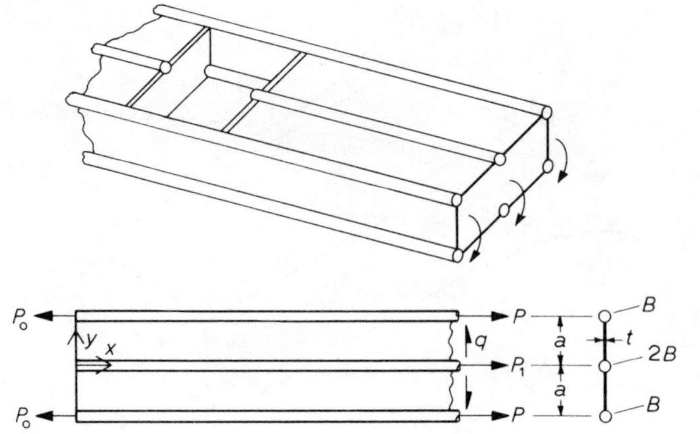

Fig. 3.37 A three-boom diffusion panel

problem around the cut-out is then idealized as part of a very long panel having three booms, one of which is cut off at the end. The equations of equilibrium are simply

$$2P + P_1 = 2P_0; \qquad q = P'$$

Choose, say, the edge load $P(x)$ as the unknown redundancy.

Then:

in the outer booms, $\sigma = P/B; \quad \bar{\sigma} = \bar{P}/B$

in the central boom, $\sigma = P_1/2B = (2P_0 - 2P)/2B\colon \quad \bar{\sigma} = -2\bar{P}/2B$

in the panels, $q = P'; \quad \bar{q} = \bar{P}'$

The PVF is:

$$\int_0^\infty \left[2\left(\frac{\bar{P}}{B}\right)\left(\frac{P}{BE}\right)B + \left(\frac{-2\bar{P}}{2B}\right)\left(\frac{2P_0 - 2P}{2BE}\right)2B + 2\left(\frac{\bar{P}'}{t}\right)\left(\frac{P'}{Gt}\right)ta \right] \mathrm{d}x = 0$$

Integrating $P'\bar{P}'$ by parts, and rearranging:

$$\int_0^\infty [-P'' + \alpha^2(P - P_0/2)]\bar{P}\,\mathrm{d}x + [P'\bar{P}']_0^\infty = 0 \tag{3.59}$$

where $\alpha^2 = 2Gt/BEa$. Solving the differential equation in the integrand,

$$P(x) = C_1 \mathrm{e}^{\alpha x} + C_2 \mathrm{e}^{-\alpha x} + P_0/2$$

If the panel is infinitely long we can discard the positive exponent, leaving the remaining boundary condition $P(0) = P_0$.

Thus

$$P = \frac{P_0}{2}[1 + \mathrm{e}^{-\alpha x}] \quad \text{and} \quad q = P' = \frac{-P_0\alpha}{2}\mathrm{e}^{-\alpha x}$$

We note that $\bar{P}(0) = 0$ because virtual applied loads are not permissible, and $P'(\infty) = 0$; therefore the last terms in (3.59) also vanish. Again the solution for large x diffuses to a simple uniform stress state $P = P_0/2$ and $P_1 = 2(P_0 - P_1) = P_0$, and zero shear stress. The shear stresses are the mechanism whereby the load is taken away from the outer booms to activate the central boom, just as it did in the previous constrained four-boom tube (or in Problem 3.12). This phenomenon, where shears move recalcitrant edge loads into or away from the centre of wide-flanged beams and boxes, is known as 'shear lag'. If we assume that the boom area B is of order $at/2$, then α is again of order $1/a$ so that the diffusion process is still rapid. There are, however, two unfortunate paradoxes in this simple analysis.

Firstly, the maximum shear occurs at $x = 0$ where its value is $-P_0\alpha/2t$; but if the panel end is free and unsupported the shear stress should be zero, so the solution is not valid. One way of resolving this contradiction is to insert a transverse boom or 'end post' across the panel and so enable the end shears to equilibrate each other, as shown in Fig. 3.38. This is good design practice anyway, as we shall shortly see. It is possible to preserve the respectability of our analysis by claiming that this end post is *rigid* and therefore, possessing no strain, will not change the virtual-work integral!

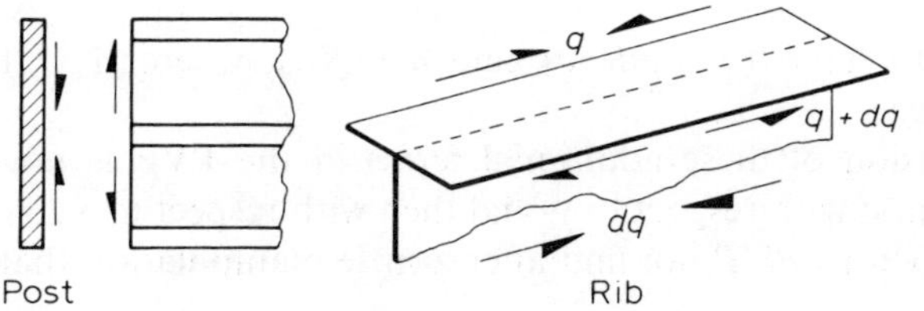

Fig. 3.38 Rigid posts and rigid ribs

Secondly, the shear stress is a function of x and consequently we have violated the equation of equilibrium

$$\frac{\partial \sigma_{yy}}{\partial y} + \frac{\partial \sigma_{xy}}{\partial x} = 0$$

in ignoring the transverse stresses σ_{yy}. By a similar ruse we could claim that the box was supported by closely spaced ribs to equilibrate this out of balance (see Fig. 3.38 again), but of course these ribs should also be rigid. This assumption is common and it can be shown that the ribs do not have to be too close or too rigid.

If neither of these two physical alternatives is available, then the presence of transverse stresses and strains must be admitted, but their inclusion in the virtual work expression is fairly straightforward, provided we can still confine the longitudinal stresses σ_{xx} to the booms – remembering that the 'effective' boom areas can always be supplemented to allow for these panel stresses. In which case, if we put $\sigma_{xx} = 0$ in the other equation of equilibrium,

$$\frac{\partial \sigma_{xx}}{\partial x} + \frac{\partial \sigma_{xy}}{\partial y} = 0$$

then σ_{xy} must be a function of x alone, and integrating the previous equation,

$$\sigma_{yy} = -\frac{\partial \sigma_{xy}}{\partial x} y + f(x)$$

Now the edge booms are attached to the panels which have relatively much higher in-plane transverse stiffness, and this effectively inhibits any tendency of the booms to bend like beams. If the booms are unable to resist transverse loads then $\sigma_{yy} = 0$ in the panel at $y = a$, so the above becomes

$$\sigma_{yy} = (a - y)\frac{\partial \sigma_{xy}}{\partial x}$$

and putting $t\sigma_{xy} = P'(x)$ as before,

$$\sigma_{yy} = (a - y)P''/t$$

This term may now be added to the work both as $\varepsilon = \sigma/E$ and $\bar{\sigma} = (a - y)\bar{P}''/t$. Having incorporated transverse strains in the panel, we can do the same in an end post. If the force in such a post is $N(y)$ then equilibrium gives us

$$\frac{\partial N}{\partial y} + tq(0) = 0$$

so $N = tq(a - y) = t(a - y)P'(0)$, and $\sigma = N/A$, where A is the post's cross-sectional area.

The inclusion of these additional terms in the PVF is now routine. After integrating first with respect to y, and then with respect to x to convert $P''\bar{P}''$ to $P'''\bar{P}'$ and then to $P''''\bar{P}$, we find after a little manipulation that

$$\int_0^\infty \left(P'''' - \frac{3E}{a^2G}P'' + \frac{6t}{a^3B}P - \frac{3t}{a^3B}P_0\right)\bar{P}\,\mathrm{d}x + \left[P''\bar{P}' - P'''\bar{P} + \frac{3E}{a^2G}P'\bar{P}\right]_0^\infty + P'(0)\bar{P}'(0)\frac{t}{A} = 0 \quad (3.60)$$

The effect of the additional terms is therefore to give us a fourth-order differential equation and extra constants of integration with which to satisfy the boundary conditions. To examine the behaviour of the solution we put $G/E = 3/8$, so that the complementary function of the differential equation behaves like $\exp(\alpha x/a)$, where

$$\alpha^4 - 8\alpha^2 + 6at/B = 0 \quad (3.61)$$

Solutions again have equal positive and negative roots. If we ignore positive exponentially increasing solutions, then two roots remain and we need two boundary conditions:

Firstly,

$$P(0) = P_0 \qquad (\text{and } \bar{P}(0) = 0 \text{ in } (3.60)) \quad (3.62)$$

Secondly, $\bar{P}'(0)$ need not be zero since the supported edge is no longer a surface, so from the coefficients of $\bar{P}'(0)$ in (3.60) we have

$$-P''(0) + \frac{t}{A}P'(0) = 0 \tag{3.63}$$

A reasonable value of B would be $at/2$. This is the value we would likely choose if the panel of area $2at$ had no actual booms but was idealized as two edge booms of area B and a central boom of area $2B$. The roots of (3.61) are now

$$\alpha = \sqrt{6}, \quad \alpha = \sqrt{2}$$

Ignoring the transverse stresses would have produced the single pair of roots $8\alpha^2 = 6at/B$ in (3.61) or $\alpha = 1.224$, which is not too far from $\alpha = \sqrt{2}$. The new root $\alpha = \sqrt{6}$ is consequently the term which allows us to satisfy the boundary condition (3.63) involving the deformation of the end post. It clearly decays more rapidly than the other term and suggests that the effect of the end post is very local indeed. We can conveniently look at the two extremes:

(a) a rigid end post, $A \to \infty$, so (3.63) says $P''(0) = 0$
(b) no post at all, $A \to 0$, so (3.63) says $P'(0) = 0$

The solutions are:

(a) $P/P_0 = 0.75 \exp\left(-\frac{\sqrt{2}x}{a}\right) - 0.25 \exp\left(-\frac{\sqrt{6}x}{a}\right) + 0.5$

(b) $P/P_0 = 1.183 \exp\left(-\frac{\sqrt{2}x}{a}\right) - 0.683 \exp\left(-\frac{\sqrt{6}x}{a}\right) + 0.5$

As it was the shear variation which revealed the inadequacies of the simple theory, it is of interest to look at it. Figure 3.39 shows the behaviour. The differences between the simple theory and that with transverse strains are clear and striking. Simple theory will overestimate the maximum shear stress by nearly 50 per cent, even if the end post is rigid. With no end post at all the simple theory overestimates the maximum stress by over 100 per cent. As expected, the differences are confined to a local region $x < a$.

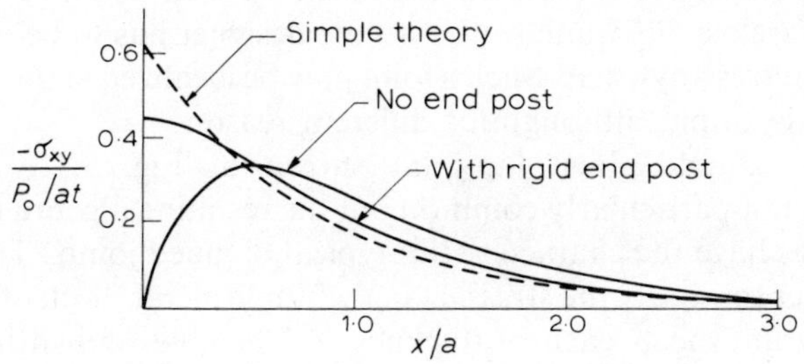

Fig. 3.39 Panel shears near an edge with post

In practice, of course, infinitely rigid end posts are not available. However, like rigid ribs they do not have to be too rigid. Figure 3.40 shows the theoretical prediction and some experimental results for a diffusion panel having a post at one end and a free edge at the other. The post is selected to have about the same cross-sectional area as the outer booms. The experimental structure is actually a honeycomb sandwich with the three booms embedded inside the core – such a configuration prevents the thin panels buckling in the region of high shear stress (see Plate 1). The panel is analysed with the boom area B supplemented by $at/2$ to account for the direct stresses σ_{xx} in the panel. The finite end post clearly behaves almost like one with infinite rigidity. The very large changes in shear stress near the free end are remarkably well predicted by the PVF solution.

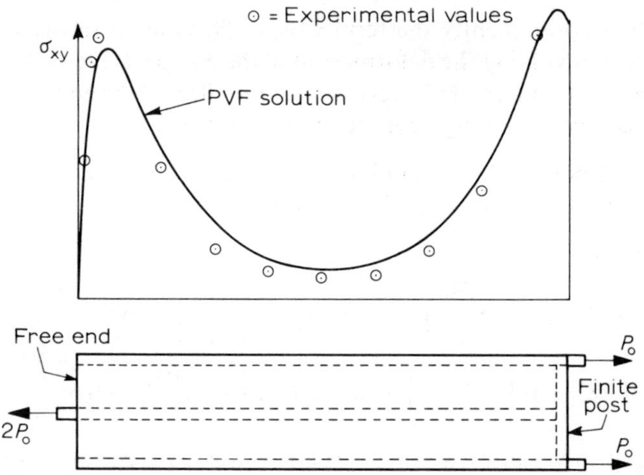

Fig. 3.40 Theoretical and experimental shear stresses in a diffusion panel

Another very common sort of diffusion problem is that which occurs when one structure is mechanically joined to another. Riveted joints and bolted joints for example exhibit characteristic exponential diffusion behaviour which should be understood for an efficient joint design. Actually, many bolted joints are not terribly efficient but rely upon highly ductile bolts yielding under abnormal stress and shedding their loads to others more favourably loaded. Adhesive-bonded joints may not be able to redistribute their stresses since the very strong epoxy resins favoured are usually quite brittle, so the designer has to be careful not to induce tensile stresses anywhere. Such a joint may be idealized in the same fashion as the previous example, although for different reasons.

Consider the double glued lap-joint shown in Fig. 3.41. This double configuration is not particularly common but the resulting algebra is simple and the solution does have the characteristics typical of glued joints. The aim of the joint is to transfer smoothly the stress σ_0 in the single metal sheet, of thickness $2t$, to the two separate sheets each of thickness t. The glue, shown hatched, is of thickness h which is usually much smaller than t; typically t/h may be of order 10. The modulus of the metal E is also much greater than the moduli E_g and G_g of the glue. For example, Young's modulus for epoxy resin is about 5 GN/m^2 compared

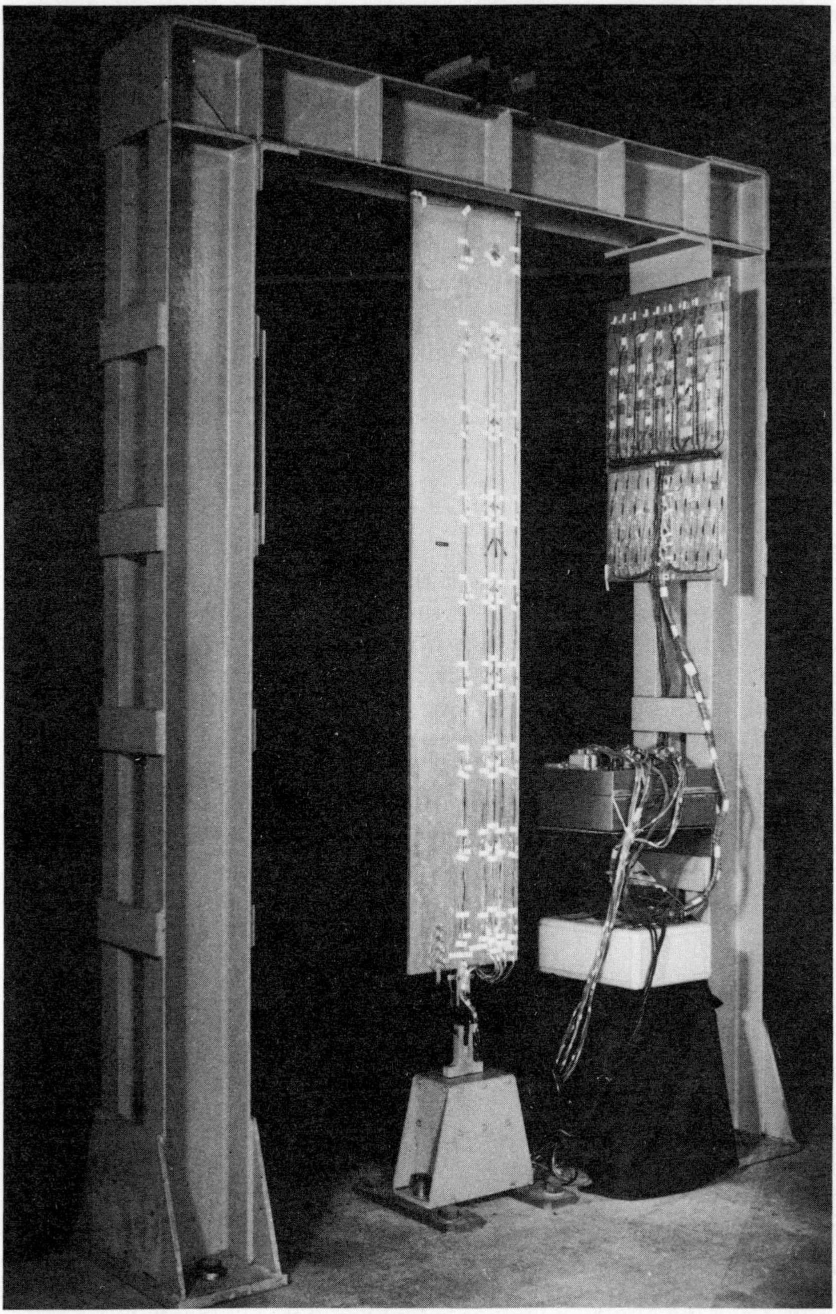

Plate 1

with 70 for light alloy and 210 for steel. Now the strain ε_{xx} must be continuous across the interfaces if the materials stretch together, so we can say that the longitudinal strains are comparable. Therefore, because $E_g \ll E$, we may assume

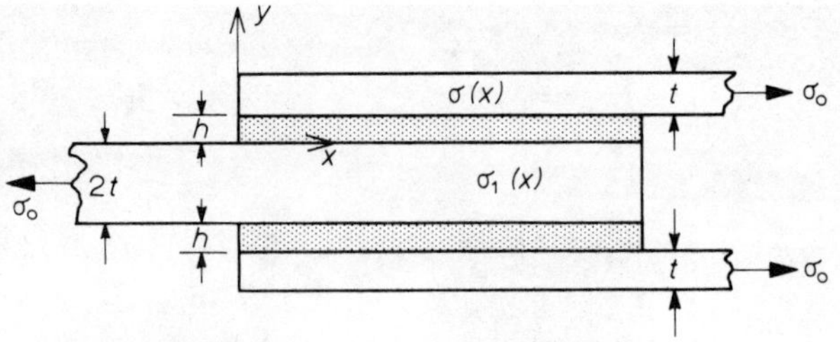

Fig. 3.41 A double glued lap joint

that the stress σ_{xx} in the glue may be ignored. This was also the starting point for analysing the previous diffusion problem, so again we may say that the shear stress in the glue varies only with x, and the transverse stress is of the form

$$\sigma_{yy} = -y\frac{\partial \sigma_{xy}}{\partial y} + f(x)$$

The stress variation in the thin metal sheets can be reasonably assumed to be linear as in simple beam theory, and in this particular symmetric configuration it is also reasonable to assume that no significant metal displacements take place in the y direction at all. In other words, there is no beam bending action in any of the metal sheets and the direct stresses therein are uniform over the thickness. If beam action is not present then we may assume that the outer sheets cannot support a direct stress along their inner edges, and hence the above equation becomes

$$\sigma_{yy} = (h - y)\frac{\partial \sigma_{xy}}{\partial x}$$

Denoting the stress in the outer sheets as $\sigma(x)$ and in the inner sheet by $\sigma_1(x)$, then from logitudinal equilibrium,

$$\sigma_1 2t + 2\sigma t = 2\sigma_0 t$$

and

$$\sigma_{xy} = \frac{t\,\partial\sigma}{\partial x}$$

Thus we have all stresses now in terms of just one unknown $\sigma(x)$, say, so

$$\sigma_1 = \sigma_0 + \sigma; \qquad \sigma_{xy} = t\sigma'; \quad \sigma_{yy} = (h - y)t\sigma''$$

To be honest, we should go a little further than this and take the direct stress $\sigma_{yy} = ht\sigma''$ across the central plate into account. Also it is not strictly accurate to ignore bending in the outer plates since the shear stress from the glue is applied with an eccentricity to these plates. In fact it is possible to add all these terms into the PVF, including the bending moments in the outer plates which are related simply to $\sigma_{yy}(h) = M''$ at the interface,[29] but they are all second-order effects as it turns out. Looking at the above three simple expressions for σ_1, σ_{xy}, and σ_{yy}, the similarity with the previous diffusion example is now clear.

On substituting into the PVF as before, we finally obtain

$$\int_V \bar{\sigma}_i^! \varepsilon \, dv = 0 = \int_0^l \left[\sigma'''' - \frac{3E_g}{h^2 G_g}\sigma'' + \frac{6E_g}{Eth^3}(\sigma - \sigma_0/2) \right] \bar{\sigma} \, dx$$

$$+ \left[\sigma''\bar{\sigma}' - \sigma'''\bar{\sigma} + \frac{3E_g}{h^2 G_g}\sigma'\bar{\sigma} \right]_0^l = 0$$

From the differential equation, the particular integral is $\sigma = \sigma_0/2$, which is the diffused solution away from both ends of the joint. Inserting $\exp(\alpha x/l)$ into the equation, the roots of α are given by

$$\alpha^4 - 3\frac{E_g}{G_g}\frac{l^2}{h^2}\alpha^2 + 6\frac{E_g}{E}\frac{h}{t}\frac{l^4}{h^4} = 0$$

or

$$\alpha^2 = \frac{3}{2}\frac{E_g}{G_g}\frac{l^2}{h^2}(1 \pm \sqrt{1 - \varepsilon}), \qquad \text{where} \quad \varepsilon = \frac{8}{3}\left(\frac{G_g}{E_g}\right)\left(\frac{G_g}{E}\right)\left(\frac{h}{t}\right)$$

Now using our quoted figures, G_g/E is of order h/t or less, so at the most ε is of order $(G_g/E_g)(h/t)^2$ which is much less than unity. We can therefore approximate $(1 - \varepsilon)^{1/2}$ as $1 - \varepsilon/2$ and the two roots for α^2 emerge as

$$\alpha_1^2 = 2\left(\frac{G_g}{E}\right)\left(\frac{t}{h}\right)\left(\frac{l}{t}\right)^2 \qquad \text{and} \qquad \alpha_2^2 = 3\left(\frac{E_g}{G_g}\right)\left(\frac{l}{h}\right)^2$$

Again if G_g/E is of order h/t, then α_1 is of order l/t, whereas α_2^2 is much larger, being of order l/h. Because both roots are large there is a rapid diffusion from the inner sheet into the outer. In fact the diffusion is so rapid that the central portion of the lap joint does nothing but support a uniform stress of $\sigma_0/2$, and no increase in load-carrying capacity is to be gained by making a longer joint. The smallest root α_1 is of the order l/t, so that if we say diffusion is mostly complete when $\exp(-\alpha x/l)$ is of order 2–3 per cent, then we find x may be only $5t$ to $10t$. Thus one half of the load $\sigma_0 t$ diffuses at one end and the balance at the other. Many attempts have been made to energize the middle of such joints by costly or clumsy tailoring of the metal thickness. The most promising avenue appears to be the development of ductile glues which obligingly yield and excite the tardy interior.

On inserting the equilibrium boundary conditions, $\sigma(0) = 0$, and $\sigma'(0) = 0$, and discarding the positive exponents for a 'long' joint $l \gg t$, the complete solution in the vicinity of the end $x = 0$, is

$$\frac{\sigma_{xy}}{\sigma_0} = \frac{\alpha_1 t}{2l}\left[\exp\left(\frac{-\alpha_1 x}{l}\right) - \exp\left(\frac{-\alpha_2 x}{l}\right) \right]$$

$$\frac{\sigma_{yy}}{\sigma_0} = \frac{\alpha_1 \alpha_2 ht}{2l^2}\left[\exp\left(\frac{-\alpha_2 x}{l}\right) - \frac{\alpha_1}{\alpha_2}\exp\left(\frac{-\alpha_1 x}{l}\right) \right]$$

where we have ignored α_1/α_2 compared with unity. Apart from the inefficient concentration of shear stress near the end, the most important feature is the corresponding concentration of the transverse tensile stress σ_{yy} whose magnitude we see is of the order σ_0, which is bad news for most glues. A sketch of these stresses is shown in Fig. 3.42 for typical values t/h of order 10, l/t of order 5; so $\alpha_1 = 5$; $\alpha_2 = 50$. The large tensile stresses spell trouble if the glue is brittle. The fact that they are confined to such a small region is no comfort if they cause a small crack to develop, since the effective end of the joint will simply move towards the centre. The tendency of bonded lap joints to fail due to tensile glue cracking has been found out the hard way in practice – it is often referred to as 'low peeling strength'. It was common in the aircraft industry to rivet just the ends of glued joints!

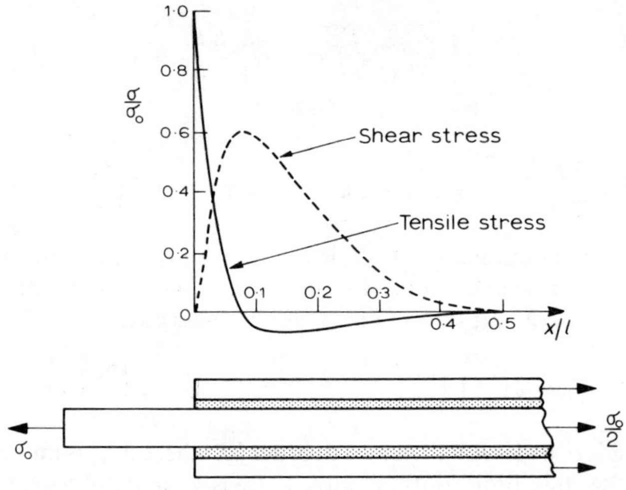

Fig. 3.42 Glue stresses near the end of a short double lap joint

3.11 Gross deformations

Chapter 2 on the PVD was concluded by extending the virtual-work formulation to large displacements by the simple expedient of looking at small incremental displacements from the position of equilibrium (equation (2.56)). Exactly the same analogous argument can be applied to virtual forces; the the PVF can be written as

$$\int_V \delta\boldsymbol{\sigma}^t\boldsymbol{\varepsilon}\,\mathrm{d}v = \int_V \delta\mathbf{p}_v^t\mathbf{u}\,\mathrm{d}v + \int_S \delta\mathbf{p}_s^t\mathbf{u}\,\mathrm{d}s$$

Again the right-hand side of this equation can be removed by insisting that the incremental stress field $\delta\boldsymbol{\sigma}$ is self-equilibrating, requiring no $\delta\mathbf{p}_s$ or $\delta\mathbf{p}_v$. This ruse will remove the displacements $\mathbf{u}$ from the PVF but unfortunately they still remain in disguised form. We have to remember that the virtual stresses must satisfy the equations of equilibrium like $\mathbf{D}\ \delta\boldsymbol{\sigma} = \mathbf{0}$ in V and $(\mathbf{D}n)\ \delta\boldsymbol{\sigma} = \mathbf{0}$ on S, where the operators in $\mathbf{D}$ are with respect to the correct geometry. But the current geometry $(x + u,\ y + v,\ z + w)$ involves the unknown displacements u, v, and w which cannot be ignored in gross deformations. Consequently, direct compatibility arguments cannot be avoided, since we cannot actually set up the PVF to satisfy compatibility indirectly. It does seem preferable therefore to face up to the compatibility problem and to use the PVD instead where gross deformations are involved.

3.12 Summary

1. The unit-load theorem: $r_i = \int_V \bar{\boldsymbol{\sigma}}^t \boldsymbol{\varepsilon}\, dv$, where $\bar{\boldsymbol{\sigma}}$ are the stresses in equilibrium with a unit load $R_i = 1$. The stress–strain law is immaterial.
2. For a linearly elastic structure, $\mathbf{r} = \mathbf{FR}$, where $f_{ij} = \int_V \boldsymbol{\sigma}_i^t \boldsymbol{\varepsilon}_j\, dv$ and the stresses due to $R_i = 1$ or $R_j = 1$ are $\boldsymbol{\sigma}_i$ and $\boldsymbol{\sigma}_j$.
3. If the stresses in a bar or beam are written in terms of their forces $\mathbf{P}_g$ as $\boldsymbol{\sigma} = \boldsymbol{\beta}\mathbf{P}_g$, the element flexibility $\mathbf{f}_g = \int_V \boldsymbol{\beta}^t \boldsymbol{\phi} \boldsymbol{\beta}\, dv$.
4. If the element forces can be written as $\mathbf{P}_g = \mathbf{b}_g\mathbf{R}$, then

$$\mathbf{F} = \sum_g \mathbf{b}_g^t \mathbf{f}_g \mathbf{b}_g$$

5. If the element forces have to be written in terms of redundancies $\mathbf{X}$ as $\mathbf{P}_g = \mathbf{b}_{0g}\mathbf{R} + \mathbf{b}_{xg}\mathbf{X}$, then

$$\mathbf{X} = -\mathbf{D}_{xx}^{-1}\mathbf{D}_{x0}\mathbf{R}$$

where

$$\mathbf{D}_{xx} = \sum_g \mathbf{b}_{xg}^t \mathbf{f}_g \mathbf{b}_{xg}; \qquad \mathbf{D}_{x0} = \sum_g \mathbf{b}_{xg}^t \mathbf{f}_g \mathbf{b}_{0g}$$

6. In non-matrix notation, the above may be written as

$$\sum_k \delta_{jk} X_k + \delta_{j0} = 0$$

where

$$\delta_{j0} = \int_V \boldsymbol{\sigma}_j^t \boldsymbol{\varepsilon}_0\, dv; \qquad \delta_{jk} = \int_V \boldsymbol{\sigma}_j^t \boldsymbol{\varepsilon}_k\, dv$$

For beams and frameworks:

$$\delta_{j0} = \int \left(\frac{M_j M_0}{EI} + \frac{N_j N_0}{AE} \right) dz; \quad \delta_{jk} = \int \left(\frac{M_j M_k}{EI} + \frac{N_j N_k}{AE} \right) dz$$

Problems

(Asterisks indicate problems with worked solution.)

P 3.1

All bars in the pinjointed framework shown have the same l, A and E. Show that the horizontal displacement of the joint J is $2Pl/3AE$ to the left. What is the vertical displacement of joint J?

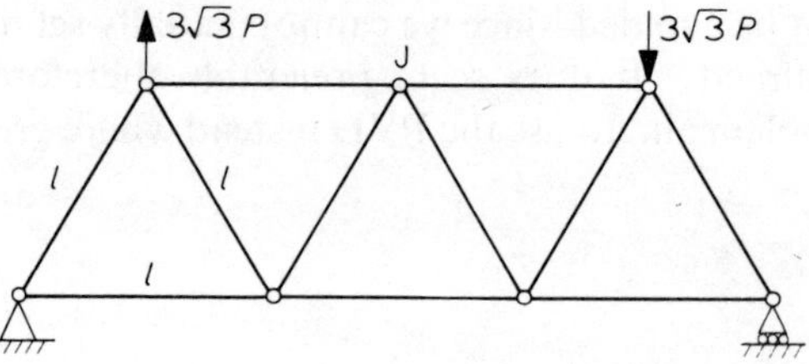

Fig. P. 3.1

P 3.2

Show that the vertical displacement and rotation of the point A in the two beams is as indicated. The temperature in Fig. P 3.2(b) varies linearly through the depth of the beam's symmetrical section as well as linearly along the length.

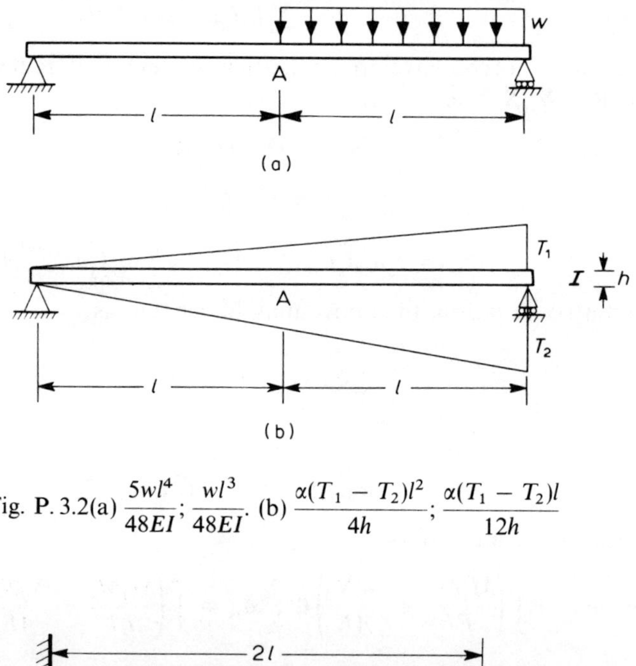

Fig. P. 3.2(a) $\dfrac{5wl^4}{48EI}$; $\dfrac{wl^3}{48EI}$. (b) $\dfrac{\alpha(T_1 - T_2)l^2}{4h}$; $\dfrac{\alpha(T_1 - T_2)l}{12h}$

P 3.3

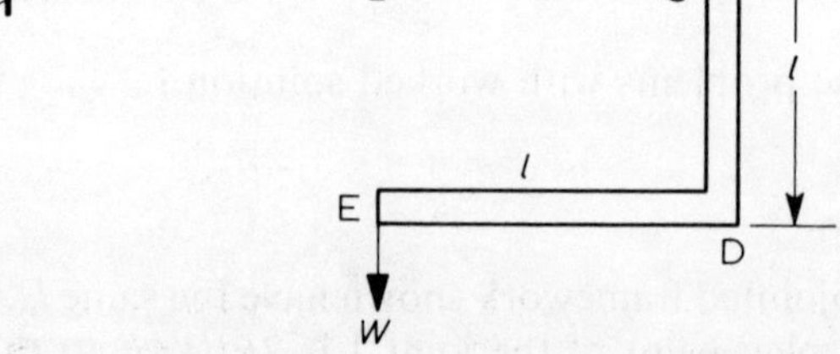

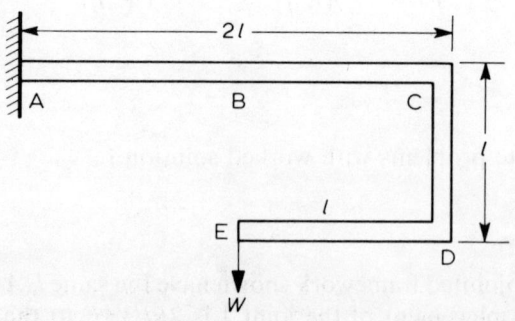

Fig. P. 3.3

The beam in the figure has a uniform flexural rigidity EI throughout, and a cross-sectional area A. Find the deflection of the loaded point E and confirm that extensional deformations contribute little to the deflection if $k/l \ll 1$ where $I = Ak^2$. Examine the contributions to the unit-load integral from the individual beam portions AB, BC, and ED and explain with the aid of a sketch why they all contribute equally to the deflection of E.

P 3.4

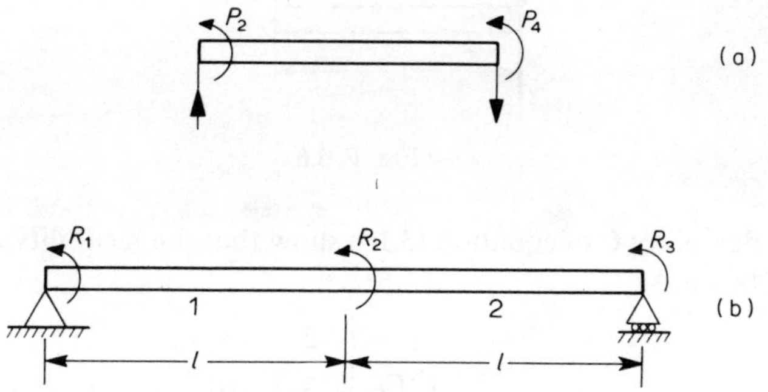

Fig. P. 3.4

(a) Show that the flexibility matrix (3.38) for a beam element subjected to end 'forces' $\mathbf{P}_g = \{P_2 \quad P_4\}$ as shown is:

$$\mathbf{f}_g = \frac{l}{6EI}\begin{bmatrix} 2 & -1 \\ -1 & 2 \end{bmatrix}$$

(b) The simply supported beam shown is subjected to couples $\mathbf{R} = \{R_1 \quad R_2 \quad R_3\}$. Show that the matrix $\mathbf{b}_g$ for the two elements is:

$$g = 1: \quad \mathbf{b}_1 = \begin{bmatrix} 1 & 0 & 0 \\ -\frac{1}{2} & \frac{1}{2} & \frac{1}{2} \end{bmatrix} \qquad g = 2: \quad \mathbf{b}_2 = \begin{bmatrix} \frac{1}{2} & \frac{1}{2} & -\frac{1}{2} \\ 0 & 0 & 1 \end{bmatrix}$$

Thence show that the total flexibility is

$$\mathbf{F} = \frac{l}{12EI}\begin{bmatrix} 8 & -1 & 4 \\ -1 & 2 & -1 \\ -4 & -1 & 8 \end{bmatrix}$$

*P 3.5**

Obtain the flexibility matrix of the previous example but using the alternative definition $\mathbf{f}_g$ of equation (3.15).

P 3.6

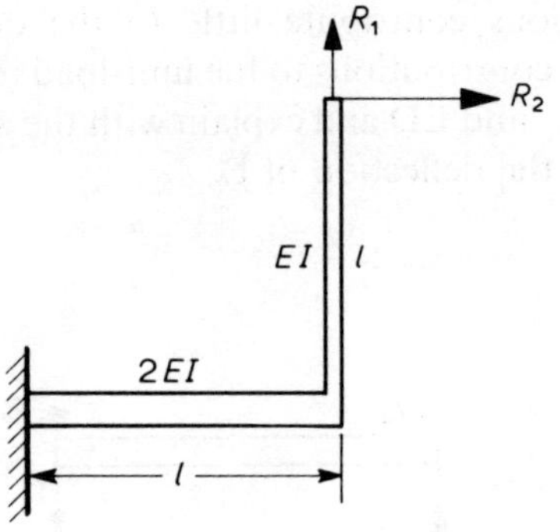

Fig. P. 3.6

Using the flexibility $\mathbf{f}_g$ of equation (3.15) show that the flexibility of the cranked cantilever beam is

$$\mathbf{F} = \frac{l^3}{12EI}\begin{bmatrix} 2 & -3 \\ -3 & 10 \end{bmatrix}$$

*P 3.7**

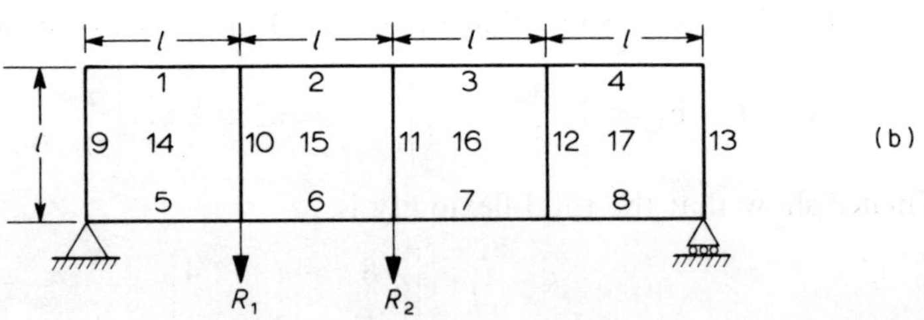

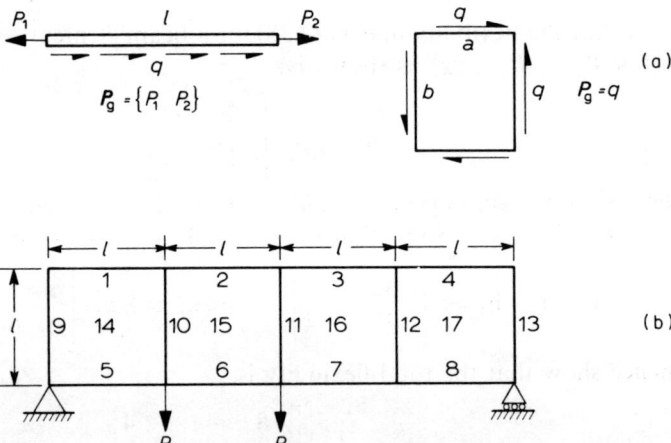

Fig. P. 3.7

The beam in the figure is an idealized plate girder in which all the direct stress-carrying material is idealized as horizontal or vertical booms. The web is idealized as four pure-shear panels sustaining only shear stress. The structure therefore has two typical components shown in Fig. P 3.7(a). Firstly, a 'boom' element loaded by two end loads P_1 and P_2, and a uniform shear flow

$q = (P_1 - P_2)/l$; and secondly, a pure shear panel. Find the flexibility of these two types of elements. Set up the $\mathbf{b}_g$ matrices for the 17 elements in the plate girder and show that the flexibility matrix for the load vector $\{R_1 \quad R_2\}$ is

$$\mathbf{F} = \frac{l}{24AE}\begin{bmatrix} 62.5 & 57 \\ 57 & 94 \end{bmatrix}$$

The boom areas A have been put equal to $lt/4$ and $E/G = 3$. Comment on the individual contributions to $\mathbf{F}$ from the vertical booms and from the shear panels.

P 3.8

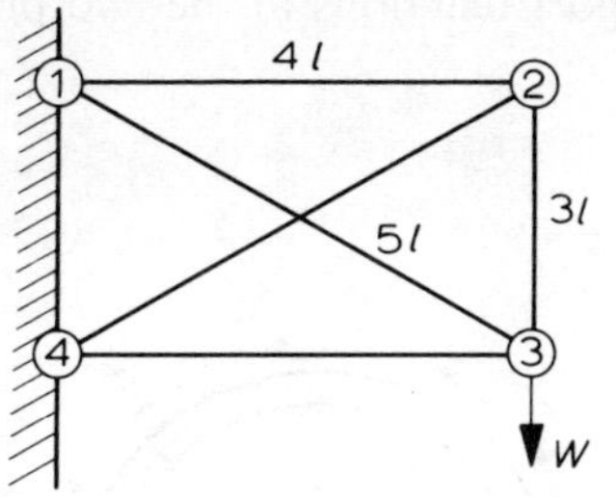

Fig. P. 3.8

Using the force method, and choosing your redundancy to take advantage of symmetry, show that the bar forces are: $N_{12} = 28W/45$; $N_{23} = 7W/15$; $N_{34} = -32W/45$; $N_{13} = 8W/9$; $N_{24} = -7W/9$. Next use the unit-load method to show that the vertical deflection of the loaded joint 3 is $11.2\,Wl/AE$, and check your answer by using two different sets of bar forces in equilibrium with the unit load.

P 3.9

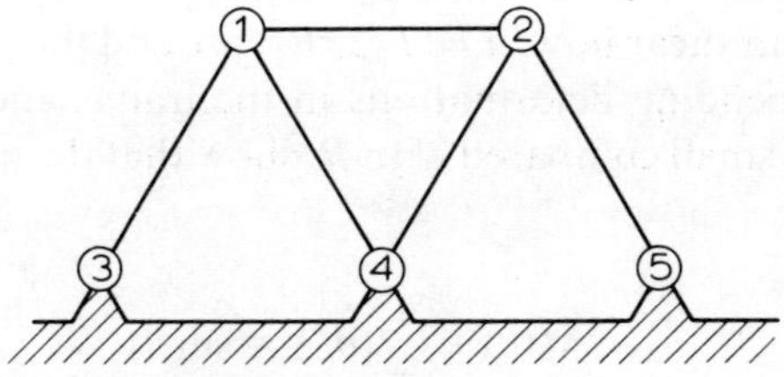

Fig. P. 3.9

All the bars shown have the same properties l, A, and E. The bar (1–3) has a coefficient of expansion α and is heated through $T°$. Using the force method show that the maximum strain in the framework is $4\alpha T/5$.

P 3.10

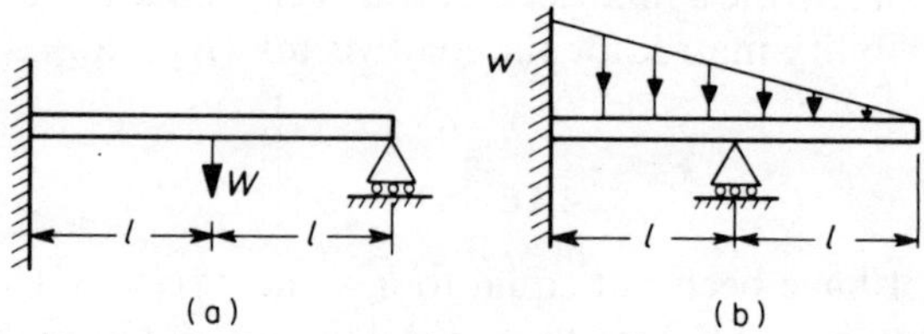

Fig. P. 3.10

Show that the rolling support reactions in the two propped cantilevers are (a) $5W/16$, and (b) $49wl/40$.

*P 3.11**

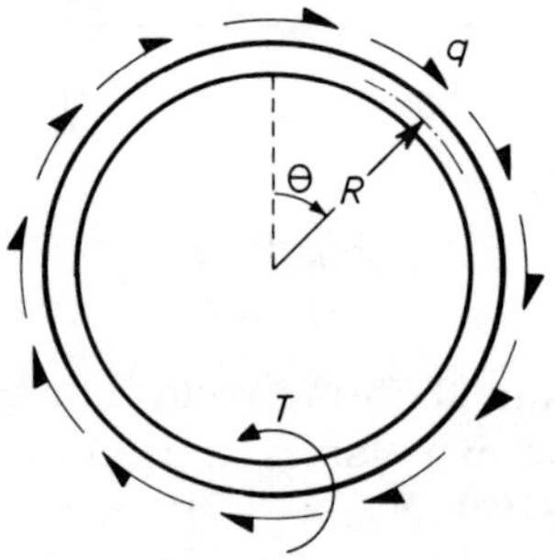

Fig. P. 3.11

A uniform circular frame is attached to a thin-walled circular fuselage, and is subjected to a pure torque T as shown in the figure. Provided the frame is reasonably stiff it can be assumed that the torque is eventually reacted in the fuselage as a uniform shear flow, $q = T/2\pi R^2$, around the periphery of the frame. Considering only bending deformations in the frame, and assuming the frame section has a depth small compared with R, show that the bending moment in the frame is

$$M(\theta) = \frac{T}{\pi}(\theta/2 - \sin\theta)$$

*P 3.12**

The figure shows the side of a building structure composed of a regular framework whose sides are 'filled in' by shear panels. Thus the bending of the

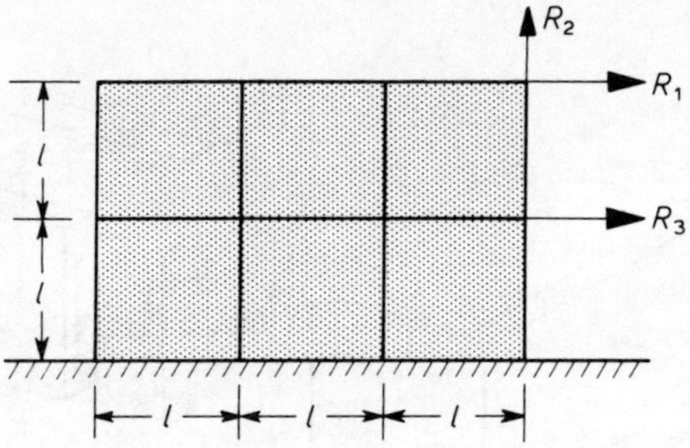

Fig. P. 3.12

framework is inhibited and we may idealize the structure as axial-load-carrying booms containing the pure-shear panels in the manner of Problem 3.7. Show that this structure has four redundancies, and set up suitable $\mathbf{b}_{0g}$ and $\mathbf{b}_{xg}$ systems for solving the redundancies

P 3.13

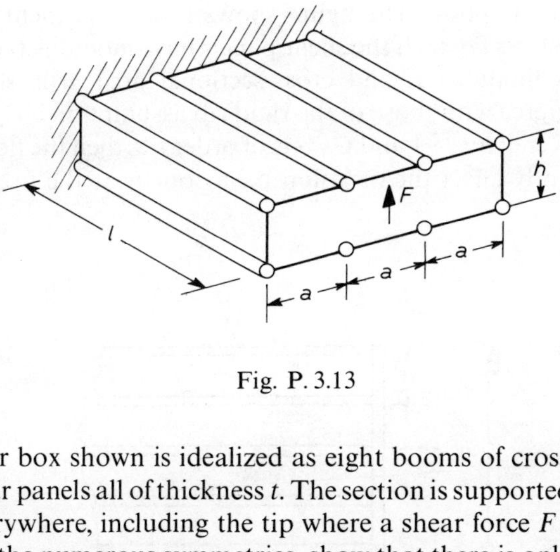

Fig. P. 3.13

The cantilever box shown is idealized as eight booms of cross-sectional area B, and pure shear panels all of thickness t. The section is supported by closely spaced rigid ribs everywhere, including the tip where a shear force F is applied. Taking advantage of the numerous symmetries, show that there is only one redundancy, and show that the boom loads at the built-in end are

$$\pm \frac{Fl}{4h}\left[1 \pm \frac{\tanh \lambda l}{\lambda l}\right]$$

where $(\lambda l)^2 = 2(G/E)(l/a)^2(at/B)$. (The larger loads are at the corner booms, hence the term 'shear lag' for the lack of diffusion from these regions.) If the boom areas are of order at, how long do you think the box need be for the shear lag to be unimportant in this problem?

P 3.14

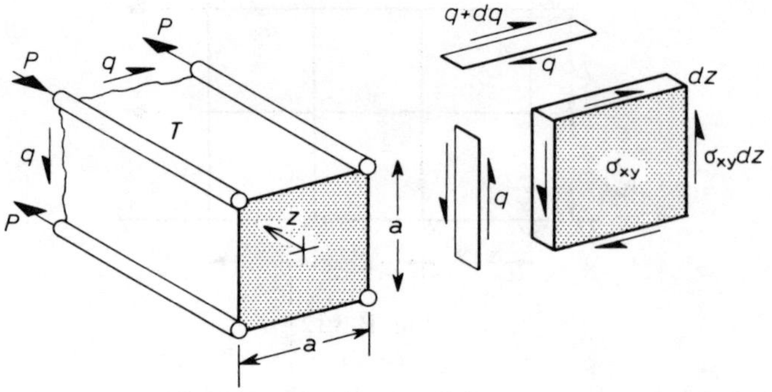

Fig. P. 3.14

The figure shows a very long four-boom tube which (for simplicity) is unloaded but has a temperature increase $T°$ applied to one boom. The self-equilibrating loads $\pm P$ and shears $q = P'(z)/2$ are shown. The tube is filled completely with a core having a finite shear modulus G_c, and it is required to see whether the consequent deformations predict a different behaviour to that of the 'closely spaced *rigid* ribs' assumption. The figure shows how an element of core must experience a shear stress $\partial q/\partial z$. If the shear panels have modulus G and thickness t, the four booms modulus E and cross-sectional area $at/2$, show that the theoretical results agree with those of the rigid rib assumption if G_c is of order G and $a/t \gg 1$. However, if $a/t \gg 1$ and G_c/G is of order t/a, then the flexibility of the core does significantly affect the diffusion behaviour near the tip, $z = 0$.

P 3.15

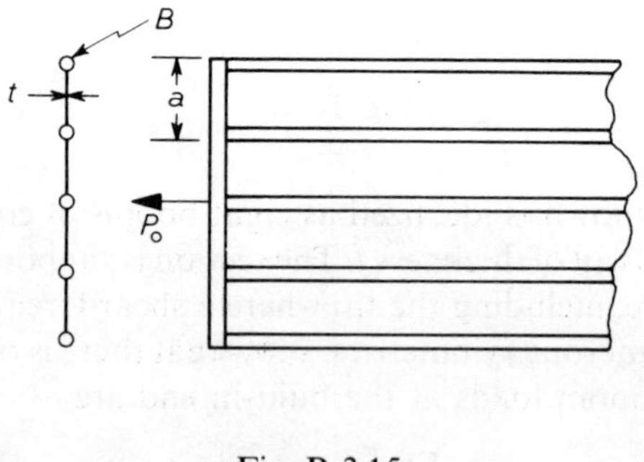

Fig. P. 3.15

A very long panel having five identical booms connected by four pure-shear panels is subjected to a central load P_0. Show that the problem has two redundancies and obtain a fourth-order differential equation for one of them. If $B = at$, $E/G = 8/3$, show that the diffusion length is not much greater than a.

4

APPROXIMATE METHODS

4.1 Introduction

So far we have used virtual work as an indirect substitute for direct equilibrium or compatibility arguments. The sceptical reader may possibly have been unconvinced by the power of the methods in solving for stress or displacements, and may prefer direct methods. Such a sceptic will also have noticed that all the solutions obtained so far have been exact – or at least appeared to be exact. No structure can be solved exactly and we have made clear from the onset that the first step in any analysis is to make assumptions. Thus the 'exact' beam theory relies on the kinematic idealization that plane sections remain plane – an *assumed displacement field.* Similarly, pinjointed frameworks have members whose ends are free to rotate at joints. The boom shear panels of the last chapter were static idealizations in that we *assumed stress fields,* and to add insult to injury we occasionally moved the lines of action of direct stresses to fictitious booms. In this chapter we take the process of approximation much further with the aim of analysing structures which have no simple kinematic or static model at all. Apart from frameworks, most structures fall into this category: the problem is entirely a geometric one as is obvious when we think of a ship, an aircraft, or a shell roof, or any of the infinite variety of three-dimensional structures which occur in mechanical engineering where a complex shape is machined from the solid. Some assumptions clearly have to be made, even when applying virtual-work arguments, otherwise we shall simply recover the continuum partial differential equations of equilibrium and compatibility ((1.52) and (1.66)).

Confronted with partial differential equations, the mathematician is likely to attempt to separate variables in some promising fashion, and then solve the consequent *total* differential equations. We shall be using this device also, but prefer to work through a gentler introduction by solving one-dimensional problems first to illustrate the fundamental use of virtual work in approximate solutions.

There is no doubt that virtual work (or variational energy methods) have evolved as the most powerful way of using approximate methods in structural mechanics. We will approach problems in stages using both the PVD and the PVF, starting by assuming displacement or stress fields in one dimension, and then broadening the scope to two- and three-dimensional geometries. Having shown how and why the two virtual-work methods deliver approximate solutions, we will then generalize the technique to employ both assumed displacement and assumed stress fields – the so-called *mixed method.* As a digression we will also relax our customary constraint that virtual displacements, or virtual forces, should be put to zero on a surface where the true displacements or true forces are prescribed.

4.2 The Rayleigh–Ritz technique[6,32,33]

The most drastic analytical idealization is to express the displacement or stress field as a *complete* series of some sort over the entire structure, and then hope that the series will converge sufficiently rapidly for us to be able to truncate it after a moderate number of terms. Having done this it merely remains to find the magnitudes of each term in the series, thus the problem is converted from one of solving differential equations to one of finding a finite number of unknown coefficients – the amplitudes of the terms in the assumed series. We have to solve for a set of discrete numbers, and the problem is said to be 'discretized'. (Our familiar frameworks are of course already in this form.) Such a procedure is well known in Fourier analysis where *known* functions are expressed as trigonometric series whose coefficients are readily found by integration, and the convergence of such series is well established. Fourier series are orthogonal and it is possible to obtain a single equation for each unknown coefficient – a ruse we shall also employ.

The usual definition of a 'complete' series is that it will eventually converge when used to represent any given function which is well behaved in some sense. This definition is not particularly helpful, and by 'complete' we will mean a series such that no single term can be expressed as an expansion in all the others. Thus duplication is not allowed. Omission of a single term can be equally fatal.

Recalling, then, that in using the PVD (or PVF) we are obliged to make the virtual displacements (or stresses) satisfy compatibility (or equilibrium) requirements, then the traditional Rayleigh–Ritz technique is to express the true displacements (or stresses) as a complete series, each term of which satisfies the necessary compatibility (or equilibrium) conditions. We then use these terms *one at a time* as possible virtual displacements or stresses in the PVD (1.69) or PVF (1.73), and so generate as many equations as the number of terms we choose to retain in the hopefully convergent series.

It is also possible to use the developed versions of the PVD (1.72) or PVF (1.75) where Gauss's identity was used to produce the equations of equilibrium and compatibility inside the integrals. In this form the procedure is attributed to Bubnov–Galerkin,[7,41] and of course it must give the same answers, although it is frequently stated that the Galerkin method requires the assumed displacement (or stress) series to satisfy also equilibrium (or kinematic) boundary conditions. This is not so provided that the surface integrals are not discarded in (1.72) or (1.75). The Galerkin form is not normally convenient nor popular since it necessarily involves finding differential equations like $\mathbf{D\sigma} + \mathbf{p}_v = \mathbf{0}$ or

$\varepsilon - \mathbf{D}^t\mathbf{u} = \mathbf{0}$, although we are not of course trying to solve them. The Rayleigh–Ritz expressions for internal work are much simpler and involve only volume integrals whereas the Galerkin form has surface integrals also. The Rayleigh–Ritz technique evaluates positive work products like $\boldsymbol{\sigma}^t\bar{\boldsymbol{\varepsilon}}$, whereas the Galerkin method uses products like $\bar{\mathbf{u}}^t(\mathbf{D}\boldsymbol{\sigma} + \mathbf{p}_v)$ and the user has to get the sign right. However, the Galerkin form does have the virtue that it can be extended as a 'weighted residual' technique, to which we shall return.

The purpose behind the Rayleigh–Ritz method is of course that we replace the solution of differential equations by that of solving simultaneous linear equations with simple constants as unknowns, a routine task since the advent of readily accessible computers. Only integrals involving the work done in stress or displacement fields are needed and the method is ideally suited to structures whose geometry is complicated, or even discontinuous, as most structures are. We will now look at a succession of simple examples and examine the separate effects of choice of series, convergence, accuracy, and the treatment of boundary conditions. We choose initially simple structures describable in terms of only one independent variable so that the principles are not obscured by algebra – most examples will examine the behaviour of the ubiquitous beam.

4.3 **Simple beams** (Using the PVD)

Connider the uniform beam of Fig. 4.1, simply supported at $z = \pm l$ and subjected to the symmetric loading shown, varying linearly from w_0 per unit length to zero at the ends. We will express the deflection $v(z)$ as a *polynomial series*, and for this reason have chosen the curious discontinuous loading to avoid obtaining an exact solution (!). The exact discontinuous solution is in fact

$$v(z) = \frac{w_0 l^4}{120EI}\left[16 - 20\left(\frac{z}{l}\right)^2 + 5\left(\frac{z}{l}\right)^4 - \frac{|z|^5}{l^5}\right]$$

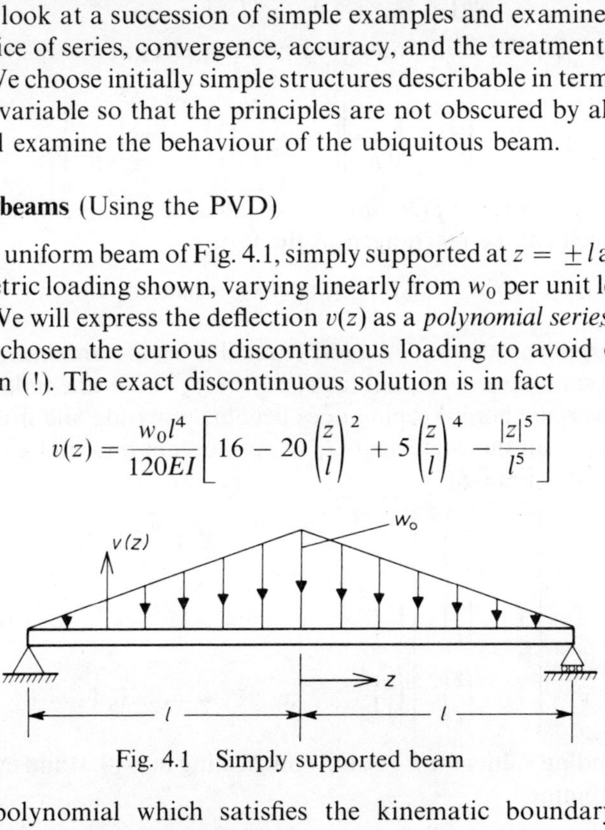

Fig. 4.1 Simply supported beam

A suitable polynomial which satisfies the kinematic boundary conditions $v(\pm l) = 0$, and which is symmetrical, is

$$v(z) = \left[1 - \left(\frac{z}{l}\right)^2\right]\left[v_1 + v_2\left(\frac{z}{l}\right)^2 + v_3\left(\frac{z}{l}\right)^4\right] \tag{4.1}$$

where v_1, v_2, and v_3 are the constant coefficients to be determined, and it remains to be seen whether three of them are sufficient. We are content to let the PVD satisfy equilibrium so we can ignore the equilibrium boundary condition at $z = \pm l$, that is,

$$EIv''(\pm l) = 0, \qquad \text{or} \qquad 2v_1 + 10v_2 + 18v_3 = 0 \tag{4.2}$$

The PVD, noting the symmetry of the problem, is

$$2\int_0^l EIv''\bar{v}''\,\mathrm{d}z = 2\int_0^l -w_0\left(1 - \frac{z}{l}\right)\bar{v}\,\mathrm{d}z$$

where there are no terms at the ends $z = \pm l$ since there is no applied moment to do work over $\bar{v}'(\pm l)$ and there is no deflection $\bar{v}(\pm l)$ for the reaction to do work over. For brevity we write the equation as

$$\int_0^l \left[l^4 v''\bar{v}'' + \beta\left(1 + \frac{z}{l}\right)\bar{v}\right]\mathrm{d}z = 0 \tag{4.3}$$

where $\beta = w_0 l^4/EI$. We could now find v'' from (4.1) and similarly find $\bar{v}''$ by putting

$$\bar{v}(z) = \left[1 - \left(\frac{z}{l}\right)^2\right]\left[\bar{v}_1 + \bar{v}_2\left(\frac{z}{l}\right)^2 + \bar{v}_3\left(\frac{z}{l}\right)^4\right]$$

The familiar argument that $\bar{v}_1$, $\bar{v}_2$, and $\bar{v}_3$ are all arbitrary and independent then expression which can be rearranged in the form:

$$f_1(v_1, v_2, v_3, \beta)\bar{v}_1 + f_2(v_1, v_2, v_3, \beta)\bar{v}_2 + f_3(v_1, v_2, v_3, \beta)\bar{v}_3 = 0$$

The familiar argument that $\bar{v}_1$, $\bar{v}_2$, and $\bar{v}_3$ are all arbitrary and independent then leads to three simultaneous equations $f_1 = f_2 = f_3 = 0$ in the unknowns v_1, v_2, and v_3. However, the book-keeping does become unwieldy and it is much more convenient to put all the coefficients $\bar{v}_n$ to zero except one, and so generate the equations one at a time. So:

$$\bar{v} = \left[1 - \left(\frac{z}{l}\right)^2\right] \qquad \text{i.e.} \quad \bar{v}_1 = 1,\ \bar{v}_2 = \bar{v}_3 = 0$$

$$\bar{v} = \left[1 - \left(\frac{z}{l}\right)^2\right]\left(\frac{z}{l}\right)^2 \qquad \text{i.e.} \quad \bar{v}_1 = 0,\ \bar{v}_2 = 1,\ \bar{v}_3 = 0$$

$$\bar{v} = \left[1 - \left(\frac{z}{l}\right)^2\right]\left(\frac{z}{l}\right)^4 \qquad \text{i.e.} \quad \bar{v}_1 = \bar{v}_2 = 0,\ \bar{v}_3 = 1$$

with corresponding values for $\bar{v}''(z)$. On substituting into (4.3) and evaluating the integrals we obtain:

$$\text{for} \quad \bar{v} = \left[1 - \left(\frac{z}{l}\right)^2\right]: \qquad v_1 + v_2 + v_3 = -0.104\beta$$

$$\text{for} \quad \bar{v} = \left[1 - \left(\frac{z}{l}\right)^2\right]\left(\frac{z}{l}\right)^2: \quad 4v_1 + 16.8v_2 + 18.63v_3 = -0.05\beta$$

$$\text{for} \quad \bar{v} = \left[1 - \left(\frac{z}{l}\right)^2\right]\left(\frac{z}{l}\right)^4: \quad 0.667v_1 + 3.105v_2 + 4.324v_3 = -0.0025\beta$$

Instead of using (4.3) we could have used the Galerkin form, which is obtained by integrating the PVD twice by parts:

$$\int_0^l \left[EIv'''' + w_0\left(1 - \frac{z}{l}\right)\right]\bar{v}\,dz + [EIv''\bar{v}']_0^l - [EIv'''\bar{v}]_0^l = 0$$

As $\bar{v}'(0) = 0$, $\bar{v}(l) = 0$, and $v'''(0) = 0$, this simply reduces to

$$\int_0^l \left[EIv'''' + w_0\left(1 - \frac{z}{l}\right)\right]\bar{v}\,dz + EIv''(l)\bar{v}'(l) = 0 \tag{4.4}$$

Identical equations are produced.

On solving the simultaneous equations we find

$$v_1 = -0.133\,33\beta; \qquad v_2 = 0.032\,55\beta; \qquad v_3 = -0.003\,42\beta$$

Several things may be said before we examine this solution in detail. Firstly, the convergence seems adequate. Secondly, although our series violated $EIv''(\pm l) = 0$, when we examine (4.2) we find

$$-l^2v'' = 2v_1 + 10v_2 + 18v_3 = -0.266\beta + 0.325\beta - 0.061\beta = -0.002\beta$$

so the PVD has left a small residual bending moment at the ends. Thirdly, we should like to know how accurate this solution is, assuming that we do not have the exact solution available. One way is to increase the number of terms in the series (4.3) and examine the effect on the answers. We should in particular examine the effect on the bending stresses which depend on EIv'' which does not converge as fast as $v(z)$ itself.

Finally, we ask ourselves: what have we tried to do? The answer is most obvious when we examine the Galerkin form (4.4). Thus our assumed solution does not *exactly* satisfy the equilibrium requirements that

$$EIv'''' + w_0\left(1 - \frac{z}{l}\right) = 0 \quad \text{for} \quad -l < z < l; \quad \text{and} \quad EIv''(\pm l) = 0$$

However, we have forced the error in these two expressions to do no work over the three available virtual displacements in $\bar{v}(z)$ and $\bar{v}'(l)$ obtained by selecting $\bar{v}_1$, $\bar{v}_2$, and $\bar{v}_3$ separately. The more items we decide to include in our series, the more virtual displacements $\bar{v}_1, \bar{v}_2, \ldots, \bar{v}_n$ there will be over which we force the residual error to do no work. Eventually there will be no room for manoeuvre in the residual error, which will have to be negligible everywhere. This residual error is the difference between $-EIv'''$ (the beam resistance) and the applied load $w_0[1 - (z/l)]$. We may imagine this residual as a continuous artificial loading (or constraint) forcing the beam to maintain our solved deflection. If we were to gradually remove this constraint, and allow the beam to move to its true position of equilibrium, then positive work must be done by the applied forces on the beam as the constraints are removed. If this were not so then the beam would not move on decreasing the constraints, thus the elastic potential energy of the beam must increase as the constraints are decreased. *We must therefore expect the strain energy in a structure, when found in terms of our approximate solution using the*

PVD, to be less than the true strain energy. Moreover, as we increase the number of terms in the series we would expect the solution to improve and the solved strain energy to increase gradually and approach the true value. It is consequently referred to as a *lower bound*, or a solution in which the structure is over-stiff. This fact is not in itself a useful check on the accuracy of a solution. Although it is often referred to as giving a lower bound on deflections and stresses 'in the mean', there is no guarantee that the stress or displacement at a particular point is a lower bound. (There are means of obtaining the bound at a point but they involve finding Green's function and are only suitable for structures whose governing differential equations and boundaries are simple.) It is considered more straightforward and systematic to check the accuracy, as we said, by increasing the degrees of freedom in the assumed solution, and seeing if significant changes in displacements and stresses occur.

In our simple beam problem, the central deflection for example is $v(0) = v_1 = -0.133\,31\beta$, compared to the exact value of $-0.133\,33\beta$. The bending stresses are not so accurate, and EIv'' is shown in the following table.

z/l	0	0.2	0.4	0.6	0.8	1.0
M (approx)/$w_0 l^2$	−0.331 75	−0.315 31	−0.269 26	−0.198 52	−0.108 01	−0.002 67
M (exact)/$w_0 l^2$	−0.333 33	−0.314 63	−0.264 00	−0.189 33	−0.098 67	0

The bending stresses are clearly not less than the true ones everywhere.

Suppose we now choose to load the beam by a single concentrated force P (see Fig. 4.2), then the deflection at the load is a direct measure of the strain energy and must therefore be a lower bound. If we choose a series which satisfies both kinematic *and* static boundary conditions, we might expect the convergence to be improved. Thus, let

$$v(z) = \sum_{n=1,3,5,\ldots} v_n \cos\frac{n\pi z}{2l} \tag{4.5}$$

and choose one term at a time for $\bar{v} = \cos(m\pi z/2l)$.

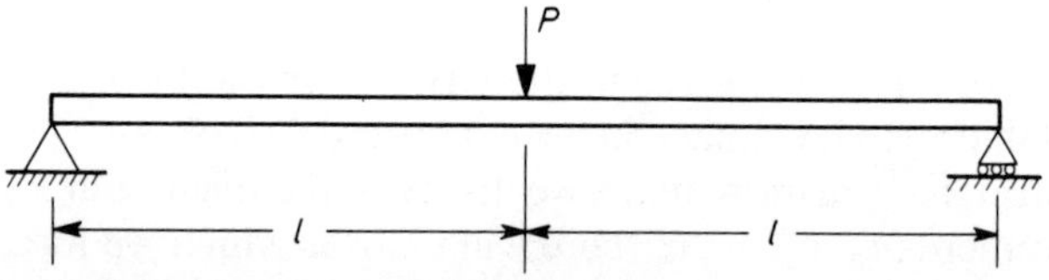

Fig. 4.2 Concentrated load

The PVD becomes, for $m = 1, 3, 5, \ldots,$

$$2\int_0^l \left(\sum_n - EI\frac{n^2\pi^2}{4l^2} v_n \cos\frac{n\pi z}{2l}\right)\left(\frac{-m^2\pi^2}{4l^2}\cos\frac{m\pi z}{2l}\right) dz = -P(1)$$

The choice of the trigonometric series (4.5) now seems even more judicious since it is an *orthogonal* series, that is,

$$\int_0^l \cos\frac{n\pi z}{2l}\cos\frac{m\pi z}{2l}\,dz = 0$$

for m and n both odd, unless $m = n$, when it is $l/2$. All terms in the summation therefore vanish except one, and we have

$$\frac{EIn^4\pi^4}{16l^3}v_n = -P, \quad \text{or} \quad v_n = -\frac{16Pl^3}{\pi^4 EIn^4}$$

We are therefore spared the effort of solving simultaneous equations, and the complete solution is

$$v(z) = \frac{-16Pl^3}{\pi^4 EI}\sum_{n=1,3,\ldots}\frac{1}{n^4}\cos\frac{n\pi z}{2l} \tag{4.6}$$

The central deflection, and therefore a measure of the stored energy, $\frac{1}{2}Pv(0)$, is

$$v(0) = \frac{-16Pl^3}{\pi^4 EI}\sum\frac{1}{n^4}$$

Obviously a truncated version of this series will be less than the infinite series, so the strain energy approaches the exact value from below as we anticipated. As it happens we can actually sum

$$\sum_{n=1,3,\ldots}\frac{1}{n^4} = \frac{\pi^4}{96}$$

whence $v(0) = -Pl^3/EI$, the exact solution. Also the bending moment

$$M = -EIv''(z) = -\frac{4Pl}{\pi^2}\sum\frac{1}{n^2}\cos\frac{n\pi z}{2l} \tag{4.7}$$

This converges like $1/n^2$, which is a slower rate than the deflection. The central moment, incidentally, can also be summed since

$$\sum_{n=1,3,\ldots}\frac{1}{n^2} = \frac{\pi^2}{8}; \quad \text{so} \quad M(0) = -\frac{Pl}{2}$$

The shear force $-EIv'''$ converges even slower, like $1/n$, and at $z = 0$ we find $-EIv'''(0) = 0$, whereas the shear force actually changes discontinuously from $-P/2$ to $+P/2$. In the manner of a Fourier series, the Rayleigh–Ritz has found the average value of zero.

4.4 Vibrations of a rod

Another problem is selected to show that the inclusion of dynamic inertia forces in the PVD is quite straightforward, and to take advantage of the Rayleigh–Ritz procedure when the structure's cross-sectional properties are not constant. We look at the longitudinal vibrations of a tapered bar, but assume that the bar is slender with small taper, so the axial stress may be uniform over the section and transverse strains may be neglected. The propagation of sound waves in such bars has been discussed by Rayleigh himself,[32] who classified their behaviour, when their length is of the order of the wavelength of sound, to be roughly independent of their cross-sectional area variation.

The problem is not without interest since ultrasonic machine tools usually consist of a tapered bar whose conveniently broad end is electromagnetically excited at high frequency, causing the narrow end to vibrate and act as a machining tool (see Fig. 4.3). Moreover, it is desirable to excite such a bar at its resonant frequency when the exciting force is only required to overcome damping which is hopefully small, so that the resonant frequency is very close to the undamped natural frequency. Bearing this application in mind, it will also be of interest to select a vibration mode which has a convenient node at which to support the tool. Consider then the steady-state vibrations, after initial transients have decayed, of the tapered fixed-free rod shown in Fig. 4.4. The axial displacement $w(z)$ at a section whose area is $A(z)$ will produce a strain $\varepsilon_{zz} = \partial w/\partial z$. We then reverse the inertia force in the manner of D'Alembert so that we may treat the problem as a quasi-static one where the body force is simply

$$p_{vz} = -\rho \frac{\partial^2 w}{\partial t^2}, \qquad \rho \text{ being the material density}$$

The PVD, assuming Hooke's law,

and using $$\int_V \boldsymbol{\sigma}^t \bar{\boldsymbol{\varepsilon}}\, dv = \int_V \mathbf{p}_v^t \bar{\mathbf{u}}\, dv$$

becomes $$\int_0^l \left(E\frac{\partial w}{\partial z}\right)\frac{\partial \bar{w}}{\partial z} A(z)\, dz = \int_0^l \left(-\rho \frac{\partial^2 w}{\partial t^2}\right) \bar{w} A(z)\, dz \tag{4.8}$$

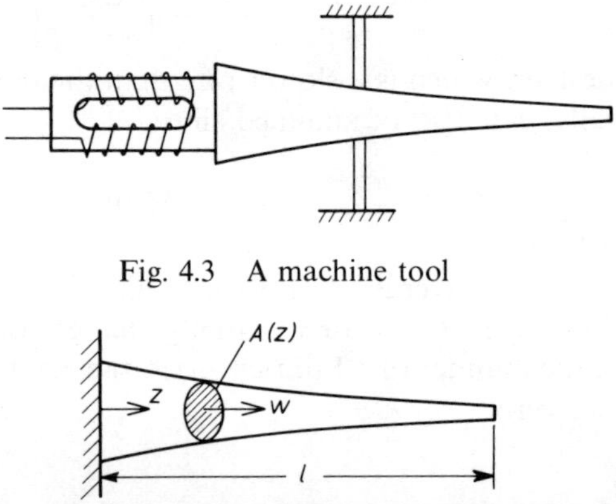

Fig. 4.3 A machine tool

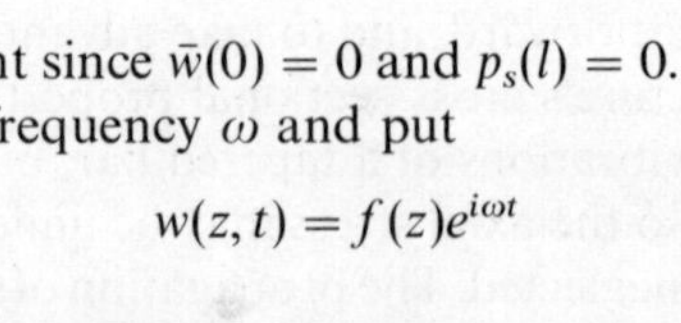

Fig. 4.4 Vibrating tapered rod

No surface term is present since $\bar{w}(0) = 0$ and $p_s(l) = 0$. We consider the natural vibrations at a circular frequency ω and put

$$w(z, t) = f(z)e^{i\omega t}$$

whence (4.8) becomes

$$\int_0^l \left[Af'(z)\bar{f}'(z) - A\frac{\omega^2}{C^2} f(z)\bar{f}(z) \right] dz = 0 \tag{4.9}$$

where $C = (E/\rho)^{1/2}$ is the speed of sound in the material. Equation (4.9) is immediately suitable for application of the Rayleigh–Ritz procedure, but before doing this let us briefly examine the exact solution by converting (4.9) to the Galerkin form by integrating,

$$-\int_0^l \left\{ [Af'(z)]' + A\frac{\omega^2}{C^2} f(z) \right\} \bar{f}(z)\,dz + [Af'(z)\bar{f}(z)]_0^l = 0 \qquad (4.10)$$

The governing differential equation of 'equilibrium' is therefore

$$[Af'(z)]' + A\frac{\omega^2}{C^2} f(z) = 0 \qquad (4.11)$$

with the kinematic boundary condition $f(0) = 0$. The second term in (4.10) provides the equilibrium boundary condition

$$Af'(l) = 0 \qquad (4.12)$$

that is, $\sigma_{zz} = Ee^{i\omega t} f'(l) = 0$ as we might have expected. Equation (4.11) has only a few simple solutions; for example, when $A(z)$ is constant or varies exponentially. (Variations of the form $A(z) = z^n$ have been examined by Tsui,[34] who obtained solutions in terms of Bessel functions of the first and second kind.) When $A(z)$ is a constant, (4.11) becomes

$$f''(z) + \frac{\omega^2}{C^2} f(z) = 0$$

With $f(0) = 0$ and $f'(l) = 0$, the solution is

$$f(z) = f_0 \sin\frac{\omega z}{C}, \qquad \text{where} \quad \cos\frac{\omega l}{C} = 0$$

Thus the natural frequencies are given by $\omega l/C = \pi/2;\ 3\pi/2;$ etc.

Now suppose we have a linearly tapering rod whose cross-sectional area $A(z)$ is given by

$$A(z) = A_0\left(1 - \frac{2z}{3l}\right)$$

the area-ratio of the two ends is 3:1. Equation (4.9) becomes

$$\int_0^l \left(1 - \frac{2z}{3l}\right)\left[f'(z)\bar{f}'(z) - \frac{\omega^2}{C^2} f(z)\bar{f}(z) \right] dz = 0 \qquad (4.13)$$

We must satisfy the kinematic boundary condition $f(0) = 0$, and it is further convenient if we can also satisfy the equilibrium condition $f'(l) = 0$. This may be done with the trigonometric series

$$f(z) = \sum_{n=1,3,\ldots} f_n \sin\frac{n\pi z}{2l}$$

Suppose we truncate this series at $n = 3$. After substituting into (4.13) and performing the integrations for the two virtual displacements $\overline{f_1} = 1$ and $\overline{f_3} = 1$ alone, we arrive at the two equations:

$$\overline{f_1} = 1, \overline{f_3} = 0: \quad f_1\left(\frac{\pi^2}{12} + \frac{1}{6}\right) + f_3\left(\frac{1}{2}\right) - \frac{\omega^2 l^2}{C^2}\left[f_1\left(\frac{1}{3} - \frac{2}{3\pi^2}\right) + f_3\left(\frac{2}{3\pi^2}\right)\right] = 0$$

$$\overline{f_3} = 1, \overline{f_1} = 0: \quad f_1\left(\frac{1}{2}\right) + f_3\left(\frac{3\pi^2}{4} + \frac{1}{6}\right) - \frac{\omega^2 l^2}{C^2}\left[f_1\left(\frac{2}{3\pi^2}\right) + f_3\left(\frac{1}{3} - \frac{2}{27\pi^2}\right)\right] = 0$$

or

$$\begin{bmatrix} 0.9891 - 0.2658\left(\dfrac{\omega l}{C}\right)^2 & 0.5 - 0.0675\left(\dfrac{\omega l}{C}\right)^2 \\ 0.5 - 0.0675\left(\dfrac{\omega l}{C}\right)^2 & 7.5689 - 0.3258\left(\dfrac{\omega l}{C}\right)^2 \end{bmatrix}\begin{bmatrix} f_1 \\ f_3 \end{bmatrix} = \begin{bmatrix} 0 \\ 0 \end{bmatrix} \tag{4.14}$$

This is a standard *eigenvalue problem* (see Appendix A) of the form

$$\mathbf{Ax} - \lambda\mathbf{Bx} = \mathbf{0}$$

where λ in this case is $(\omega l/C)^2$. However, we can also resort to earlier arguments and state that if $\{f_1 \quad f_3\}$ is to have a nonzero solution, then the coefficient matrix must be singular. Putting its determinant to zero, we obtain a quadratic for the (eigen) values of $(\omega l/C)^2$ with positive roots,

$$\frac{\omega l}{C} = 1.919 \qquad \text{and} \qquad 4.893$$

These compare with $\pi/2 = 1.571$ and $3\pi/2 = 4.712$ for the uniform rod. The corresponding mode shapes (eigenvectors) are given by substitution back into (4.14).

Thus $$f_3/f_1 = -0.0394 \text{ or } -4.811$$

The first natural mode is therefore

$$f(z) = f_1\left(\sin\frac{\pi z}{2l} - 0.0394\sin\frac{3\pi z}{2l}\right)$$

and the second is

$$f(z) = f_1\left(\sin\frac{\pi z}{2l} - 4.811\sin\frac{3\pi z}{2l}\right)$$

The first term dominates the fundamental mode while the second term dominates the second harmonic, confirming Rayleigh's suggestion that the behaviour will be close to that of a uniform bar. The first mode has no intermediate node while the second has a node at $z/l = 0.707$, compared with $z/l = 0.667$ for the uniform rod. Had we retained only one term f_1, we would have simply obtained, instead of (4.14),

$$\left(\frac{\omega l}{C}\right)^2 = \left(\frac{\pi^2}{12} + \frac{1}{6}\right)\Big/\left(\frac{1}{3} - \frac{2}{3\pi^2}\right), \qquad \text{or} \qquad \frac{\omega l}{C} = 1.929$$

This frequency is slightly higher than 1.919, obtained with two terms in the series, and confirms that over-stiff PVD solutions become less stiff as the approximation is improved.

4.5 A four-boom tube

The Rayleigh-Ritz method is equally useful in obtaining approximate solutions using the PVF. It is not, however, convenient to demonstrate this fact on simple beams since the unknown force redundancies are already simple discrete quantities, and a series approximation is unnecessary. We choose therefore to re-examine the constraint stresses in the idealized four-boom tube of Fig. 3.35 for which we obtained an exact solution (3.58). We recall that equilibrium requirements lead to the force pattern shown in Fig. 4.5, and to the equations

$$q_1 = \frac{T}{2ab} + \frac{P'}{2}; \qquad q_2 = \frac{T}{2ab} - \frac{P'}{2}$$

The PVF, $\int_V \bar{\boldsymbol{\sigma}}_i^t \boldsymbol{\varepsilon}\, dv = 0$, becomes

$$\int_0^l \left\{ \frac{8GtP\bar{P}}{EB} + \left[\frac{T}{ab}(a-b) + P'(a+b) \right] \bar{P}' \right\} dz = 0 \tag{4.15}$$

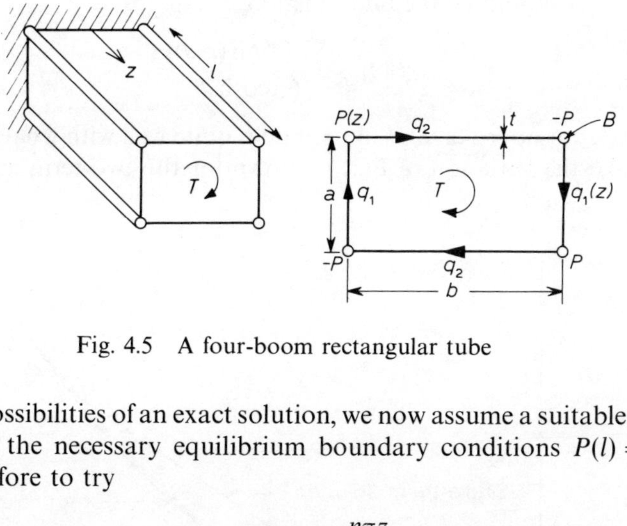

Fig. 4.5 A four-boom rectangular tube

Ignoring the possibilities of an exact solution, we now assume a suitable series for $P(z)$ satisfying the necessary equilibrium boundary conditions $P(l) = 0$. It is tempting therefore to try

$$P(z) = \sum_n P_n \sin \frac{n\pi z}{l}$$

In fact this turns out to give us the exact solution (3.58), since the exact hyperbolic terms are convertible to trigonometric ones! We will therefore be contrite and choose a polynomial, and for this it is more convenient to measure z from the tip.

Thus

$$P(z) = \sum_n P_n (z/a)^n \tag{4.16}$$

where P_n are coefficients having the dimension of force. To solve the imminent set of simultaneous equations, it is necessary to invent some dimensions, so we choose typically

$$B = at; \qquad b = 2a; \qquad G/E = 3/8; \qquad l = b = 2a$$

This choice represents a fairly short box and it is to be expected that the constraint effects at the root will be significant everywhere. Equation (4.15) is now

$$\int_0^{2a} \left\{ \frac{3}{a} P\bar{P} + \left[3aP' + \frac{T}{2a} \right] \bar{P}' \right\} \mathrm{d}z = 0 \tag{4.17}$$

We bravely terminate the series at $n = 2$, and obtain two equations from (4.17):

for $\quad \bar{P} = z/a_1 \quad \bar{P}' = 1/a$: $\qquad 14P_1 + 24P_2 = -T/a$

for $\quad \bar{P} = (z/a)^2, \quad \bar{P}' = 2z/a^2$: $\qquad 60P_1 + 128P_2 = -5T/a$

Thus

$$P_1 = -0.0227T/a \qquad \text{and} \qquad P_2 = -0.0284T/a$$

This is hardly a converging series and does not look a promising start. We would normally try a few more terms before examining the suitability of the series, but if we did this we would find that P_3 is small, and P_2 almost vanishes, the reason being that $P(z)$ should be the odd function

$$P(z) = \frac{-T\sinh(z/a)}{6a\cosh 2}$$

Nevertheless, our approximate solution does quite well with what we have given it. In Fig. 4.6 the variation of $P(z)$ is shown for the two-term and single-term

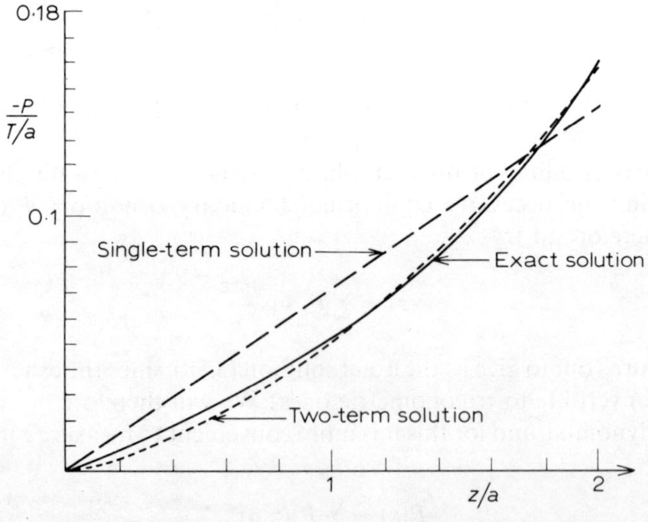

Fig. 4.6 Approximate solutions for four-boom tube loads

approximations. The two-term solution is very successful. Both linear and quadratic approximations appear to steer a mean path and produce maximum errors of roughly equal and opposite sign. This 'mean fit' is characteristic of PVD and PVF solutions and we will return to this aspect later.

It is convenient to take this example and again discuss exactly what our approximation has done. We return therefore to the Galerkin form obtainable by integrating (4.15) by parts:

$$\int_0^l \left\{\frac{8Gt}{BE}P - (a+b)P''\right\}\bar{P}\,\mathrm{d}z + \left\{\left[\frac{T(a-b)}{ab} + (a+b)P'\right]\bar{P}\right\}_0^l = 0$$

The governing differential equation of compatibility is the term in the integrand and so our approximate solution will attempt to satisfy

$$\frac{8Gt}{BE}P - (a+b)P'' = 0$$

But again there will be a residual error (a displacement) over which our virtual forces $\bar{P}$ in the series will be made to do no work. The general error in the equation $\boldsymbol{\varepsilon} - \mathbf{D}^t\mathbf{u}$ is not always easy to visualize, although for specific structures it is clear why $\boldsymbol{\varepsilon}$ may not be equal to $\mathbf{D}^t\mathbf{u}$. For instance, if a bar in a pinjointed framework is 'cut', compatibility is destroyed for that member and we cannot put $\boldsymbol{\Delta} = \mathbf{ar}$ (1.27). Similarly, if a hinge is introduced into a beam, compatibility is destroyed and we can no longer say $\varepsilon_{zz} = -yv''(z)$. The introduction of excessive kinematic freedom will not (unless the result is a mechanism) prevent us from satisfying equilibrium as the PVF requires, but the structure will be 'overflexible' and the applied forces will do more work than they would over the exact kinematically compatible displacements. In contrast to the PVD, *we must expect the strain energy in a structure, when found from an approximate PVF solution,* to be *greater than the true strain energy.* For this reason such solutions are referred to as an *upper bound.* It is a bound on energy or work, and does not guarantee that the stress or displacement at a particular point will be greater than the exact value. In our example the structure is loaded only by the tip torque so the tip rotation happens to be a true measure of work and will be an upper bound. Using the unit-load method with our solved stresses the tip rotation is readily found to be

$$\theta(l) = 0.790T/Ga^2$$

whereas the exact solution (see section 3.9) is $0.735T/Ga^2$.

Finally, it must be conceded that a polynomial series is not totally satisfactory for a long box, $l \gg b$, since the constraints we know are confined to a region of order b near the root. The exact solution is an exponential variation and no other series behaves like this over the entire length. We must later face reality and admit that in many problems it is not possible to attempt a series expansion covering the entire structure, particularly if local variations are expected.

4.6 Comparisons of PVD and PVF solutions

Having shown the use of the Rayleigh–Ritz method for both displacement and force analyses, it is of interest to apply them both to the same problem, selecting a single applied load so that its deflection is a true measure of the strain energy and provides the bounds we have mentioned. We should also like a fairly simple orthogonal series solution so that the effect of increasing the degrees of freedom is readily apparent, so a beam has been chosen which is supported on an elastic foundation or continuous spring (Fig. 4.7). The internal forces are therefore continuous and can be expanded as a series; moreover, we may vary the ratio of the beam stiffness (EI) to the spring stiffness (k) to test the ability of the assumeed series to cope with diverse variations in stress and displacement. Both kinematic and equilibrium boundary conditions are satisfied by expanding

$$v(z) = \sum_{n=1}^{N} v_n \sin\frac{n\pi z}{l} \tag{4.18}$$

thus

$$\bar{v}(z) = \sin\frac{m\pi z}{l}, \qquad m = 1, 2, 3, \ldots, N$$

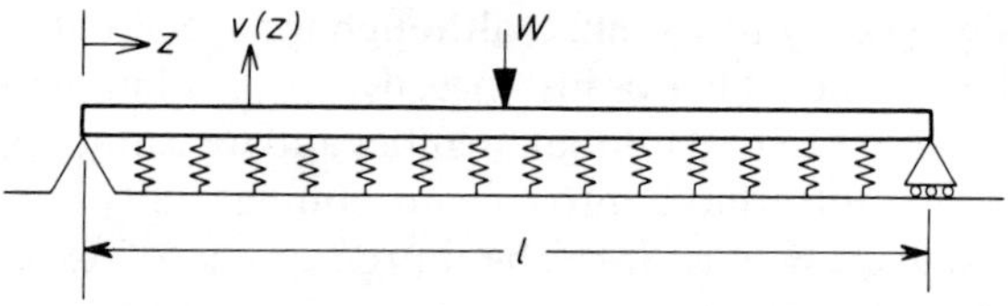

Fig. 4.7 Beam on continuous spring

The PVD becomes

$$\int_0^l EIv''\bar{v}''\,\mathrm{d}z + \int_0^l kv\bar{v}\,\mathrm{d}z = -W\bar{v}(l/2)$$

Substituting (4.18),

$$\int_0^l \left\{\left[\sum -EI\frac{n^2\pi^2}{l^2}v_n \sin\frac{n\pi z}{l}\right]\left[-\frac{m^2\pi^2}{l^2}\sin\frac{m\pi z}{l}\right] + k\left[\sum v_n \sin\frac{n\pi z}{l}\right]\sin\frac{m\pi z}{l}\right\}\mathrm{d}z = -W\sin\frac{m\pi}{2}$$

Because the integrals of $\sin(n\pi z/l)\sin(m\pi z/l)$ are zero, unless $m = n$ when it is $l/2$, only one term survives in the expansion, and we find

$$v_n = 0 \qquad \text{for } n \text{ even}$$

$$v_n = \frac{-2W(-1)^{(n-1)/2}}{kl + EI\pi^4 n^4/l^4} \qquad \text{for } n \text{ odd} \tag{4.19}$$

The central deflection

$$v(l/2) = \sum v_n(-1)^{(n-1)/2} = -\frac{2W}{kl}\sum_{1,3,\ldots}^{N}\frac{1}{1+n^4\beta} \tag{4.20}$$

where $\beta = \pi^4(EI/l^4)/k$ is a measure of the relative stiffness of the beam and the spring. If the series (4.20) is truncated we obviously get a lower bound, the value of $v(l/2)$ increasing as we increase the number of terms. The central bending moment at $z = l/2$ is

$$M(l/2) = -EIv''(l/2) = \frac{-2Wl}{\pi^2} \sum_{1,3,\ldots}^{N} \frac{n^2\beta}{1 + n^4\beta} \tag{4.21}$$

This also converges, more slowly, from a lower bound. We examine the two extreme limits of spring stiffness.

A weak spring

Let $k \to 0$, keeping EI finite, that is $\beta \to \infty$. We should expect to recover the solution for a simply supported beam with no spring. From (4.20)

$$v(l/2) \to \frac{-2W}{kl\beta}\sum \frac{1}{n^4} = \frac{-2Wl^3}{EI\pi^4}\sum \frac{1}{n^4}$$

This converges very rapidly, and can in fact be summed since the sum of $(1/n^4)$ is $\pi^4/96$, whence $v(l/2) = -Wl^3/48EI$, the exact solution. Similarly, $M(l/2) = Wl/4$ when we sum $(1/n^2)$ as $\pi^2/8$. We might have anticipated a fairly rapid convergence since the exact solution for $v(z)$ is two cubic polynomials joined at $z = l/2$ with continuity up to the second derivative – not a difficult shape for a Fourier expansion.

A stiff spring

Let $EI \to 0$, keeping k finite, that is $\beta \to 0$. The concentrated load W is eventually resisted entirely by the spring as the beam disappears, so a singularity $v(l/2) \to \infty$ appears, and $v(z)$ is zero elsewhere. Not surprisingly this series will not converge at all in the limit, so as $\beta \to 0$,

$$v(l/2) \to -(2W/kl)(1 + 1 + 1 + \cdots) = -\infty$$

and $M(z) \to 0$. We should therefore expect a gradual deterioration in the convergence as β decreases. The behaviour of both $v(l/2)$ and $M(l/2)$ is shown in Fig. 4.8 and confirms this. With seven terms in the series, $v(l/2)$ converges to within 3 per cent for the smallest $\beta = 0.001$, while the bending moment is only 60 per cent of its true value; although $\beta = 0.001$ does represent an extremely flexible beam. (Also both graphs have been non-dimensionalized with respect to the exact solutions and this exaggerates the deficiency in the stresses. The absolute error, at $N = 13$, is almost independent of β.) The central deflection approaches the true value monotonically from below only because it is a measure of the strain energy; the convergence elsewhere is less predictable. For example, at $z = l/4$ the value, for $\beta = 0.1$, is shown in the following table. Convergence is not monotonic, and indeed is initially *from above*.

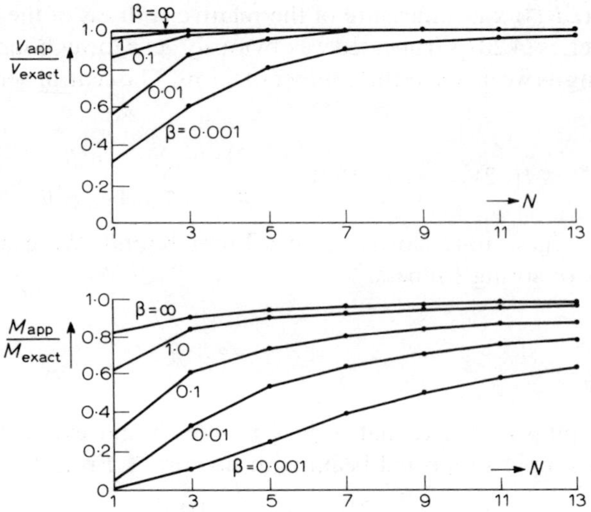

Fig. 4.8 Convergence of PVD for beam on spring $\beta = EI\pi^4/kl^4$

Convergence of $v(l/4)$

N	1	3	5	7	9	11	13
$\dfrac{v_{approx}}{v_{exact}}$	1.153	1.014	0.993	0.998	1.001	0.999	1.000

The convergence of the bending moment is naturally worse, so we will now re-examine this problem using the PVF and so obtain stresses directly without recourse to differentiating a series. It was also a little unfair to expect accurate stresses near a concentrated load where the third derivative of $v(z)$ was discontinuous, so this time we will represent the discontinuity exactly.

Equilibrium must be satisfied for the beam loaded by W at $z = l/2$ and by the continuous spring whose 'stress–strain' relationship is

$$p(z) = kv(z) \tag{4.22}$$

where $p(z)$ is the induced force per unit length. There are no other applied loads except the concentrated weight W. To avoid treating an element under this load as a special case, we write the applied load per unit length as $W\delta(z - l/2)$, where $\delta(z - l/2)$ is the Dirac delta function which is zero everywhere except at $z = l/2$ where it is infinite, but whose enclosed area is unity. The equilibrium requirements are therefore, noting Fig. 4.9,

$$\frac{dF}{dz} = W\delta(z - l/2) + p(z); \qquad \frac{dM}{dz} = F$$

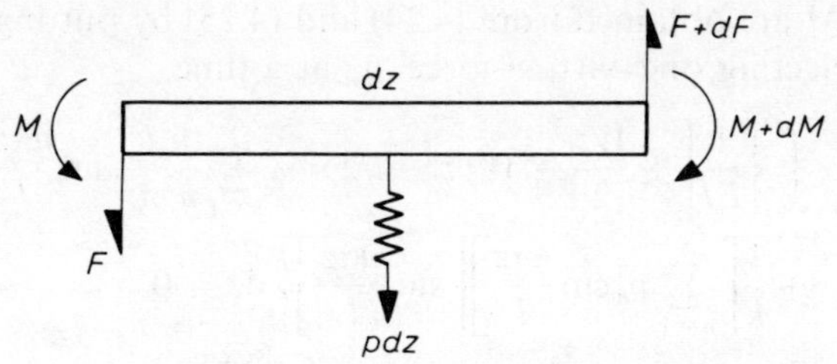

Fig. 4.9 Elemental forces

Therefore

$$M''(z) = W\delta(z - l/2) + p(z) \tag{4.23}$$

To apply the PVF we expand the spring force as

$$p(z) = \sum_{n=1}^{N} p_n \sin \frac{n\pi z}{l} \tag{4.24}$$

This expansion happens also to satisfy the kinematic conditions that $v = p/k = 0$ at $z = 0$ and l. Substituting (4.24) into (4.23) and integrating,

$$M'(z) = C_1 + W\langle z - l/2\rangle^0 - \sum_{n=1}^{N} \frac{p_n \cos(n\pi z/l)}{n\pi/l}$$

For convenience we have introduced McCauley[35] brackets $\langle\ \ \rangle$ to denote a quantity which takes zero value when its argument is negative, thus

$$\int_0^z W\delta(z - l/2)\,\mathrm{d}z = W\langle z - l/2\rangle^0 = \begin{matrix} W \text{ for } z > l/2 \\ 0 \text{ for } z < l/2 \end{matrix}$$

Similarly,

$$M(z) = C_1 z + C_2 + W\langle z - l/2\rangle^1 - \sum_{n=1}^{N} \frac{p_n \sin(n\pi z/l)}{(n\pi/l)^2}$$

To satisfy the equilibrium requirements that $M(0) = M(l) = 0$, we have $C_2 = 0$ and $C_1 l + Wl/2 = 0$.

Therefore,

$$M(z) = -\frac{Wz}{2} + W\langle z - l/2\rangle - \sum_{n=1}^{N} \frac{p_n \sin(n\pi z/l)}{(n\pi/l)^2} \tag{4.25}$$

We now have all the stresses in terms of the unknown constants p_n and the applied load W. The PVF is then

$$\int_V \bar{\boldsymbol{\sigma}}_i^t \boldsymbol{\varepsilon}\,\mathrm{d}v = \int_0^l \left[\bar{M}\left(\frac{M}{EI}\right) + \bar{p}\left(\frac{p}{k}\right) \right] \mathrm{d}z = 0$$

where $\bar{p}$ and $\bar{M}$ are obtained from (4.24) and (4.25) by putting the applied load $W = 0$ and selecting one 'virtual force' $\bar{p}_m$ at a time.

$$\text{For } \bar{p}_m = 1: \quad \int_0^l \left\{ \frac{1}{EI}\left[-\frac{Wz}{2} + W\langle z - l/2\rangle - \sum_{n=1}^{N} \frac{p_n l^2}{n^2\pi^2} \sin\frac{n\pi z}{l} \right]\left[\frac{-l^2}{m^2\pi^2} \sin\frac{m\pi z}{l} \right] + \frac{1}{k}\left[\sum_{n=1}^{N} p_n \sin\frac{n\pi z}{l} \right]\left[\sin\frac{m\pi z}{l} \right] \right\} dz = 0$$

Once again, only the single term for $m = n$ emerges from the trigonometric products, so after some reduction we find

$$p_n = 0 \qquad \text{for } n \text{ even,} \quad \text{and} \quad p_n = \frac{-2W(-1)^{(n-1)/2}}{l(1 + 3n^4)} \tag{4.26}$$

for $n = 1, 3, 5, \ldots$; where $\beta = \pi^4 EI/kl^4$ as before. On substituting back into (4.25), the moment at $z = l/2$ becomes

$$M(l/2) = \frac{-Wl}{4} + \frac{2Wl}{\pi^2} \sum_{1,3,\ldots} \frac{1}{n^2 + \beta n^6} \tag{4.27}$$

If $\beta \to \infty$, $M(l/2) \to -Wl/4$ as expected for no spring. If $\beta \to 0$, putting $\sum 1/n^2 = \pi^2/8$, we find $M(l/2) = 0$ (which is fairly obvious if the beam disappears.)

Knowing the beam stresses, the deflection can be found using the unit-load theorem or in this case simply integrating $EIv'' = -M$, whence

$$EIv = \frac{Wz^3}{12} - \frac{W}{6}\langle z - l/2\rangle^3 - \sum \frac{p_n \sin(n\pi z/l)}{n^4\pi^4/l^4} + C_3 + C_4 z$$

Putting $v(0) = v(l) = 0$; $C_3 = 0$ and $C_4 = -Wl^2/16$. The central deflection is therefore given by

$$EIv(l/2) = \frac{-Wl^3}{48} + \frac{2Wl^3}{\pi^4} \sum_{1,3,\ldots} \frac{1}{n^4(1 + \beta n^4)} \tag{4.28}$$

This result converges much faster than the previous PVD solution (4.20) as was intended. It clearly gives an upper bound when truncated. In fact, even for small β, it converges so rapidly that it is not worth plotting the behaviour. The moment $M(l/2)$ is shown in Fig. 4.10 and it is apparent that the PVF stresses converge from above more rapidly than the PVD stresses from below.

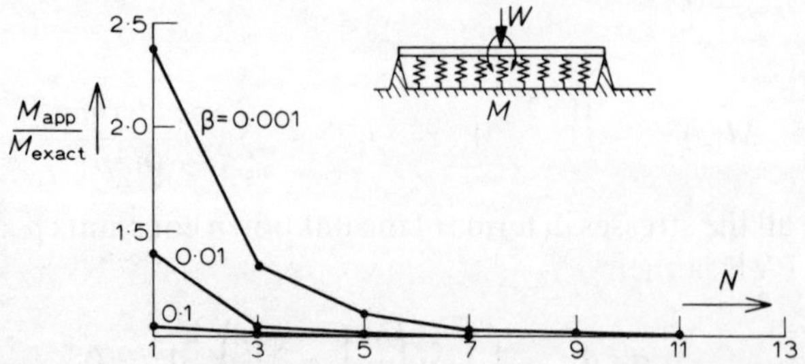

Fig. 4.10 Convergence of moments in PVF solution

Armed with the knowledge that the PVD overestimates the stiffness while the PVF underestimates it, it is tempting to speculate whether a combination of the two – *a mixed method* – could produce an approximate solution which is better than either the PVD or the PVF. This turns out to be possible, and as is often the case there is also a bonus in using a method in which neither compatibility nor equilibrium are satisfied exactly.

4.7 Mixed methods

We could demonstrate the technique immediately on the previous beam–spring model, but the general formulation is quite simple and there is little point in losing this simplicity in algebra. We briefly recapitulate and repeat the general PVD and PVF.

PVD

$$\int_V \boldsymbol{\sigma}^{t}\bar{\boldsymbol{\varepsilon}}\,dv - \int_V \mathbf{p}_v^{t}\bar{\mathbf{u}}\,dv - \int_{S_p} \mathbf{p}_s^{t}\bar{\mathbf{u}}\,ds = 0 \qquad (1.69)$$

where $\bar{u}$ has been put to zero on the surface S_u where $\mathbf{u} = \tilde{\mathbf{u}}$ is prescribed.

PVF

$$\int_V \bar{\boldsymbol{\sigma}}^{t}\boldsymbol{\varepsilon}\,dv - \int_V \bar{\mathbf{p}}_v^{t}\mathbf{u}\,dv - \int_{S_u} \bar{\mathbf{p}}_s^{t}\tilde{\mathbf{u}}\,ds = 0 \qquad (1.73)$$

where $\bar{\mathbf{p}}_s$ has been put to zero on S_p where th forces $\mathbf{p}_s$ are prescribed.

The above forms are the most convenient to use in the Rayleigh–Ritz method, but we did also convert them when proving that they delivered the equations of equilibrium and compatibility. The alternative (Galerkin) forms were:

PVD

$$\int_V \bar{\mathbf{u}}^{t}(\mathbf{D}\boldsymbol{\sigma} + \mathbf{p}_v)\,dv - \int_{S_p} \bar{\mathbf{u}}^{t}[(\mathbf{D}n)\boldsymbol{\sigma} - \mathbf{p}_s]\,ds = 0 \qquad (1.72)$$

provided compatibility is satisfied exactly, i.e. $\bar{\boldsymbol{\varepsilon}} = \mathbf{D}^{t}\bar{\mathbf{u}}$.

PVF

$$\int_V \bar{\boldsymbol{\sigma}}^{t}(\boldsymbol{\varepsilon} - \mathbf{D}^{t}\mathbf{u})\,dv + \int_{S_u} [(\mathbf{D}n)\bar{\boldsymbol{\sigma}}]^{t}(\mathbf{u} - \tilde{\mathbf{u}})\,ds = 0 \qquad (1.75)$$

provided equilibrium is satisfied exactly, i.e. $\mathbf{D}\bar{\boldsymbol{\sigma}} + \bar{\mathbf{p}}_v = \mathbf{0}$ in V and $(\mathbf{D}n)\bar{\boldsymbol{\sigma}} = \mathbf{0}$ on S_p.

Now we interpreted the Rayleigh–Ritz method on beams as a device to reduce the residual errors in the equations of equilibrium or compatibility in the manner suggested by Galerkin. This argument clearly applies to the general equations (1.72) and (1.75) also. Thus:

1. In the PVD the residual errors in the equilibrium equations $\mathbf{D}\boldsymbol{\sigma} + \mathbf{p}_v$ in V and $(\mathbf{D}n)\boldsymbol{\sigma} - \mathbf{p}_s$ on S_p, are reduced to negligible significance by making the integrals in (1.72) zero over as many virtual displacements $\bar{\mathbf{u}}$ as we deem necessary.
2. In the PVF the residual errors in the compatibility equations, $\boldsymbol{\varepsilon} - \mathbf{D}^t\mathbf{u}$ in V and $\mathbf{u} - \bar{\mathbf{u}}$ on S_u, are reduced to negligible significance by making the integrals in (1.75) zero over as many virtual stresses $\bar{\boldsymbol{\sigma}}$ and tractions $(\mathbf{D}n)\bar{\boldsymbol{\sigma}}$ as possible.

In the mixed method we abandon trying to satisfy either equilibrium or compatibility exactly, and instead use *both* equations (1.72) and (1.75) to do this approximately for both requirements. The two equations are quite separate, and the only physical link (apart from the virtual weights) is the stress–strain law which links $\boldsymbol{\varepsilon}$ in (1.75) to $\boldsymbol{\sigma}$ in (1.72). There seems little point in pretending that the two equations are unified; however, this has been traditionally done in the extremum energy version known as the Hellinger–Reissner principle[38] which is shown in Appendix B to be equivalent to (1.72) + (1.75). This further generalization of virtual work, by interpreting it as a Galerkin procedure for satisfying both equilibrium and compatibility equations approximately, is itself a special case of the technique of 'weighted residuals', discussed admirably by Crandall[36] and expanded in a review article by Finlayson and Scriven.[37] It is worth while digressing a little and discussing the philosophy behind these methods since it is pertinent to our use of the virtual work approach as opposed to the more common use of extremum energy principles.

The idea behind the method of weighted residuals (MWR) is in essence very simple. It consists of making a series approximation (piecewise or global) to some function $\psi(x, y, z)$ which is supposed to satisfy a differential field equation in a given region V. When the assumed series is substituted into the differential equation, unless it is fortuitously the exact solution, an error will result. The MWR consists of integrating the product of this error, over the region V, with some suitable weighting function, and putting the result equal to zero. The aim is to so choose the approximation and the weighting function as to ensure that the error everywhere is smaller than some acceptable value. The differential equation need not even be linear, but if it is then the MWR is easier to discuss. Suppose the function is governed by the differential equation

$$L(\psi) = f(x, y, z) \tag{4.29}$$

where L is a linear differential operator and $f(x, y, z)$ is a known function prescribed in V. The unknown $\psi(x, y, z)$ is then expanded as a complete series of 'likely' functions $\phi_n(x, y, z)$, that is, the ϕ_n satisfy required boundary conditions, continuity requirements, etc. We write

$$\psi = \sum_n a_n \phi_n$$

where the a_n are constants to be determined, as in the Rayleigh–Ritz procedure. The error or 'residual' everywhere in V is simply

$$L\left(\sum_n a_n\phi_n\right) - f$$

and we wish to minimize it in some fashion.

One method is that of *least squares*, whereby we choose the a_n to minimize

$$\int_V \left[L\left(\sum_n a_n\phi_n\right) - f\right]^2 \mathrm{d}v$$

Differentiating with respect to a_n we obtain

$$\int_V \left[L\left(\sum_n a_n\phi_n\right) - f\right] L(\phi_n)\,\mathrm{d}v = 0 \qquad \text{for all } \phi_n.$$

We therefore have n simultaneous linear equations in the unknown a_n. The weight on the residual in this case is $L(\phi_n)$.

Another, more brutal, alternative is to equate to zero the integral of the error itself,

$$\int_{V_n} \left[L\left(\sum_n a_n\phi_n\right) - f\right] \mathrm{d}v = 0$$

over a series of subdomains V_n, and so generate the required number of equations for a_n. In the limit if we shrink these domains to single points we are simply ensuring zero error at a discrete number of points. This 'collocation' technique is equivalent to selecting delta functions as the weighting functions.

Finally, the Galerkin procedure simply selects the individual terms, in the series for ψ, as the weighting functions; thus

$$\int_V \left[L\left(\sum_n a_n\phi_n\right) - f\right] \phi_n\,\mathrm{d}v \tag{4.30}$$

The Galerkin procedure seems to be the most popular and systematic of all these MWR techniques. It is straightforward to apply, once we know the differential equation, or series of differential equations, in V. Of course the series must converge, and it can be shown that if the operator L is self-adjoint, then equation (4.30) can be converted to an extremum problem – that is, the minimization of some 'functional' in terms of which the convergence can be examined. It may be this latter aspect which has led to the PVD and PVF usually being stated as the principles of minimum potential or minimum complementary potential energy (see Appendix B). However, we do not feel this approach is necessary or fruitful. The theorem of minimum potential energy is the only one which has a real physical basis, and the others are as fictional as virtual displacements or forces. They can only claim utility if they either throw light on the problem or are easier to apply. We feel that we have already shown that the PVD and PVF are more concise and illuminating than their extremum versions. Now that we have generalized their use (and more is to come), the search for equivalent functionals to minimize – if they exist – seems pointless.

We now summarize the mixed method as the two equations:

$$\int_V \bar{\mathbf{u}}^t(\mathbf{D}\boldsymbol{\sigma} + \mathbf{p}_v)\,dv - \int_{S_p} \bar{\mathbf{u}}^t[(\mathbf{D}n)\boldsymbol{\sigma} - \mathbf{p}_s]\,ds = 0 \tag{4.31}$$

and

$$\int_V \bar{\boldsymbol{\sigma}}^t(\boldsymbol{\varepsilon} - \mathbf{D}^t\mathbf{u})\,dv + \int_{S_u} [(\mathbf{D}n)\bar{\boldsymbol{\sigma}}]^t(\mathbf{u} - \tilde{\mathbf{u}})\,ds = 0 \tag{4.32}$$

It is often convenient to convert the above surface integrals $\mathbf{u}^t(\mathbf{D}n)\boldsymbol{\sigma}$; for example, where signs are tricky, or in finite elements where volume integrals lead to a more symmetric form. Using Gauss, equation (1.71), they can therefore be written in an alternative form:

$$\int_V \boldsymbol{\sigma}^t\mathbf{D}^t\bar{\mathbf{u}}\,dv = \int_V \mathbf{p}_v^t\bar{\mathbf{u}}\,dv + \int_{S_p} \mathbf{p}_s^t\bar{\mathbf{u}}\,ds \tag{4.33}$$

and

$$\int_V (\bar{\boldsymbol{\sigma}}^t\boldsymbol{\varepsilon} + \mathbf{u}^t\mathbf{D}\bar{\boldsymbol{\sigma}})\,dv = \int_{S_u} [(\mathbf{D}n)\bar{\boldsymbol{\sigma}}]^t\tilde{\mathbf{u}}\,ds \tag{4.34}$$

Equation (4.33) of course looks remarkably like the PVD, but we must emphasize that we do not now put $\bar{\boldsymbol{\varepsilon}} = \mathbf{D}^t\bar{\mathbf{u}}$. Similarly, (4.34) does look like the PVF until we note that we are not making $\bar{\boldsymbol{\sigma}}$ self-equilibrating, that is, $\mathbf{D}\bar{\boldsymbol{\sigma}}$ is not zero in V.

The algebraic link between the equations (4.31) and (4.32), or the pair (4.33) and (4.34), occurs because we expand both $\boldsymbol{\sigma}$ and $\mathbf{u}$ in some series fashion and then select $\bar{\boldsymbol{\sigma}}$ and $\bar{\mathbf{u}}$ as the individual terms in the same series. The residuals in stress equilibrium are coupled with weights in displacements, and vice versa.

One of the attractions of the MWR approach is that we are not obliged to solve differential equations $\mathbf{D}\bar{\boldsymbol{\sigma}} = \mathbf{0}$ as in the PVF. The freedom from satisfying $\boldsymbol{\varepsilon} - \mathbf{D}^t\mathbf{u} = \mathbf{0}$ is not so liberating since it is easier to differentiate than solve differential equations. Another advantage is that we are separately expanding $\boldsymbol{\sigma}$ and $\mathbf{u}$, and do not have to solve for $\boldsymbol{\sigma}$ through $\boldsymbol{\varepsilon}$ by forming $\mathbf{D}^t\mathbf{u}$ and so lose accuracy by differentiating a series. We must also admit that the mixed solution is neither an upper or lower bound – it *may* be even better – but we have already discussed the limited usefulness of 'mean' bounds.

The mixed technique is now illustrated on the beam–spring for which we already have 'pure' PVD and PVF solutions in section 4.6. The structure consists of two components, the beam and the spring. The first term in (4.33) therefore consists of bending stresses and spring stresses, thus

$$\int_0^l [M(-\bar{v}'') + p\bar{v}]\,dz = w[-\bar{v}(l/2)] \tag{4.35}$$

The second equation involves $\boldsymbol{\varepsilon}$ which is written in terms of $\boldsymbol{\sigma}$ using the stress–strain law. For the beam the curvature is written as M/EI, and for the spring the displacement is written as p/k. Thus

$$\int_0^l \left[\bar{M}\frac{M}{EI} + \bar{p}\frac{p}{k} + v\bar{M}'' + v(-\bar{p}) \right] \mathrm{d}z = 0 \tag{4.36}$$

If the reader is in any doubt as to the signs or the formulation of the products in the above equation, it is always possible to write the MWR, Galerkin, version (4.31) and (4.32), which can, of course, be derived from (4.35) and (4.36) by integrating by parts:

$$\int_0^l (-M'' + p)\bar{v}\,\mathrm{d}z + [-M\bar{v}' + M'\bar{v}]_0^l + W\bar{v}(l/2) = 0 \tag{4.37}$$

and

$$\int_0^l \left[\left(\frac{M}{EI} + v''\right)\bar{M} + \left(\frac{p}{k} - v\right)\bar{p} \right] \mathrm{d}z + [v\bar{M}' - v'\bar{M}]_0^l = 0 \tag{4.38}$$

The integrals contain the expected differential equations to be satisfied approximately, and the other terms arise if neither $M(z)$ nor $v(z)$ satisfy boundary conditions exactly.

We now assume *separate* expansions for $M(z)$, $p(z)$ and $v(z)$, and will satisfy equilibrium and kinematic boundary conditions simply to improve the convergence. The simple nature of this problem now becomes a slight disadvantage since it leaves us little room to manoeuvre in selecting the series. For example, it is tempting to select

$$M = \sum M_n \sin\frac{n\pi z}{l}; \qquad p = \sum p_n \sin\frac{n\pi z}{l}; \qquad v = \sum v_n \sin\frac{n\pi z}{l}$$

If we do this, the series are orthogonal, and on inserting $\sin(n\pi z/l)$ as the weights $\bar{M}$, $\bar{p}$, and $\bar{v}$, we obtain single equations for the unknowns. In fact each term in the series then satisfies exactly the compatibility equations $M/EI + v'' = 0$ and $p/k - v = 0$ in (4.38), so we simply retrieve the PVD solution. We will therefore be deliberately obdurate and select different expansions for the forces and the displacements. The lack of orthogonality now makes the algebra tedious, but it will suffice to take single-term expansions to illustrate the nature of the solution.

We assume

$$M = M_1 \sin\frac{\pi z}{l}; \qquad p = p_1 \sin\frac{\pi z}{l} \tag{4.39}$$

but

$$v = v_1 z(z - l) \tag{4.40}$$

All boundary conditions are satisfied;

$$M(0) = M(l) = v(0) = v(l) = 0$$

Substituting into (4.35) with $\bar{v} = z(z - l)$, we have

$$\int_0^l \left[\left(M_1 \sin\frac{\pi z}{l}\right)(-2) + \left(p_1 \sin\frac{\pi z}{l}\right) z(z - l) \right] \mathrm{d}z = \frac{Wl^2}{4}$$

or

$$p_1 + M_1\frac{\pi^2}{l^2} = \frac{-W\pi^3}{16l} \tag{4.41}$$

Equation (4.36), using $\bar{M} = \sin(\pi z/l)$ and $\bar{p} = \sin(\pi z/l)$ separately, produces

$$\int_0^l \left[\frac{M_1}{EI}\sin^2\frac{\pi z}{l} + v_1 z(z-l)\left(\frac{-\pi^2}{l^2}\sin\frac{\pi z}{l}\right)\right]\mathrm{d}z = 0$$

and

$$\int_0^l \left[\frac{p_1}{k}\sin\frac{\pi z}{l} - v_1 z(z-1)\right]\sin\frac{\pi z}{l}\,\mathrm{d}z = 0$$

giving

$$M_1 = -8EIv_1/\pi \qquad \text{and} \qquad p_1 = -8l^2kv_1/\pi^3 \tag{4.42}$$

Substituting these values back into (4.41), we obtain finally

$$v_1 = \frac{W\pi^6}{128kl^3(1+\beta)}, \qquad \text{where} \qquad \beta = \frac{\pi^4 EI}{kl^4} \text{ again}$$

Thus the maximum deflection $v(l/2) = -v_1 l^2/4$, is given by

$$\frac{v_{\max}}{(-2W/kl)} = \frac{0.939}{1+\beta} \tag{4.43}$$

Because of the single load W this deflection is still a measure of the structure's strain energy, so it is of interest to compare this simple result with that obtained by the lower-bound PVD and upper-bound PVF. Assuming $v = v_1 z(z-l)$ and inserting into the PVD

$$\int_0^l (EIv''\bar{v}'' + kv\bar{v})\,\mathrm{d}z = -W\bar{v}(l/2)$$

we find

$$\frac{v_{\max}}{(-2W/kl)} = \frac{0.937}{1+1.232\beta} \tag{4.44}$$

For a PVF solution we must use (4.25) and select the single term $p = p_1 \sin \pi z/l$, so

$$M = -\frac{Wz}{2} + W\left\langle z - \frac{l}{2}\right\rangle - \frac{l^2}{\pi^2}p_1 \sin\frac{\pi z}{l} \tag{4.45}$$

with the self-equilibrating system $\bar{\boldsymbol{\sigma}}$ generated by putting $p_1 = 1$ and $W = 0$. We find

$$\frac{v_{\max}}{(-2W/kl)} = \frac{1}{\beta}\left[\frac{\pi^4}{96} - \frac{1}{1+\beta}\right] \tag{4.46}$$

The results (4.43), (4.44), and (4.46) are shown in the following table.

Maximum deflection at point load W

β	Exact	PVF	Mixed	PVD
2	0.3406	0.3407	0.3130	0.2708
1	0.5145	0.5147	0.4695	0.4200
0.1	1.0418	1.0559	0.8535	0.8347
0.01	1.7510	2.4579	0.9296	0.9261

$$\beta = \frac{EI\pi^4}{kl^4}$$

The mixed solution does indeed lie between the PVF and the PVD although the PVF is easily the more accurate because the discontinuity at the load W is more accurately represented by (4.45). For small β and finite k the stiffness of the beam, EI, is so small that very rapid changes in $v(z)$ must occur under the load. It is then being rather optimistic to expect much more accuracy from a single-term approximation.

4.8 Relaxed boundary conditions

The extremum energy versions of virtual work, discussed in Appendix B, all imagine small arbitrary increments instead of virtual alternatives. Thus the principle of minimum potential energy imagines small actual increments $\delta\mathbf{u}$ which must necessarily be zero wherever the true displacements $\mathbf{u}$ are prescribed on S_u. In contrast, the PVD does not have to be so restrictive. We have found it so far convenient to exclude virtual displacements $\bar{\mathbf{u}}$ on S_u to avoid introducing unknown forces into the work equations: these forces are the reactions (say $\mathbf{p}_R$) present over the surface S_u. But this restriction is only a convenience; it is not a necessity. There are some circumstances in which it will be convenient to introduce these reactions and allow nonzero virtual displacements. For example, in choosing a suitable approximate expansion for $\mathbf{u}$ it is fairly straightforward to insist that the *total* series for $\mathbf{u}$ is equal to some prescribed $\tilde{\mathbf{u}}$ (say zero) on S_u, but it might be tedious – or impossible – to make every single term zero. And it is every single term we select to be the $\bar{\mathbf{u}}$. Also, it is possible that we may have to evaluate the reactions from the solved displacements by performing

$$\mathbf{p}_R = (\mathbf{D}n)\boldsymbol{\sigma} = (\mathbf{D}n)\boldsymbol{\kappa}\boldsymbol{\varepsilon} = (\mathbf{D}n)\boldsymbol{\kappa}\mathbf{D}^t\mathbf{u}$$

There is therefore the usual loss of accuracy when differentiating an approximate $\mathbf{u}$.

The relaxed PVD (1.69) now becomes

$$\int_V \boldsymbol{\sigma}^t\bar{\boldsymbol{\varepsilon}}\,dv = \int_V \mathbf{p}_v^t\bar{\mathbf{u}}\,dv + \int_{S_p} \mathbf{p}_s^t\bar{\mathbf{u}}\,ds + \int_{S_u} \mathbf{p}_R^t\bar{\mathbf{u}}\,ds \tag{4.47}$$

On using the usual Gauss transformation we produce an additional term

$$\int_{S_u} \bar{\mathbf{u}}^t[(\mathbf{D}n)\boldsymbol{\sigma} - \mathbf{p}_R]\,ds$$

and clearly there is no reason why the PVD should not deliver equilibrium conditions $(\mathbf{D}n)\boldsymbol{\sigma} = \mathbf{p}_R$, even though the $\mathbf{p}_R$ are unknown reactions.

In using (4.47) the unknowns are the displacements $\mathbf{u}$ and the forces $\mathbf{p}_R$ which happen to be mixed, but we must emphasize that *this is not a mixed method*, it is still the 'pure' PVD. The literature often confuses the choice of variable with the choice of method. The relaxed PVD (4.47) does have its equivalent in extremum energy theorems, but the energy functionals have to be further supplemented by including constraints in which the reactions appear as Lagrange multipliers. However, we must not digress by discussing less elegant formulations.

The PVF may be similarly relaxed by not insisting that the virtual applied forces $\bar{\mathbf{p}}_s$ are zero on S_p, where the true forces $\mathbf{p}_s$ are prescribed. The consequence is that now unknown displacements $\mathbf{u}$ are introduced into the PVF (1.74) which takes the relaxed form

$$\int_V \bar{\boldsymbol{\sigma}}_i^t \boldsymbol{\varepsilon}\, dv = \int_{S_u} \bar{\mathbf{p}}_s^t \bar{\mathbf{u}}\, ds + \int_{S_p} \bar{\mathbf{p}}_s^t \mathbf{u}\, ds \tag{4.48}$$

The reason for adopting this form may again be a local difficulty in choosing stress series, each term of which produces boundary tractions $(\mathbf{D}n)\bar{\boldsymbol{\sigma}}$ which are zero on S_p. There is no reason to use (4.48) as a device for introducing $\bar{\mathbf{u}}$ in a more acurate fashion, since the *direct* determination of $\mathbf{u}$ from $\mathbf{D}^t\mathbf{u} = \boldsymbol{\varepsilon} = \boldsymbol{\phi\sigma}$ involves integration which improves the accuracy. The *indirect* way of obtaining $\mathbf{u}$ from $\boldsymbol{\sigma}$ by using the unit-load theorem is of course nothing more than (4.48) anyway. In discussing whether the product $\mathbf{p}^t\mathbf{u}$ is retained or put to zero on a surface, the distinction may not always be that clear cut. The product has several compnents and they may not all be of the same nature. For example, a simply supported plate has prescribed zero displacement but prescribed zero moments.

We now solve two examples to illustrate the advantages, in approximate PVD and PVF solutions, of relaxing constraints on either $\mathbf{u}$ or $(\mathbf{D}n)\boldsymbol{\sigma}$ on a boundary.

Example

Figure 4.11 shows a uniform beam, both ends clamped, loaded by a uniformly distributed load. Normally in using the Rayleigh–Ritz method we would try a series for $v(z)$ each term of which satisfied the necessary boundary conditions $v(0) = v'(0) = v(l) = v'(l) = 0$. A series of the form $v(z) = z^2(z - l)^2 f(z)$ will satisfy these boundary conditions whatever the expansion for $f(z)$. However, it is tempting to choose an orthogonal series in order to avoid solving simultaneous

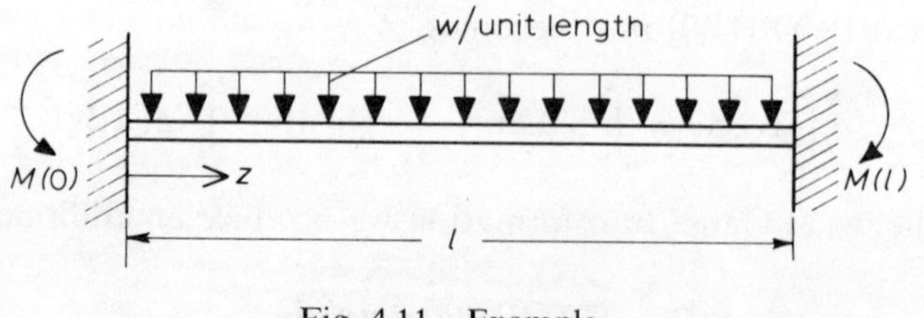

Fig. 4.11 Example

equations, and here we have an impasse: no such simple series is available. We therefore compromise and expand

$$v(z) = \sum_n v_n \sin\frac{n\pi z}{l} \tag{4.49}$$

Although $v(0) = v(l) = 0$, each term in (4.49) violates the kinematic boundary conditions $v'(0) = v'(l) = 0$. However, we are still free to insist that the complete series satisfies $v'(0) = v'(l) = 0$. Notice that (4.49) also violates equilibrium since it produces zero moments, EIv'', at $z = 0$ and l, when the moments almost certainly will not be zero. We are not, of course, obliged to satisfy equilibrium; nevertheless we expect poor moment convergence near $z = 0$ and $z = l$.

The relaxed PVD (4.47) for this problem becomes

$$\int_0^l (-EIv'')(-\bar{v}'')\,dz = \int_0^l (-w)\bar{v}\,dz + M(0)\bar{v}'(0) - M(l)\bar{v}'(l)$$

where $M(0)$ and $M(l)$ are the support reaction moments. By anticipating symmetry we may put

$$v(z) = \sum_{n=1,3,\ldots} v_n \sin\frac{nz}{l}, \qquad \text{and} \qquad M(0) = M(l)$$

Putting $\bar{v}(z) = \sin(m\pi z/l)$, the relaxed PVD becomes

$$\int_0^l \left\{\left[\sum_{1,3,\ldots} EI\frac{n^2\pi^2}{l^2} v_n \sin\frac{n\pi z}{l}\right]\frac{m^2\pi^2}{l^2}\sin\frac{m\pi z}{l} + w\sin\frac{m\pi z}{l}\right\}dz$$
$$- M(0)\frac{m\pi}{l} + M(0)\frac{m\pi}{l}\cos m\pi = 0$$

Then

$$\left(\frac{EI\pi^3}{4l^4 w}\right) n^4 v_n - \frac{n}{wl^2}M(0) = -\frac{1}{n\pi^2} \qquad \text{for all } n \text{ odd} \tag{4.50}$$

together with, putting $v'(0) = v'(l) = 0$,

$$\sum_{1,3,5,\ldots} n v_n = 0 \tag{4.51}$$

Equations (4.50) and (4.51) may be solved for the variables v_n and $M(0)$. Eliminating v_n from (4.50) and substituting into (4.51), we find

$$\left(\frac{EI\pi^3}{4l^3 w}\right) v_n = \frac{1}{n^3}\left(\frac{M(0)}{wl^2}\right) - \frac{1}{\pi^2 n^5}; \qquad \text{and} \qquad \frac{M(0)}{wl^2}\sum_{1,3,\ldots}\frac{1}{n^2} = \frac{1}{\pi^2}\sum_{1,3,\ldots}\frac{1}{n^4}$$

The infinite series can actually be summed as $\sum 1/n^2 = \pi^2/8$; $\sum 1/n^4 = \pi^4/96$; so we find

$$\frac{M(0)}{wl^2} = \frac{1}{12}; \qquad \text{and} \qquad v_n = \frac{-wl^4}{3EI\pi^3}\left(\frac{12}{\pi^2 n^5} - \frac{1}{n^3}\right)$$

The displacement series converges rapidly; but suppose we attempted to find the end moments from the displacements,

$$M(z) = -EIv'' = \frac{-wl^2}{3\pi} \sum_{1,3,\ldots} \left[\frac{12}{\pi^2}\left(\frac{1}{n^3}\right) - \frac{1}{n} \right] \sin\frac{n\pi z}{l}$$

At $z = l/2$ the convergence is reasonable; indeed we can go to the limit anyway since the sums with $1/n^3$ and $1/n$ are $\pi^3/32$ and $\pi/4$, so $M(l/2) = -wl^2/24$. But at $z = 0$ and l, M is zero instead of the value $wl^2/12$ obtained above. The series is consequently incapable of predicting the (maximum) moments at the ends. The moment

$$M(z) = \frac{-wl^2}{3\pi} \sum_{n=1}^{N} \left(\frac{12}{\pi^2 n^3} - \frac{1}{n} \right) \sin\frac{n\pi z}{l}$$

is plotted in Fig. 4.12, for $N = 7$ and 51. The behaviour near $z = 0$ is typical of the attempt of a Fourier series to represent a discontinuity.

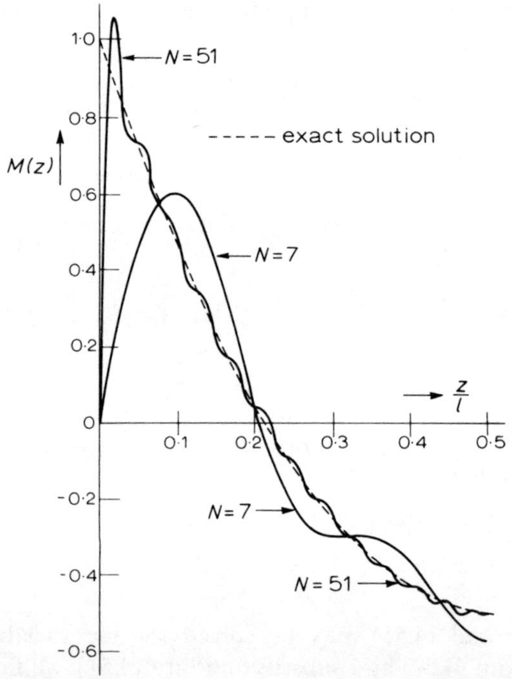

Fig. 4.12 Failure of series to predict end moment

Example

For completeness we demonstrate the use of relaxed boundary conditions on (for the last time) the four-boom tube using the PVF. In section 4.5 we moved the origin to the free tip in order to be able to satisfy the condition that $\bar{P}(z)$ was zero there. This is not necessary if we abandon this constraint, so assume

$$P(z) = \sum_{n=0}^{N} P_n(z/a)^n \tag{4.52}$$

The term P_0 is necessary because there will be a nonzero boom load at $z = 0$. Equation (4.17) now becomes, using (4.48),

$$\frac{1}{Et}\int_0^{2a}\left[\frac{3}{a}P\bar{P} + \left(\frac{-T}{2a} + 3aP'\right)\bar{P}'\right]\mathrm{d}z = \bar{P}(l)w_l$$

where w_l is the axial displacement of each of the four booms at the tip. The right-hand side represents the work done by *virtual* end loads over real displacements, whereas the *actual* end loads $P(l)$ are zero! Truncating (4.52) as before at the quadratic term, we obtain:

$$\bar{P} = 1: \qquad 2P_0 + 2P_1 + \tfrac{8}{3}P_2 - \beta = 0$$

$$\bar{P} = z/a: \qquad 2P_0 + \tfrac{14}{3}P_1 + 8P_2 - 2\beta = \frac{T}{3a}$$

$$\bar{P} = (z/a)^2: \qquad \tfrac{8}{3}P_0 + 8P_1 + \tfrac{256}{15}P_2 - 4\beta = \frac{2T}{3a}$$

where $\beta = Etw_l$. Also putting

$$P(2a) = 0: \qquad P_0 + 2P_1 + 4P_2 = 0$$

The solution is:

$$\begin{aligned}\beta &= -0.1212T/a\\ P_0 &= -0.1591T/a\\ P_1 &= 0.1364T/a\\ P_2 &= -0.0284T/a\end{aligned}$$

These figures agree with the previous solution after allowing for the change in origin to $z = 2a$. The value of the tip deflection given by $Etw_l = -0.1212T/a$ compares with the exact value of $-0.1223T/a$. This reminds us that not all deflections are overestimated by the PVF.

4.9 Buckling

The PVD has already been used in Chapter 2 to set up and solve both initial buckling problems and nonlinear snap-through problems. It is quite straightforward therefore to apply the Rayleigh–Ritz method to obtain approximate solutions, and we now select two such problems to which there is no simple closed solution.

Firstly, we examine the instability of a uniform vertical cantilever under its own weight, as shown in Fig. 4.13. We have already shown, in equation (2.66), that external work is done by a strut's end load, during an incremental displacement $\delta v(z)$, by moving over an axial shortening

$$\int_0^l v'(z)\,\delta v'(z)\,\mathrm{d}z$$

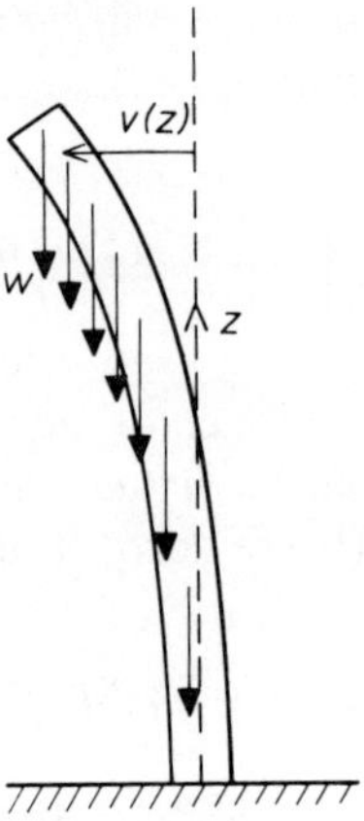

Fig. 4.13 Instability of a heavy vertical cantilever

In this problem there is a distributed weight w per unit length so the total virtual work must be

$$\begin{aligned}\int_0^l\left[\int_0^z v'\,\delta v'\,\mathrm{d}z\right]w\,\mathrm{d}z &= \left[\int_0^z v'\,\delta v'\,\mathrm{d}z\,wz\right]_0^l - \int_0^l v'\,\delta v' wz\,\mathrm{d}z\\ &= wl\int_0^l v'\,\delta v'\,\mathrm{d}z - \int_0^l wzv'\,\delta v'\,\mathrm{d}z\\ &= w\int_0^l (l-z)v'\,\delta v'\,\mathrm{d}z\end{aligned}$$

The PVD now becomes

$$\int_0^l (EIv''\,\delta v'' - w(l-z)v'\,\delta v')\,\mathrm{d}z = 0$$

or by integrating by parts the first term,

$$-\int_0^l (EIv''' + w(l-z)v')\,\delta v'\,\mathrm{d}z + [EIv''\,\delta v']_0^l = 0 \tag{4.53}$$

The exact solution is obtained from the integrand, and is the second-order differential equation in v':

$$EI(v')'' + w(l-z)(v') = 0, \qquad \text{subject to} \quad v'(0) = 0,\ v''(l) = 0$$

The equation is tractable in terms of Bessel functions. The smallest, critical, value of the weight is found to be[39]

$$w_{\text{crit}} = 7.84EI/l^3$$

To apply the Rayleigh–Ritz method to this problem, a suitable series would be

$$v'(z) = \sum_{n=1,3,\ldots} v_n' \sin\frac{n\pi z}{2l} \tag{4.54}$$

This satisfies the necessary kinematic boundary condition, $v'(0) = 0$; and as a bonus the equilibrium condition $EIV''(l) = 0$ also, so the last term in definite limits in (4.53) vanishes. On substituting (4.54) in to (4.53) and putting $\delta v'(z) = \sin m\pi z/l$:

$$\sum_n \int_0^l \left[-EIv_n'\left(\frac{n\pi}{2l}\right)^2 \sin\frac{n\pi z}{2l} + w(l-z)v_n' \sin\frac{n\pi z}{2l}\right] \sin\frac{m\pi z}{2l}\,\mathrm{d}z = 0$$

Only the simple trigonometric product is orthogonal. The linear term produces contributions from all terms in the series, so limiting ourselves to just two, we obtain

$$\begin{matrix}\text{for } m=1: \\ \\ \text{for } m=3:\end{matrix} \quad \begin{bmatrix} \dfrac{-EI\pi^2}{8wl^3} + \dfrac{1}{4} - \dfrac{1}{\pi^2} & \dfrac{1}{\pi^2} \\ \dfrac{1}{\pi^2} & \dfrac{-EI9\pi^2}{8wl^3} + \dfrac{1}{4} - \dfrac{1}{9\pi^2} \end{bmatrix} \begin{bmatrix} v_1' \\ \\ v_3' \end{bmatrix} = \begin{bmatrix} 0 \\ \\ 0 \end{bmatrix}$$

Thus buckling problems lead to eigenvalue determination in exactly the same way as did the vibrating rod. This is true of all initial-buckling and natural-frequency problems, whatever the discretized structure may be. Setting the determinant to zero, we obtain two roots

$$\frac{EI}{wl^3} = 0.0144 \qquad \text{and} \qquad 0.1276$$

The second root gives the smallest buckling load $wl^3/EI = 7.839$. Had we retained only the single term, $v' = v_1' \sin \pi z/2l$, the value would have been 8.298 which is 6 per cent too high. Once again the PVD has produced an over-stiff result.

4.10 A shallow arch

In all approximate solutions the eigenvalues are more accurate than the vectors and it is quite common to obtain a fairly accurate buckling load or natural frequency with a poor representation of the eigenvector or mode shape. Lest we become overconfident about solving buckling loads with crude displacement assumptions, we look again at the snap-through of a shallow arch. This was examined briefly in section 2.11, but this time we apply a central concentrated load W as in Fig. 4.14. The shallow shape will be taken as $v_0(z) = h_0 \sin \pi z/l$, where $h_0/l \ll 1$.

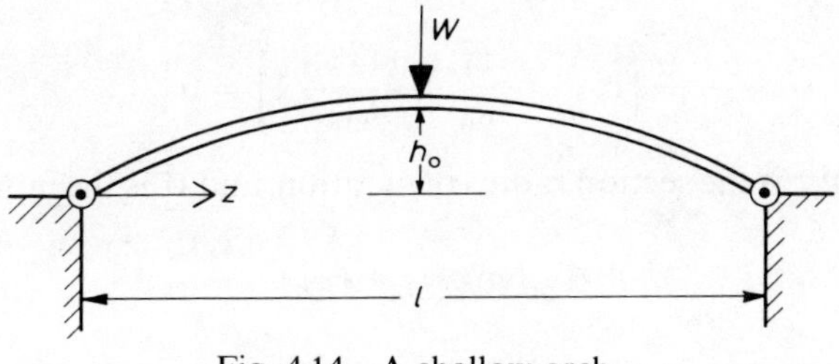

Fig. 4.14 A shallow arch

Previously we assumed that such a structure would displace symmetrically, and yet a little thought would suggest that this may not be so for an arch whose bending rigidity is small. As the load W does work this will be stored as strain energy in both axial compression and bending, and yet if the compressive energy much exceeds the bending, we might expect the arch to relieve itself of the former by moving and stretching into an S shape (Fig. 4.15), and so exchanging its axial energy into a little bending. Such a progression might be possible without the arch compressing at all – a so-called inextensional deformation. It therefore seems prudent to choose an approximate series which has both symmetric and antisymmetric freedoms. The simplest possible choice*, yet again, which satisfies both kinematic and static boundary conditions is:

$$v(z) = v_1 \sin\frac{\pi z}{l} + v_2 \sin\frac{2\pi z}{l} \tag{4.55}$$

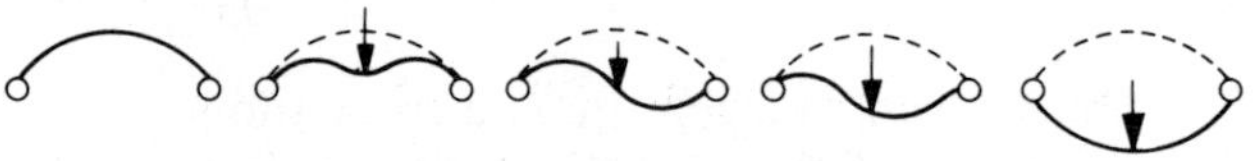

Fig. 4.15 An asymmetrical snap-through

Previously, in equation (2.62), it was shown that the axial strain in a shallow arch can be approximated as the uniform value

$$\varepsilon_{zz} = \frac{1}{2l}\int_0^l [(v')^2 - (v_0')]\,\mathrm{d}z$$

The incremental strain due to a small displacement $\delta v(z)$ is therefore

$$\delta\varepsilon_{zz} = \frac{1}{l}\int_0^l v'\,\delta v'\,\mathrm{d}z$$

The PVD for gross deformations (2.56) then becomes

$$\int_0^l EI(v'' - v_0'')\,\delta v''\,\mathrm{d}z + \frac{AE}{2l}\int_0^l [(v')^2 - (v_0')^2]\,\mathrm{d}z \int_0^l v'\,\delta v'\,\mathrm{d}z = -W\delta v(l/2)$$

Substituting $\delta v = \sin \pi z/l$ and then $\delta v = \sin 2\pi z/l$, the above equation delivers, after evaluating the integrals, two equations for v_1 and v_2:

$$4\frac{k^2}{h_0^2}\left(\frac{v_1}{h_0} - 1\right) + \left(\frac{v_1^2}{h_0^2} + 4\frac{v_2^2}{h_0} - 1\right)\frac{v_1}{h_0} = -\tilde{W} \tag{4.56}$$

and

$$\frac{v_2}{h_0}\left(16\frac{k^2}{h_0^2} + \frac{v_1^2}{h_0^2} + 4\frac{v_2^2}{h_0^2} - 1\right) = 0 \tag{4.57}$$

where $k = (I/A)^{1/2}$ is the section radius of gyration, and $\tilde{W}$ is the non-dimensional loading

$$\tilde{W} = 8Wl^3/\pi^4 AEh_0^3$$

* Even this form dangerously ignores the third harmonic.

Now had we assumed a symmetric deformation, with $v_2 = 0$, the deflection v_1 would be given by (4.56) as

$$4\frac{k^2}{h_0^2}\left(\frac{v_1}{h_0} - 1\right) + \left(\frac{v_1^2}{h_0^2} - 1\right)\frac{v_1}{h_0} = -\tilde{W} \tag{4.58}$$

This is a similar cubic to the previous solution in Fig,. 2.37, and again exhibits snap-through instability. The turning points are given by

$$\frac{\partial \tilde{W}}{\partial(v_1/h_0)} = 0, \qquad \text{or} \qquad 3\left(\frac{v_1}{h_0}\right)^2 = 1 - 4\left(\frac{k}{h_0}\right)^2$$

There are therefore no turning points at all if $k/h_0 > 1/2$, and therefore no snap-through. This is no surprise since an arch for which $h_0 < 2k$ is so shallow that its behaviour is entirely beam-like. The other extreme is when k/h_0 is vanishingly small: the arch now *appears* to behave quite differently and has a symmetric cubic behaviour

$$\left(\frac{v_1^2}{h_0^2} - 1\right)\frac{v_1}{h_0} = -\tilde{W}$$

with turning points and snap-through at $v_1/h_0 = \pm 1/\sqrt{3} = \pm 0.58$ (see Fig. 4.16). Turning now to the possibility of an asymmetric behaviour for a nonzero value of v_2, from equation (4.57) v_1 and v_2 must be coupled by

$$16\frac{k^2}{h_0^2} + \frac{v_1^2}{h_0^2} + 4\frac{v_2^2}{h_0^2} - 1 = 0 \tag{4.59}$$

and substituting this back into (4.56), we obtain

$$4\frac{k^2}{h_0^2}\left(3\frac{v_1}{h_0} + 1\right) = \tilde{W} \tag{4.60}$$

and thence from (4.59),

$$4\frac{v_2^2}{h_0^2} = 1 - 16\frac{k^2}{h_0^2} - \frac{v_1^2}{h_0^2} \tag{4.61}$$

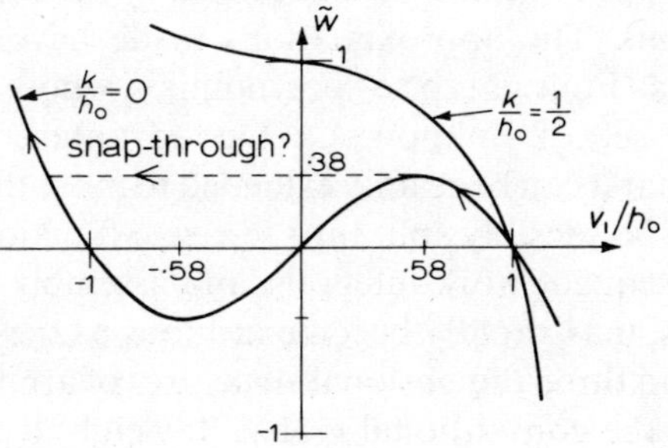

Fig. 4.16 Symmetric snap-through

This last equation shows that there is no antisymmetrical mode possible if $(k/h_0) > 1/4$. But this is a very shallow beam-like arch, so what happens if k/h_0 is much smaller? Well, equation (4.60) shows that if v_2 exists, then there is an alternative solution in which v_1 is a linear function of $\tilde{W}$, but that this solution cannot arise immediately as $\tilde{W}$ is increased from zero. It therefore becomes an alternative equilibrium solution at the intersection of (4.60) and the previous (4.58). Equating these two, we find the points

$$\frac{v_1}{h_0} = 0; \qquad \text{or} \qquad \frac{v_1}{h_0} = \pm\left(1 - 16\frac{k^2}{h_0^2}\right)^{1/2}$$

Figure 4.17 shows the two curves for a small value of $k/h_0 \ll 1$; remember that v_1 is the deflection at the load W. If k/h_0 is small (see equation (4.60)) the mode change at point 1 occurs very close to $v_1 = h_0$, and the value of $\tilde{W}$ is close to $16(k/h_0)^2$, which is much smaller than unity. The mixed-mode snap-through is therefore likely to occur almost immediately that a load is applied to a slender arch. The dangers of missing modes in nonlinear problems is every present. Of course the mixed-mode snap-through could be avoided by either increasing the bending stiffness (k/h_0) or by putting one end of the arch on rollers and never allowing the destabilizing compressive axial load to develop.

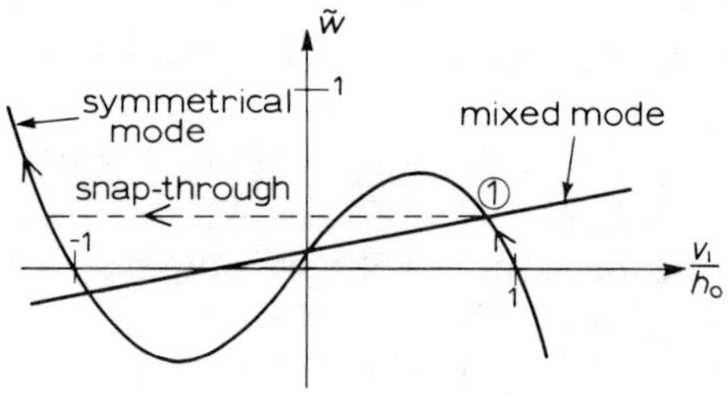

Fig. 4.17 Symmetric and asymmetric snap-through

4.11 Two-dimensional problems – the Krylov–Kantorovitch method[40,41]

So far all our approximate methods have been applied to one-dimensional structures such as rods or beams, or to idealized versions such as the boom–shear panel configurations. The approximations made have been to assume some global series, such as Fourier type or polynomials, whose coefficients constituted the new discrete set of unknowns. One advantage of this fairly drastic representation is that it can be readily extended to more than one dimension if the field variables can be sensibly split into separate functions (of x, y, and z, say). Naturally the consequent work integrals, and assembly of the equations in the discrete unknowns, may rapidly become arduous as the degrees of freedom are increased. Two- and three-dimensional structures of arbitrary shape are unlikely to be tractable by the conventional global Rayleigh–Ritz method since we are obliged to satisfy either kinematic or static boundary conditions and this can make choosing a series very difficult if the geometry is complex. For this reason, most formidable problems are tackled by a *piecewise* Rayleigh–Ritz method in which expansions are limited to small domains or *finite elements*, and the aid of the computer is sought in both performing the element integrals, summing the virtual work of all elements, and finally solving the simultaneous equations. This particular technique is described fully in Chapters 5 and 6.

However, there are a variety of problems, usually in two rather than three dimensions, for which a completely numerical description is not necessary. If a global approximation can be made, it often conveniently separates out the important parameters or groups of parameters, which govern the behaviour of the structure, and this is a bonus not forthcoming in completely numerical solutions which are usually performed as 'numerical' experiments to test the effect of various parameters. To simplify the issue, we would probably use a numerical package like finite elements to solve a geometrically complex structure like a ship, aircraft, or building, etc.; then, having the local stress fields, use a closed solution to test a particular element for strength, buckling, stress-intensity, and so on.

There is a wealth of literature on the behaviour of flat plates or shells subjected to stretching, bending, and buckling, and virtual work or energy methods are to the fore. The classical texts of Stephen Timoshenko[21,39,42] contain many such investigations. As a quick illustration we could look at the problem of a simply supported rectangular flat plate, $0 \leqslant x \leqslant a$, $0 \leqslant y \leqslant b$, loaded by a single weight W in the middle. This is a trivially simple problem, and yet it possesses no closed solution. The obvious expansion in the displacement $w(x, y)$, which satisfies both kinematic and moment boundary conditions, would be

$$w(x, y) = \sum_m \sum_n w_{mn} \sin\frac{n\pi x}{a} \sin\frac{m\pi y}{b} \tag{4.62}$$

The Rayleigh–Ritz and Galerkin forms of the PVD have been derived in (2.39) and (2.42). The latter is here the simplest because there are no terms on S_p in either $\bar{w}$ or $\partial^2 w/\partial x^2$ and $\partial^2 w/\partial y^2$. Thus, substituting (4.62) into (2.42) and putting

$$\bar{w}(x, y) = \sin\frac{q\pi x}{a} \sin\frac{r\pi y}{b}$$

we have

$$\int_0^a \int_0^b \left\{ \sum_{m,n} D w_{mn} \left[\left(\frac{n\pi}{a}\right)^4 + 2\left(\frac{n\pi}{a}\right)^2 \left(\frac{m\pi}{b}\right)^2 + \left(\frac{m\pi}{b}\right)^4 \right] \sin\frac{n\pi x}{a} \sin\frac{m\pi y}{b} \right\} \times$$
$$\sin\frac{q\pi x}{a} \sin\frac{r\pi y}{b}\, dy\, dx = -W \sin\frac{q\pi}{2} \sin\frac{r\pi}{2}$$

Both series are orthogonal, and only integrals for $q = n$ and $r = m$ survive, whence the general solution is

$$w_{mn} = -\frac{4W \sin\dfrac{n\pi}{2} \sin\dfrac{m\pi}{2}}{\pi^4 Dab\,[(n/a)^2 + (m/b)^2]^2}$$

Convergence for the deflection is rapid, behaving like $1/n^4$; the bending moments converging like $1/n^2$.

This plate-bending example works well because the assumed series satisfies all the boundary conditions and not just the kinematic, and the orthogonal series are simple to integrate over the rectangular domain. In many problems the variations in two dimensions may not be so predictable, and it is useful to turn to the powerful method developed by Kantorovitch[40] in 1933 and which falls neatly between the exact solution and the Rayleigh–Ritz–Galerkin technique of using discrete unknowns and assumed expansions in all directions. The Krylov–Kantorovitch method[41] is capable of very accurate solutions for structures with a single set of field equations. It is not very suitable for discontinuous structures, or assemblies of structures, each of which may have different governing differential equations and boundary conditions. We shall restrict ourselves to simple illustrative examples in two dimensions, using both the PVD and PVF.

The method consists simply of making a reasonable assumption of the likely variation of a stress or displacement field (as in the Ritz method) but *in one independent variable only*, and then allowing the PVD or PVF to generate the governing *ordinary* differential equation in the remaining independent variable. The initial guesstimate may be either inspired, based on known boundary conditions, inferred from similar but simpler problems, or just a shot in the dark. After solving the differential equation, this function of one variable may itself be used as a guess and the variation in the original variable now solved instead of assumed – and hence the reliance on inspiration somewhat reduced. The success of the method depends not only on the initial guess but also on whether the problem naturally separates into two variables anyway, and in which precise fashion this separation should be made. It is safe to say that if the geometry of the structure is simple then so may be the analysis, but if the geometry is complex then judicious separation of variables may be well nigh impossible.

We can do no better than choose as a first example[40] the bending of a flat clamped rectangular plate, for which no closed solution exists. The loading we take as a uniform pressure p, and the notation for deflections, etc., is as first used in section 2.9. The plate of dimensions $a \times b$ has boundary conditions that are purely kinematic; that is, on $x = \pm a/2$, $w = \partial w/\partial x = 0$; and on $y = \pm b/2$, $w = \partial w/\partial y = 0$. It is therefore sensible to use the PVD, since we should be able to hazard a guess at a likely displacement field. It should be noted that this problem is singularly tedious when analysed by the Rayleigh–Ritz technique; for example, we could choose a polynomial expansion, satisfying all boundary conditions, as

$$w(x, y) = \left(x^2 - \frac{a^2}{4}\right)^2 \left(y^2 - \frac{b^2}{4}\right)^2 \sum_{m=0}^{\infty} \sum_{n=0}^{\infty} w_{mn} x^m y^n \tag{4.63}$$

and

$$\bar{w}(x, y) = \left(x^2 - \frac{a^2}{4}\right)^2 \left(y^2 - \frac{b^2}{4}\right)^2 x^m y^n \tag{4.64}$$

The Galerkin form of the PVD for this problem follows immediately from (2.42) and is

$$\int_{-b/2}^{b/2}\int_{-a/2}^{a/2}(D\nabla^4 w - p)\bar{w}\,\mathrm{d}x\,\mathrm{d}y \tag{4.65}$$

Substituting (4.63) and (4.64) into (4.65) produces a set of equations for the unknowns w_{mn}. But (4.63) is not an orthogonal series, so that all equations contain all unknowns w_{mn}; moreover, the convergence of w_{mn} as m and n are increased is not wonderful, and of course the integrations are lengthy. For example, if we restrict ourselves to one term w_{00} we find, for a square plate, $a = b$, that the central deflection and bending moment (using the moment expression derived in (2.40)) are

$$w = 0.001\,33pa^4/D; \qquad M = -0.002\,76pa^2 \quad (\text{for } \nu = 0.3)$$

These values are respectively 6 and 21 per cent in error compared with the accurate solutions[21] of

$$w = 0.001\,26pa^4/D; \qquad M = -0.002\,28pa^2$$

Moreover, the deflected shape cannot be a fair approximation since its maximum value is *too large* and consequently it must err considerably on the small side elsewhere, since we know that the work – $\iint pw\,\mathrm{d}x\,\mathrm{d}y$ – must be an underestimate. It seems then that a quartic is a pretty poor description of the deflection. Bearing this in mind let us now use the Kantorovitch method.

We must first make a reasonable guess for the variation of $w(x, y)$ in the x variable, satisfying necessary kinematic boundary conditions. If the plate were infinitely long ($b \to \infty$) the problem would be one-dimensional and the exact solution of

$$D\frac{\mathrm{d}^4 w}{\mathrm{d}x^4} - p = 0, \qquad \text{with} \quad w = \frac{\mathrm{d}w}{\mathrm{d}x} = 0 \text{ on } x = \pm\frac{a}{2}$$

is

$$w(x) = \frac{p}{24D}\left(x^2 - \frac{a^2}{4}\right)^2$$

We therefore try

$$w = \frac{p}{24D}\left(x^2 - \frac{a^2}{4}\right)^2 f(y) \tag{4.66}$$

and

$$\bar{w} = \frac{p}{24D}\left(x^2 - \frac{a^2}{4}\right)\bar{f}(y)$$

where $f(y)$ is the unknown (dimensionless) function to be determined. (This seems at first a somewhat inauspicious beginning since – at least for the doubly symmetrical square plate – we would expect the best solution when the variation with both x and y was similar; in which case we are back with the unsatisfactory Ritz solution using (4.63).) Substituting (4.66) into the PVD (4.65) we obtain, after performing the integrations with respect to x,

$$\int_{-b/2}^{b/2} [a^4 f'''' - 24a^2 f'' + 504(f-1)]\bar{f}\,\mathrm{d}y \tag{4.67}$$

where primes denote differentiation with respect to y.

For arbitrary $\bar{f}(y)$ the expression in square brackets must vanish, and so we have the differential equation in $f(y)$ whose solution, discarding unsymmetrical terms, is

$$f(y) = C_1 \cosh\frac{\alpha y}{a}\cos\frac{\beta y}{a} + C_2 \sinh\frac{\alpha y}{a}\sin\frac{\beta y}{a} + 1 \tag{4.68}$$

where

$$\alpha = (504)^{1/4}\cos\frac{\phi}{2}; \qquad \beta = (504)^{1/4}\sin\frac{\phi}{2}; \qquad \tan^2\phi = \tfrac{5}{2}$$

C_1 and C_2 are found from the boundary conditions $w = \partial w/\partial y = 0$ on $y = \pm b/2$. Comparing this solution with the previous Ritz solution, *for a square plate*, we find for the central deflection and moment,

$$\begin{aligned} w &= 0.001\,297 pa^4/D && \text{(2.4 per cent error)} \\ M_x &= 0.0261 pa^2 && \text{(14 per cent error)} \\ M_y &= 0.0241 pa^2 && \text{(6 per cent error)} \end{aligned}$$

(These values do not agree with the answer in reference 40 which seems to be in error.)

Although this solution is not symmetrical in x and y, when compared with the previous one-term Ritz solution with a symmetrical polynomial, the PVD has clearly been able to take advantage of the extra freedom in not specifying *a priori* the form of $f(y)$. The variation in M_y is also better than that of M_x. We can now improve this answer by using our derived solution for $f(y)$ as our guess, and repeating the above procedure to solve the variation with x. Not surprisingly, this turns out to be of the same form as (4.68), so for a square plate symmetry is restored. This process may be repeated indefinitely although in this problem, because the variations with x and y are of the same form, the process merely refines the values of the constants α, β, C_1, and C_2. In fact the process converges extremely rapidly for any value of a/b. Examples are detailed by Kerr and Alexander[43] where it is shown that the second iterative values of the functions are within 10^{-4} of their asymptotic values. For the square plate the values of both deflection and moments are everywhere indistinguishable from the exact values.

The Kantorovitch method as used above, or extended by iteration, will succeed to a greater or lesser degree depending on whether the original separation of variables is justified for the particular problem. Unfortunately, one finds this out only after obtaining the solution – by substituting the answer in the governing partial differential equation and examining the size of the residual error. If simple separation of variables is restrictive we can allow further degrees of freedom in the same manner adopted in the conventional Ritz procedure. For this plate example we could, for instance, let

$$w(x, y) = \frac{p}{24D}\left(x^2 - \frac{a^2}{4}\right)^2 [f(y) + x^2 g(y)]$$

and generate two simultaneous differential equations for $f(y)$ and $g(y)$ from the separate coefficients of the virtual functions $\bar{f}(y)$ and $\bar{g}(y)$. However the resulting algebraic computation increases explosively when we do this.

The attentive reader may have noticed that we carefully chose an example (albeit one with no simple closed solution) for which all the necessary boundary conditions were kinematic, and hence we were off to a flying start when applying the PVD. We therefore now choose a second plate example for which most of the boundary conditions are of the equilibrium type. We also deliberately choose to use the 'weighted residual' Galerkin form of the plate equations (2.42) rather than the Rayleigh–Ritz (2.39) type, simply to show how the surface equilibrium residuals must be incorporated if they are nonzero – an automatic process in our formulation.

We choose to evaluate the initial buckling stress of a rectangular plate $a \times b$ subjected to a uniform edge stress, on $x = 0$, a, of $\sigma_{xx} = N/t$. Having already shown, equation (2.66), that for a simple strut the additional work done by the end load is

$$P \int_0^l v' \, \delta v' \, \mathrm{d}z$$

then for the plate there must be a similar term

$$N \int_0^a \int_0^b \frac{\partial w}{\partial x} \frac{\partial}{\partial x} (\delta w) \, \mathrm{d}y \, \mathrm{d}x$$

Suppose that the plate is simply supported on $x = 0$, $x = a$, and $y = 0$; but that the edge $y = b$ is free. This is the situation met in an axially loaded symmetrical angle-section which can buckle by rotating about the corner edge which remains straight in the axial direction and becomes a point of inflexion in the transverse sense (see Fig. 4.18). This problem is solved exactly in Timoshenko[39] where the

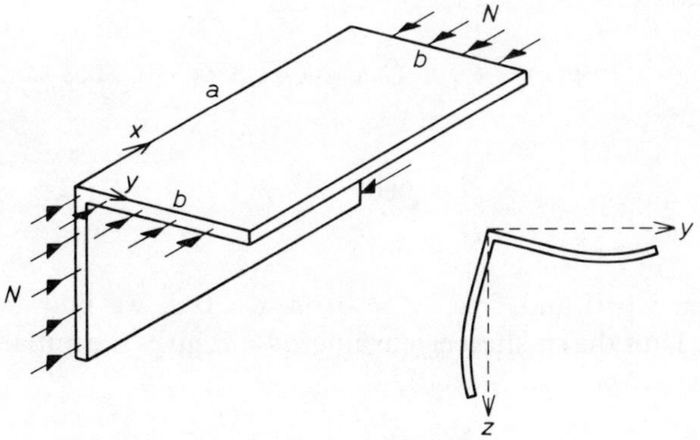

Fig. 4.18 Buckling of an angle

deflection is shown to vary sinusoidally with x but the transverse variation is a mixture of trigonometric and hyperbolic terms. We will make the drastic assumption that

$$w(x, y) = yf(x) \tag{4.69}$$

This impertinence is possible because it is strictly necessary to satisfy only kinematic boundary conditions and in this problem there is only one—$w = 0$ on $y = 0$. Assumption (4.69) actually also satisfies the equilibrium boundary condition on $y = 0$, namely that

$$M_y = D(\partial^2 w/\partial y^2 + \nu\, \partial^2 w/\partial x^2) = 0$$

However, it violates both equilibrium conditions on $y = b$ (zero moment and zero shear) and so the corresponding residuals in the Galerkin equation (2.42) must be retained. This equation (with $\theta = \pi/2$, $s = -x$ on $y = b$) becomes

$$\int_0^a \int_0^b \left(D\nabla^4 w\, \delta w - N \frac{\partial w}{\partial x} \frac{\partial \delta w}{\partial x} \right) \mathrm{d}y\, \mathrm{d}x + \int_0^a D\left[\left(\frac{\partial^2 w}{\partial x^2} + \nu \frac{\partial^2 w}{\partial y^2} \right) \frac{\partial \delta w}{\partial y} \right]_{y=b} \mathrm{d}x$$
$$- \int_0^a D\left[\frac{\partial^3 w}{\partial y^3} + (2 - \nu) \frac{\partial^3 w}{\partial x^2\, \partial y} \right]_{y=b} \delta w\, \mathrm{d}x = 0 \tag{4.70}$$

We have excluded boundary residuals on $x = 0, a$, by presupposing that $f(x)$ may eventually be chosen so as to satisfy both kinematic and equilibrium boundary conditions on $x = 0$ and $x = a$. On substituting $w = yf(x)$ and $\delta w = y\, \delta f(x)$ into (4.70), then integrating with respect to y, and converting $\delta f'$ to δf, we are left with

$$\int_0^a \left[\frac{b^3}{3} \left(f'''' + \frac{N}{D} f'' \right) - 2(1 - \nu) b f'' \right] \delta f\, \mathrm{d}x = 0 \tag{4.71}$$

where again $\delta f = 0$ on $x = 0, a$, if we are able to choose $f(x)$ such that the displacement boundary conditions, $w = 0$ on $x = 0, a$, are satisfied exactly.

The solution to (4.71) is

$$f(x) = C_1 + C_2 x + C_3 \sin \alpha x + C_4 \cos \alpha x$$

where

$$\alpha^2 = \frac{N}{D} - \frac{6(1 - \nu)}{b^2}$$

On putting $w = 0$ and $\partial^2 w/\partial x^2 = 0$ on $x = 0, a$, we find $C_1 = C_2 = C_4 = 0$, $\sin \alpha a = 0$. Thus the smallest eigenvalue $\alpha a = \pi$, gives the initial buckling stress as

$$N = \frac{\pi^2 D}{b^2} \left[\frac{b^2}{a^2} + \frac{6(1 - \nu)}{\pi^2} \right]$$

The above value is compared with Timoshenko's exact solution, for $\nu = 1/4$, in the following table:

a/b	0.5	1.0	1.6	2.0	3.0	4.0
$\frac{b^2 N}{\pi^2 D}$ (approx.)	4.456	1.456	0.847	0.706	0.567	0.518
$\frac{b^2 N}{\pi^2 D}$ (exact)	4.400	1.440	0.835	0.698	0.564	0.516

The agreement is remarkable for such a crude approximation. As we have come to expect, the approximate buckling loads are all too high.

The Kantorovitch technique can naturally also be employed in the PVF. We now analyse a (slightly) more complex structure and show how the algebra can quickly get out of hand.

4.12 Shear lag in a box girder

Consider the problem of a cantilever whose section is a thin-walled rectangular box – a simplified version of an aircraft wing-box or a box-girder bridge. The cantilever is subjected to a symmetrically placed shear force F at the tip, and completely built-in at the root $z = 0$ (see Fig. 4.19). The box profile is maintained by closely spaced ribs which are assumed to be so stiff that they can be ignored in summing the virtual work – certainly a rigid rib is assumed to exist at the tip to receive the load F. (This structure is analysed in Problem 3.13 by drastically idealizing it as an assembly of eight booms and shear panels. It is shown that a shear-lag feature exists near the constrained root where the centre region of the flanges is not excited sufficiently by the two webs.) We will now look at the flange stress field and see how inaccurate the simple boom–shear panel predictions can be.

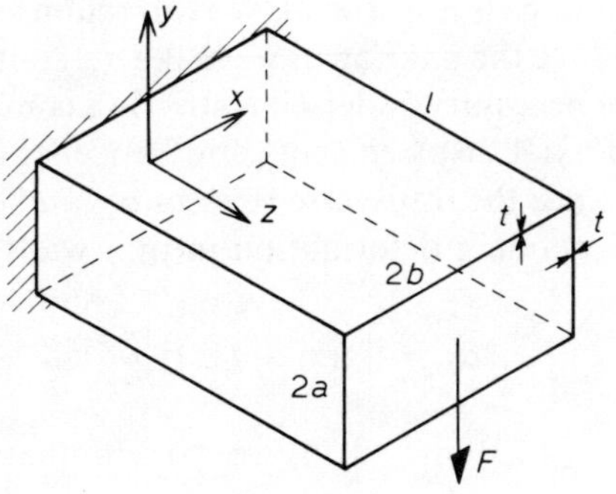

Fig. 4.19 Cantilever box with root constraints

We will use the PVF, so the necessary equilibrium requirements (1.52), noting the symmetry about $x = 0$ and $y = 0$, are as follows:

In $y = a$:

$$\frac{\partial \sigma_{zz}}{\partial z} + \frac{\partial \sigma_{xz}}{\partial x} = 0 \tag{4.72}$$

$$\frac{\partial \sigma_{xx}}{\partial x} + \frac{\partial \sigma_{xz}}{\partial z} = 0 \tag{4.73}$$

In $x = b$:

$$\frac{\partial \sigma_{zz}}{\partial z} + \frac{\partial \sigma_{yz}}{\partial y} = 0 \tag{4.74}$$

$$\frac{\partial \sigma_{yy}}{\partial y} + \frac{\partial \sigma_{yz}}{\partial z} = 0 \tag{4.75}$$

Moment equilibrium:

$$2a \int_{-b}^{b} \sigma_{zz}\Big|_{y=a} t\,dx + 2\int_{-a}^{a} \sigma_{zz}\Big|_{x=b} yt\,dy = F(l - z) \tag{4.76}$$

Shear equilibrium:

$$2\int_{-a}^{a} \sigma_{yz}\Big|_{x=b} t\,dy = -F \tag{4.77}$$

Equations (4.72) to (4.77) are all the equilibrium requirements. We now separate variables by assuming likely variations with x and y, guided by symmetry. Simple beam theory would predict a constant bending stress σ_{zz} across the flanges $y = \pm a$, and a linear variation through the vertical webs $x = \pm b$. We will assume a quadratic variation in the flanges, thus:

In $y = a$,

$$\sigma_{zz} = g(z) + \frac{x^2}{b^2} f(z) \tag{4.78}$$

If the box is wide ($b \gg a$) most of the shear lag effects are confined to the flanges and the variation in the webs is not crucial. We will retain a linear variation in the webs, and to further reduce the unknowns we make σ_{zz} continuous at the corners. (This assumption is not necessary, indeed it is strictly a compatibility argument to insist on continuity of axial displacements and hence strain ε_{zz}. It then follows that σ_{zz} is continuous since the transverse stresses σ_{xx} and σ_{yy} are zero at the box corners if we preclude bending deformation in thin walls.)

Thus, in $x = b$,

$$\sigma_{zz} = [g(z) + f(z)]\frac{y}{a} \tag{4.79}$$

Substituting (4.78) and (4.79) into (4.76) we can eliminate $g(z)$:

$$g(z) = \frac{3F(l - z)}{4at(a + 3b)} - \frac{(a + b)}{(a + 3b)} f(z) \tag{4.80}$$

Equations (4.78) and (4.79) for the bending stresses now become:
In $y = a$:

$$\sigma_{zz} = \frac{3F(l - z)}{4at(a + 3b)} + \left[\frac{x^2}{b^2} - \frac{(a + b)}{(a + 3b)}\right] f(z) \tag{4.81}$$

In $x = b$:

$$\sigma_{zz} = \left[\frac{3F(l - z)}{4at(a + 3b)} + \frac{2b}{(a + 3b)} f(z)\right] \frac{y}{a} \tag{4.82}$$

Substituting (4.81) and (4.82) into (4.72) and (4.74), and integrating with respect to x and y, we get the shear stresses:
In $y = a$:

$$\sigma_{xz} = \frac{3Fx}{4at(a + 3b)} - \left[\frac{x^3}{3b^2} - \frac{(a + b)}{(a + 3b)} x\right] f'(z) \tag{4.83}$$

In $x = b$:

$$\sigma_{yz} = \frac{3F}{8at(a + 3b)}\left[\frac{y^2}{a} - (a + 3b)\right] + \frac{ab}{(a + 3b)}\left[\frac{1}{3} - \frac{y^2}{a^2}\right] f'(z) \tag{4.84}$$

where the functions of integration with respect to x and y have been found by antisymmetry reasoning in (4.83) and an equilibrium argument in (4.84) – namely that $\sigma_{xz} = -\sigma_{yz}$ at the corner (b, a). It is easily verified that (4.84) satisfies the equilibrium requirement (4.77). On substituting (4.83) and (4.84) into (4.73) and (4.75), we can likewise find the transverse stresses:
In $y = a$:

$$\sigma_{xx} = \left[\frac{x^4}{b^4} - \frac{(a + b)6\,x^2}{(a + 3b)\,b^2} + \frac{(5a + 3b)}{(a + 3b)}\right] \frac{b^2 f''(z)}{12} \tag{4.85}$$

In $x = b$:

$$\sigma_{yy} = \left[\frac{y^3}{a^3} - \frac{y}{a}\right] \frac{abf''(z)}{3(a + 3b)} \tag{4.86}$$

where this time the functions of integration have been deduced from the equilibrium requirements that the transverse direct stresses are zero at the corners.

All the necessary preparatory work for the PVF is now complete. The stresses everywhere have been found in terms of the applied load F and a single (redundancy) function $f(z)$. Notice that in all our expressions the stress system statically equivalent to F (i.e. $f(z) = 0$) is none other than simple beam theory. The solution $f(z)$ therefore represents a self-equilibrating correction to simple theory.

We now apply the PVF which is no more difficult in principle that in all previous examples. We select $\bar{\sigma}_{zz}$, $\bar{\sigma}_{xx}$, $\bar{\sigma}_{yy}$, and $\bar{\sigma}_{xz}$ from equations (4.81) to (4.86) by putting $F = 0$, and denote their virtual nature by bars on $\bar{f}(z)$, $\bar{f}'(z)$, and $\bar{f}''(z)$. The true elastic strains $\boldsymbol{\varepsilon}$ are found from the same equations using

$$\varepsilon_{xx} = \frac{1}{E}(\sigma_{xx} - \nu\sigma_{xx}) \quad \text{etc.}$$

The virtual work integral $\int \bar{\boldsymbol{\sigma}}'\boldsymbol{\varepsilon}\, \mathrm{d}v$ with respect to x and y is a fearsome algebraic integral equation in $f, \bar{f}, f', \bar{f}', f''$, and $\bar{f}''$ which is then converted to an integral equation with the integrand a coefficient purely of $\bar{f}(z)$ in the now familiar fashion. To simplify the discussion, and for economy of paper, we will choose the particular values $\nu = 1/3$ and $b = 3a$ – this is a fairly wide box and the shear lag effects in the flanges should be noticeable. We also take the length $l = 4b$, a fairly short cantilever so that, for the purpose of illustration, the root constraint effects are not dwarfed by large bending stresses. The final equation then becomes:

$$\begin{aligned} \int_0^{12a} (1.448a^4 f'''' - 0.873a^2 f'' + 0.40f)\bar{f}\,\mathrm{d}z \\ + [0.313F/at + 1.019af' - 1.448a^3 f''']_0^{12a}\bar{f} \\ + [-0.536F/at + 0.146f + 1.448a^2 f'']_0^{12a}\, a\bar{f}' = 0 \quad (4.87) \end{aligned}$$

The differential equation extracted from the above integral has the solution, for a 'long' box,

$$f(z) = \mathrm{e}^{-cz/a}\left(C_1 \sin\frac{dz}{a} + C_2 \cos\frac{dz}{a}\right), \qquad \text{where} \qquad c = 0.643;\ d = 0.335$$

Notice that at the tip, $z = 12a$, $\mathrm{e}^{-cz/a} = 0.0004$, so that the root constraint effects have essentially diffused completely and we are justified in ignoring the positive exponent in the above 'long box' solution. The boundary conditions, necessary to determine C_1 and C_2, are found from the last two terms in square brackets in (4.87) since at $z = 0$ virtual stresses represented by $\bar{f}(0)$ and $\bar{f}'(0)$ are permissible and nonzero.

The stresses at the root section are shown in Fig. 4.20. Also shown are the simple beam theory, a six-boom solution, and the shear stresses predicted by Argyris and Dunne.[44] The bending stresses σ_{zz} show the characteristic shear lag effects, the values at the corners being almost double those in the middle of the flanges. The crude six-boom idealization, which predicts stresses only in the middle and at the corners, is a useful quick approximation between the Krylov–Kantorovitch solution and simple beam theory; but it does underestimate the peak stresses. The shear stresses in the webs are predictable by almost any theory, but not so in the flanges where reference 44, for example, predicts zero shear. The transverse stresses are not predicted by any other theory and are seen here not to be negligible. We can also see – with hindsight – how we could have halved our algebra by making the common assumption for shallow boxes that in the webs the shears are constant and the transverse stresses are negligible.

We have, of course, chosen a geometrically simple shape. Any other shape than rectangular – like a wing box – would have resulted in some fairly heavy algebra.

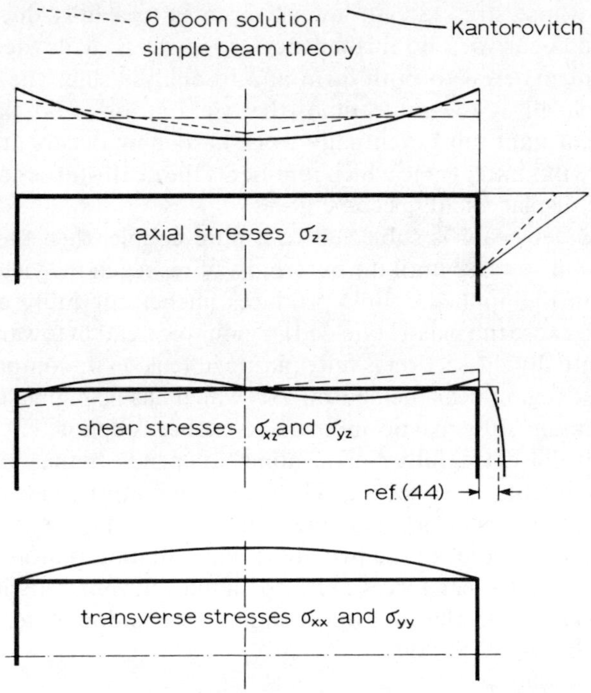

Fig. 4.20 Shear lag at the root of a cantilever; ——— Kantorovitch, – – – beam theory, - - - - - - six-boom solution

In practice, a structure could also be discontinuous at internal ribs or diaphragms, and, although the Kantorovitch technique and the PVF would produce all the equations together with boundary and continuity conditions, it is clear that the algebra would continue to multiply. As we have said before, we really have no choice but to use a numerical technique like finite elements to analyse complex fabricated structures, but analytical methods like the above do give us a useful tool to predict likely behaviour and serve as a check on finite-element modelling. They also isolate important structural and geometric parameters, like b/a in this example, and the relative unimportance of the stress variation through the webs.

4.13 Plastic collapse of frameworks

This chapter concludes with a fairly brief treatment of the prediction of plastic collapse of frameworks – mostly stiff-jointed frameworks. The PVD will prove to be a powerful way of marshalling equilibrium statements but it will also employ approximate stress fields, so it is another example of a mixed method where both stresses and displacement fields are approximate.

Collapse analysis, or limit analysis, is commonly used for civil engineering frameworks which are made of mild steel, although the philosophy can be extended to predict the eventual collapse of any structure which is made from a material whose stress–strain law asymptotes to a fully ductile state at which strain can occur with no further increase in stress. Mild steel happens to be a convenient material to both form and to analyse since the dislocations occur pretty well all at the same time – the 'yield stress' – and thereafter the stress remains constant until eventually work-hardening occurs at very large strains. We ignore this latter stage which represents the death-throes of most frameworks when the displacements become gross.

If a slender beam is subjected to a pure couple, then the resisting bending moment will increase until the outermost fibres reach the yield stress σ_y. Further increase in the applied couple produces further curvature and bending strain which will cause the plastic (yielded) region to spread in towards the neutral axis, until eventually the section is fully plastic in tension or compression except for a tiny elastic region near the neutral axis which has negligible moment resistance. The maximum 'fully plastic' moment resistance is denoted as M_p, shown in Fig. 4.21. The value of this fully plastic moment depends on the yield stress and on the shape of the beam cross-section. The neutral axis must divide the section into two equal areas, A, resulting in a pure couple $M_p = A\sigma_y h$ where h is the distance between the centroids of the two areas. The situation is more complicated when axial forces and shear forces are also significant, but we will assume that our frameworks are slender and that bending strains dominate.

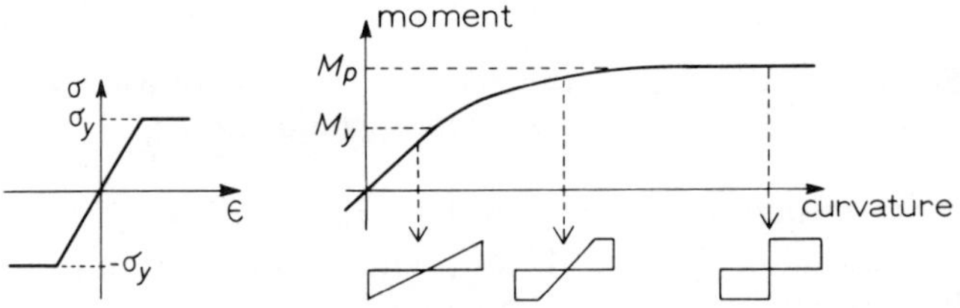

Fig. 4.21 Stress–strain variation across a mild steel beam

Once a beam section develops M_p, no further resistance to bending is possible at that section, and the behaviour is thereafter analogous to a pinjoint as far as further loading is concerned. It is possible that the beam may then become a mechanism and collapse. Most frameworks, though, are very redundant and can tolerate the successive development of many plastic hinges as the applied load is increased, until final collapse may occur in a mechanism with multiple hinges. One way of analysing this sort of problem would be to apply all the loads **R** and obtain an elastic solution to the redundant structure, then locate the maximum bending moment and, by putting it equal to M_p, obtain a value of the load vector **R** which produces the first plastic hinge. It is now convenient to represent the *magnitude* of the total load pattern **R** by some scalar λ, and call the loading $\lambda\mathbf{R}$. A further increment in λ can now be applied and the elastic solution found with the new hinge in place. A higher value of λ is then found at which the next peak bending moment reaches a local M_p. Such an analytical reconstruction of the gradual development of hinges with increasing load is naturally a time-consuming and expensive business, and we now demonstrate a more direct method of finding the collapsed state and avoiding intermediate stages. We will assume that λ controls the magnitude of a fixed load pattern $\mathbf{R}$; in fact, if the load pattern can be changed or cycled then other loading/unloading phenomena can occur (and be analysed by the PVD), but the reader is referred to the many standard texts[45,46] which treat incremental collapse and shakedown.

Suppose then we attempt to guess where the plastic hinges might occur in the simple beam of Fig. 4.22, and insert just enough for the beam to collapse. Because the shear force in the beam is constant between the supports and the central load, the bending moment must vary linearly and so any maxima must occur at the ends or in the middle. Inserting these hinges as shown in the figure, we clearly have a collapse mechanism. It is now merely a matter of using equilibrium on the beam which is now statically determinate as a consequence of prescribing the moments to be M_p at the three hinges. However, having drawn the mechanism it would be unforgivable to use direct equilibrium arguments when we have a ready-made virtual displacement pattern with which to relate the moments M_p to the applied load. The moments M_p and the rotations $\bar{\theta}$ will clearly correspond in sign, so all the products in the internal work are positive. The PVD is simply

$$M_p\bar{\theta} + M_p\bar{\theta} + M_p 2\bar{\theta} = W(l/2)\bar{\theta}; \qquad \text{or} \qquad W = 8M_p/l$$

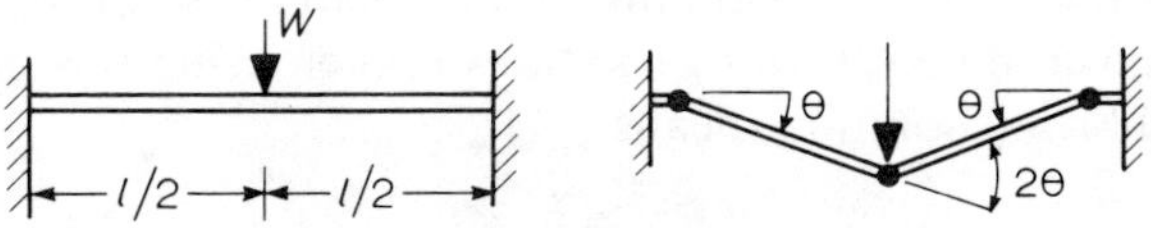

Fig. 4.22 Collapse of a single beam

(In this particular example the load happens to be also that value which produces the first plastic hinge since the elastic moments at the three hinges all have the same magnitude.)

This part of the analysis – the equilibrium relationship between W and the hinge moments – is exact. We should be perfectly clear about this: the assumed virtual displacement pattern can be anything. The choice of discrete hinges and 'rigid' links happens to deliver equations containing only the moments at those hinges, but the equation would still be valid even if the true moments at those points were not M_p. As we have said before, the stress–strain law is irrelevant to the principle of virtual work. But there are, of course, several approximations and uncertainties. The assumed mechanism may not be that in which the structure does finally collapse, so our assumption that M_p is achieved at all our chosen hinges may be erroneous. Further, we may find that there are moments elsewhere in excess of M_p. Some guesswork therefore appears to be necessary; nevertheless we can be systematic and produce two very useful bounds as a consequence of these speculative answers.

Firstly, suppose we assume a likely collapse mechanism involving a set of kinematically possible (compatible) hinge rotations $\bar{\theta}$, and let the load pattern $\mathbf{R}$ have a magnitude $\lambda\mathbf{R}$. Then the PVD becomes

$$\lambda\mathbf{R}^t\bar{\mathbf{r}} = \sum M_p\bar{\theta} \qquad (4.88)$$

Now this may not be the true collapse mechanism or the true 'critical' collapse load $\lambda_c\mathbf{R}$, in which case the hinges will have moments M, some of which may be less than M_p, and there may be fully plastic moments developed elsewhere which we have not anticipated. However, it is still true that the PVD will provide an equilibrium statement between $\lambda_c\mathbf{R}$ and these moments M. Thus,

$$\lambda_c\mathbf{R}^t\bar{\mathbf{r}} = \sum M\bar{\theta} \tag{4.89}$$

This time any M can be less than M_p (indeed M and $\bar{\theta}$ may not even be of the same sign so the right-hand side of (4.89) is less than that of (4.88).

Thus

$$\lambda > \lambda_c \tag{4.90}$$

The consequence of guessing the wrong collapse mechanism is therefore to produce a collapse load which will be greater than the true value. The solution is an *upper bound*. Statement (4.90) is known as the *kinematic* or *upper-bound theorem*.

Secondly, suppose we can find a system of internal bending moments which are in equilibrium with – or statically equivalent to – the applied loads $\lambda\mathbf{R}$, and in a statically indeterminate structure there may be many such alternatives. We choose to limit the maximum bending moment *anywhere* to M_p. If the actual collapse mechanism involves hinge rotations θ, then using this in the PVD we know our equilibrium solution means

$$\lambda\mathbf{R}^t\bar{\mathbf{r}} = \sum M\bar{\theta} \tag{4.91}$$

But if the assumed moment distribution is not the correct collapse distribution then some or all of the values of M will be less than M_p. The exact collapse load would of course lead to

$$\lambda_c\mathbf{R}^t\bar{\mathbf{r}} = \sum M_p\bar{\theta} \tag{4.92}$$

From (4.91) and (4.92) we can therefore say

$$\lambda < \lambda_c \tag{4.93}$$

This, the *static theorem*, leads to a lower bound on λ_c.

If we can find a set of kinematically possible hinges with bending moments equal to M_p and which also satisfy equilibrium and lead to moments everywhere else which are less than M_p, then *both* (4.93) and (4.90) must hold, so λ must equal λ_c. This correct solution must therefore be unique.

As an example, consider the portal framework of Fig. 4.23, with component elements having fully plastic moments and lengths as indicated. Three likely mechanisms are shown in Figs 4.23(a), (b), and (c). Applying the PVD to:

(a) $2M_p\bar{\theta} + M_p\bar{\theta} + M_p2\bar{\theta} + M_p2\bar{\theta} = Wl2\bar{\theta} \qquad \therefore \quad W = 3.5M_p/l$

(b) $M_p\bar{\theta} + M_p2\bar{\theta} + M_p\bar{\theta} = 2W(l/2)\bar{\theta} \qquad \therefore \quad W = 4M_p/l$

(c) $2M_p\bar{\theta} + M_p2\bar{\theta} + M_p3\bar{\theta} + M_p2\bar{\theta} = 2W(l/2)\bar{\theta} + Wl2\bar{\theta} \qquad \therefore \quad W = 3M_p/l$

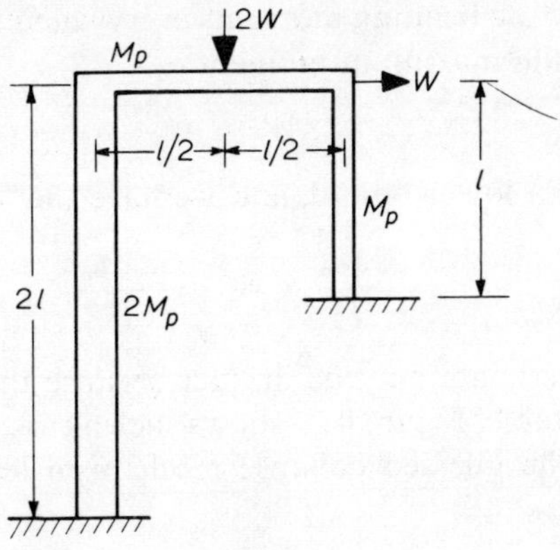

Fig. 4.23 Portal framework

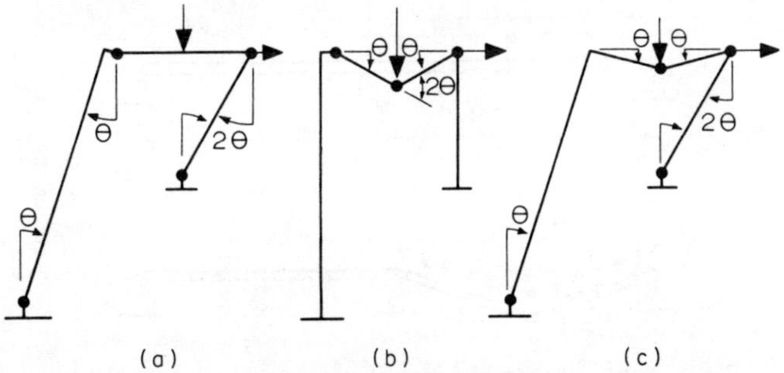

Fig. 4.23 (a), (b), and (c). Possible collapse mechanisms

There are no better alternatives, so the collapse load from (c) is $W_c = 3M_p/l$. Suppose we had chosen only (a) and (b), then a lower bound can be found by looking in detail at the bending moments in (say) (a) which, by virtue of the PVD, is an equilibrium solution statically equivalent to the applied loads. The bending moment distribution with $W = 3.5M_p/l$ (or $M_p = 2Wl/7$) is shown in Figure 4.24.

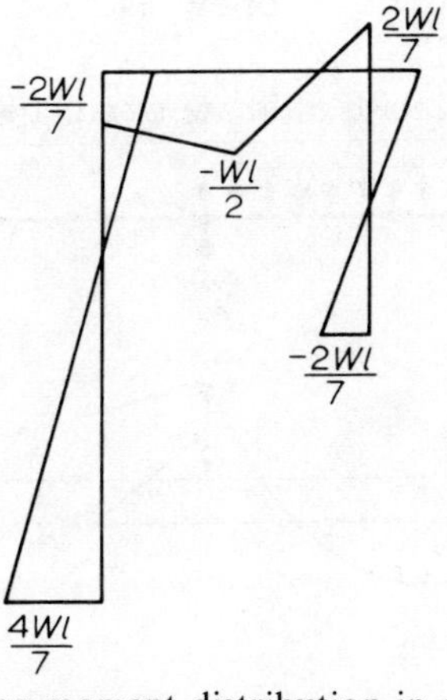

Fig. 4.24 Bending moment distribution in collapse mode (a)

We can now restrict the bending moments everywhere to be no greater than M_p, by scaling so that the maximum moment

$$\tfrac{1}{2}Wl = M_p$$

Thus $W = 2M_p/l$ is a lower bound, and we have the fairly crude estimate

$$\frac{2M_p}{l} < W_c < \frac{3.5M_p}{l}$$

Sometimes our estimate may be inexact because the precise location of the hinges is not predictable. Figure 4.25 shows such a case in a uniform beam with a distributed load. The guessed collapse mechanism leads to the upper-bound solution:

$$M_p 2\bar{\theta} + M_p\bar{\theta} = (wl)\tfrac{1}{2}(\tfrac{1}{2}\bar{\theta}) \qquad \text{or} \qquad wl^2 = 12M_p \tag{4.94}$$

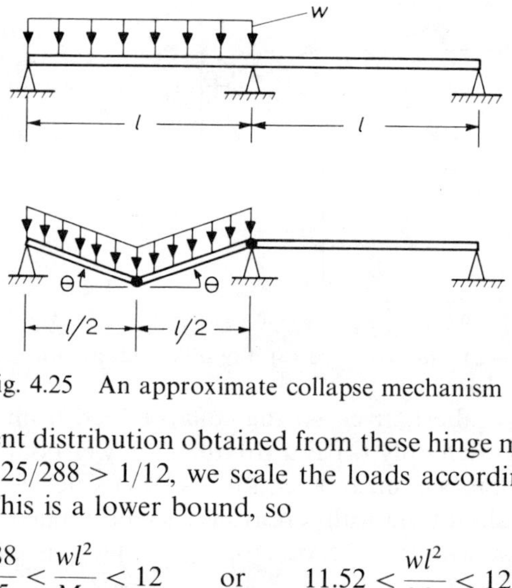

Fig. 4.25 An approximate collapse mechanism

The bending moment distribution obtained from these hinge moments is shown in Fig. 4.26. Since $25/288 > 1/12$, we scale the loads accordingly so that now $25wl^2/288 = M_p$. This is a lower bound, so

$$\frac{288}{25} < \frac{wl^2}{M_p} < 12 \qquad \text{or} \qquad 11.52 < \frac{wl^2}{M_p} < 12$$

This time we have a fairly precise estimate of collapse.

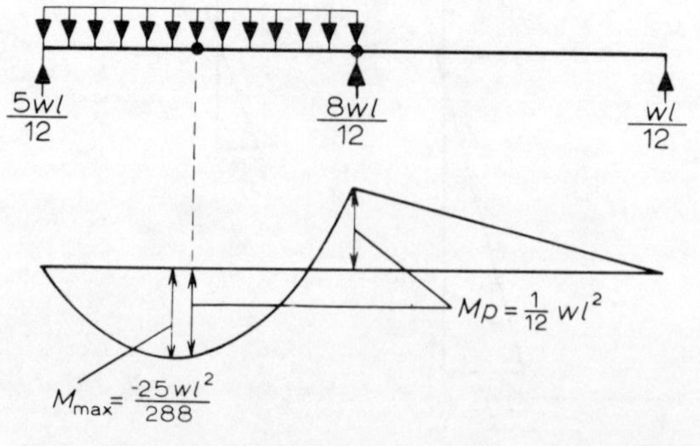

Fig. 4.26 Moment distribution in assumed collapse

4.14 Summary

Rayleigh–Ritz or Galerkin method

PVD Assume a likely displacement series (satisfying kinematic boundary conditions) and obtain strains by satisfying compatibility. Use the PVD with one term at a time as the virtual displacement and so generate simultaneous equations in the unknown coefficients. The solution will be an over-stiff bound on strain energy.

PVF Assume a likely self-equilibrating series for stress, satisfying both internal equilibrium and static boundary conditions. Use the PVF with one term at a time for the virtual stresses, and so generate simultaneous equations for the unknown coefficients. The solution will be an over-flexible bound on strain energy.

Mixed method

Assume a series for both displacements and for stresses. Do not satisfy compatibility or equilibrium exactly, but employ the weighted residual form to do so approximately 'in the mean', using (4.31) and (4.32) or alternatively (4.33) and (4.34).

Relaxed boundary conditions

When using the PVD (or PVF) the virtual displacements (or forces) need not satisfy kinematic (or static) boundary conditions provided the unknown reactions (or displacements) are retained in the formulation.

Krylov–Kantorovitch method

For two-dimensional problems. Assume field variations in one variable and solve the consequent ordinary differential equations in the remaining variable.

Plastic collapse of frameworks

Assume kinematically consistent collapse mechanisms, use the PVD to find the smallest collapse load which must be an upper bound. Any statically equivalent distribution of moments, which nowhere exceed M_p, can provide a lower bound.

Problems

(Asterisks indicate problems with worked solutions.)

P 4.1

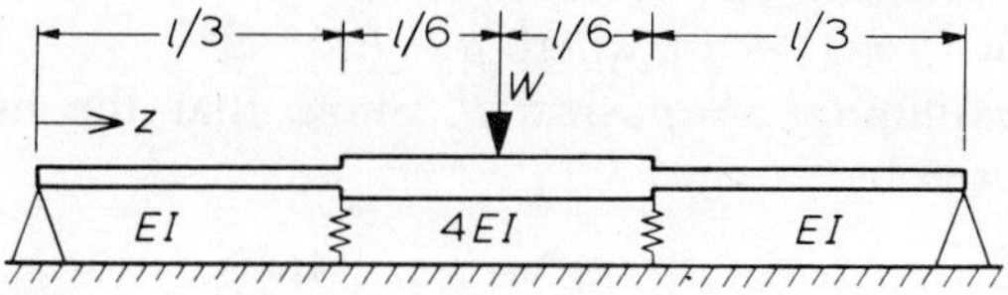

Fig. P. 4.1

A beam of length l is simply supported at its ends and supported on two identical springs, a distance $l/6$ from the centre. The flexural rigidity of the two outer portions of the beam is EI, and of the central portion, $4EI$, as shown. The extensional stiffness of each ring is $100EI/l^3$, and the beam is loaded by a central weight W.

Assuming the deflected shape is

$$v(z) = v_1 \sin\frac{\pi z}{l} + v_3 \sin\frac{3\pi z}{l}$$

show that $v_1 = -0.719\tilde{W}$; $v_3 = -0.0247\tilde{W}$; where $\tilde{W} = Wl^3/2EI\pi^4$. Show also that the spring forces are $0.324W$. (The exact solution is $0.334W$.)

*P 4.2**

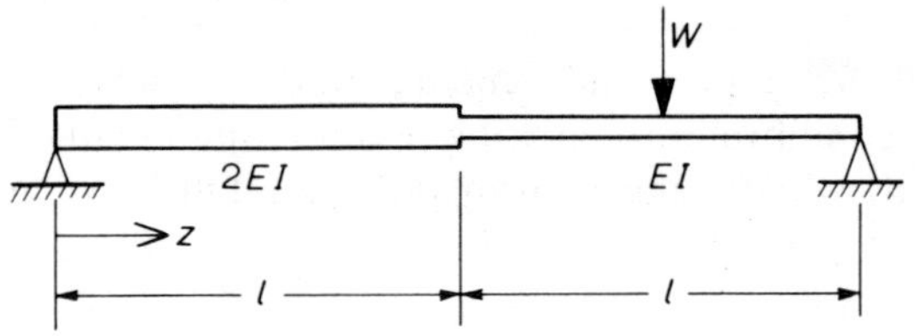

Fig. P. 4.2

A simply supported beam has discontinuous stiffness as shown, and a concentrated load W placed at $z = 3l/2$. Assuming a three-term series for

$$v(z) = \sum_n v_n \sin\frac{n\pi z}{2l}$$

examine the convergence of the bending moment, and sketch its variation along the beam, comparing it with the exact solution (which, of course, is not discontinuous).

P 4.3

A simply supported beam of length l and stiffness EI is supported by a continuous elastic spring of stiffness k per unit length (spring force $p = kv$/unit length), and loaded by a uniformly distributed load w/unit length. Obtain a one-term Rayleigh–Ritz solution by:

(a) The PVD: assuming $v = v_1 \sin \pi z/l$. Show that the maximum bending moment is $4wl^2\beta/\pi^3(1+\beta)$, where $\beta = EI\pi^4/kl^4$.

(b) The PVF: assuming $p = p_1 \sin \pi z/l$. Show that the maximum bending moment is given by

$$M_{\max} = \frac{wl^2}{8} - \frac{4wl^2}{\pi^3(1+\beta)}$$

Sketch the variation, with β, of the two solutions as β varies from 0 to ∞, and deduce that the exact solution is fairly accurately bounded.

P 4.4

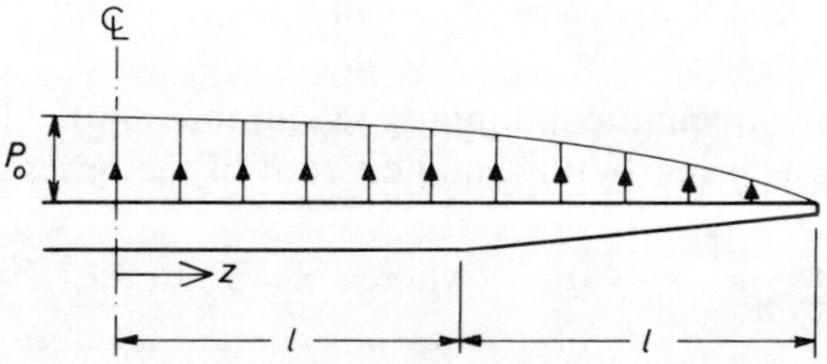

Fig. P. 4.4

The figure shows one half of an idealized aircraft wing structure subjected to a distributed lifting force $p(z) = p_0(1 - z^2/4l^2)$ per unit length. The tapered beam has the properties:

$$0 < z < l\colon I = I_0; \qquad l < z < 2l\colon I = I_0\left(\frac{3}{2} - \frac{z}{2l}\right)$$

It is required to obtain a two-term polynomial approximation to the deflected shape $v(z)$. Suggest why the simplest possible series may be of the form

$$v(z) = \frac{z^2}{l^2}\left(v_1 + v_2\frac{z}{l}\right)$$

Show that $v_1 = 0.376\beta$; $v_2 = -0.083\beta$; where $\beta = p_0 l^4/EI_0$. Sketch the bending moment distribution and compare with the readily obtained exact solution.

P 4.5

A uniform open tube of length $2l$ is rigidly clamped at both ends $z = 0$ and $z = 2l$, and has a torque T_0 applied at the centre $z = l$. The Saint-Venant torsion rigidity is GJ, and the torsion-bending rigidity $E\Gamma = \frac{1}{4}GJl^2$ (see section 2.10 for these definitions). Solve this problem by assuming that the rate of twist

$$\theta'(z) = z(z - l)(\theta_1 + \theta_2 z/l) \qquad \text{for} \quad 0 < z < l$$

Compare the value of the maximum twist with that obtained using just the first term in the series, and the exact solution of $5T_0 l/42GJ$.

P 4.6

The buckling weight of a heavy vertical cantilever was predicted in section 4.9 by assuming a sine series for $v(z)$ (equation (4.54)). Solve this problem by assuming

$$v(z) = v_1(z/l)^2 + v_2(z/l)^3$$

and show that the buckling load $wl^3/EI = 7.89$.

*P 4.7**

A uniform strut of length l, flexural stiffness EI, is fully clamped at $z = 0$ and simply supported at $z = l$. Assuming a deflection

$$v = \sum_{n=1}^{N} v_n \sin\frac{n\pi z}{l}$$

which violates the kinematic boundary condition $v'(0) = 0$, show that the buckling load P_{CR} is given by the smallest root of the series

$$\sum_{n=1}^{N} \frac{1}{n^2 - \lambda} = 0, \qquad \text{where} \quad \lambda \neq P_{cR}/(\pi^2 EI/l^2).$$

Confirm that four terms ($N = 4$) are needed to get within 7 per cent of the exact solution, $\lambda = 2.05$.

P 4.8

A uniform cantilever of mass m/unit length undergoes flexural vibrations at a natural frequency ω, so that the deflected shape $v(z)$ is of the form $v(z)\sin\omega t$. This problem may be imagined as a beam in 'equilibrium' under applied inertia forces $p(z)$/unit length,

$$p(z) = -m\frac{\partial^2 v}{\partial t^2} = m\omega^2 v$$

(a) Assuming $v = z^2(v_1 + v_2 z/l)$ find the natural frequency using the PVD, and compare the answer with that predicted by $v = v_1 z^2$.

(b) To use the PVF we must satisfy 'equilibrium' requirements, $M'' = -p$. Then

$$\int_V \bar{\boldsymbol{\sigma}}^t \boldsymbol{\varepsilon}\, dv = \int_V \bar{\mathbf{p}}_v^t \mathbf{u}\, dv \qquad \text{becomes} \qquad \int_0^l \frac{\bar{M}M}{EI}\, dz = \int_0^l (\bar{M}'')\left(\frac{M''}{m\omega^2}\right) dz$$

Find the natural frequency by assuming $M(z) = M_1(1 - z/l)^2$ which satisfies equilibrium conditions $M(l) = M'(l) = 0$. Improve this unsatisfactory guess by satisfying also the kinematic conditions $v(0) = v'(0) = 0$, by noting that $v = p/m\omega^2 = -M''/m\omega^2$, so putting

$$M(z) = M_1\frac{z^4}{l^4}\left(1 - \frac{z}{l}\right)^2$$

Compare all four predictions to see whether the frue frequency can be estimated from the bounds in (a) and (b).

P 4.9

Solve Problem 4.8 by the mixed method, assuming

$$M = M_1\left(1 - \frac{z}{l}\right)^2 \qquad \text{and} \qquad v = v_1\left(\frac{z}{l}\right)^2$$

and compare the answer with those of Problem 4.8(a) and (b). The equations to be used are (4.33) and (4.34), which become

$$\int_0^l M\bar{v}''\,\mathrm{d}z = \int_0^l p\bar{v}\,\mathrm{d}z, \qquad p = m\omega^2 v; \quad \text{and} \quad \int_0^l \left(\bar{M}\frac{M}{EI} + v\bar{M}''\right)\mathrm{d}z = 0$$

P 4.10

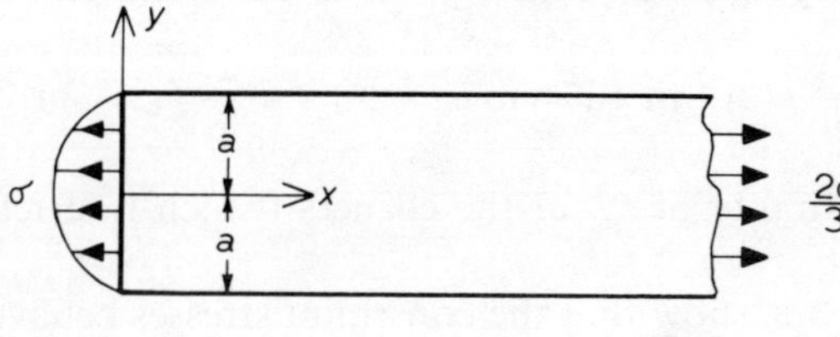

Fig. P. 4.10

The very long plate shown may be the wide flange of an I-beam which due to shear lag has a parabolic stress distribution across the section $x = 0$ given by $\sigma_{xx} = \sigma(1 - y^2/a^2)$. It is required to apply the Kantorovitch technique to see how rapidly the stress diffuses to the constant value.

The shear stresses must vary in an antisymmetrical fashion about the centre $y = 0$, so assume

$$\sigma_{xy} = f'(x)(a^2 - y^2)y$$

which satisfies necessary boundary conditions on $y = \pm a$. Thence show that all other equilibrium requirements will be satisfied if

$$\sigma_{yy} = \tfrac{1}{4}f''(x)(a^2 - y^2)^2; \qquad \sigma_{xx} = \sigma(1 - y^2/a^2) - f(x)(a^2 - 3y^2)$$

Obtain a fourth-order differential equation for $f(x)$ and show that diffusion is virtually complete at $x = 2a$.

*P 4.11**

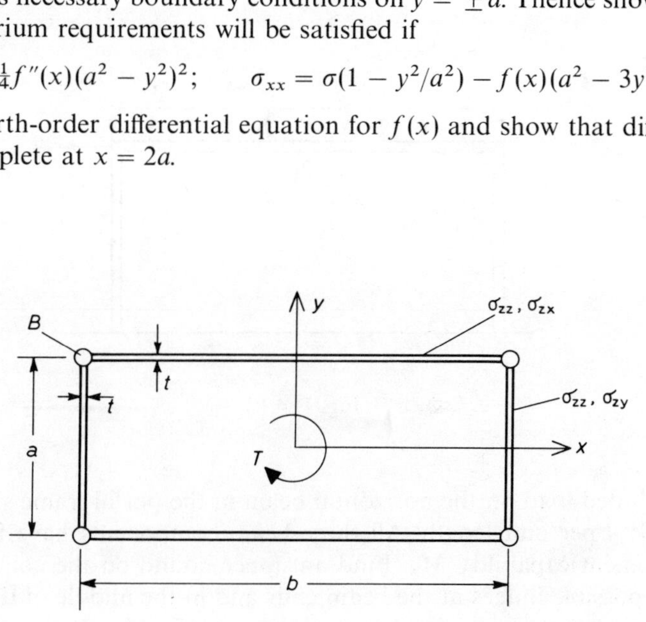

Fig. P. 4.11

A long four-boom rectangular box is subjected to a constant torque T and fully clamped at $z = 0$. Its shape is maintained by closely spaced ribs which inhibit any transverse strains ε_{xx} or ε_{yy}. We wish to investigate, using the PVF, the constraint stresses. However, the crude idealization of section 3.9 (Fig. 3.35) is to be discarded and the 'shear panels' are to be allowed to take direct stresses. We may then look at the complete range of small and large boom areas, B, to test the validity of the boom-panel idealization.

Taking advantage of antisymmetry, assume a variation

$$\sigma_{zz} = \frac{2x}{b} f(z) \quad \text{in} \quad y = a/2; \qquad \sigma_{zz} = \frac{2y}{a} f(z) \quad \text{in} \quad x = b/2$$

This ensures continuity of σ_{zz} at the corners (which is strictly a compatibility requirement).

Taking $G/E = 3/8$, show that the constraint stresses behave like $\exp(-\alpha z/a)$, where

$$\frac{1}{\alpha^2} = \frac{\mu}{6} + \beta(1 + \mu)^2 (\tfrac{1}{3} + \tfrac{1}{2}\beta) + \frac{(1 + \mu^3)}{1 + \mu}$$

where $\mu = b/a$, $\beta = 4B/2(a + b)t$. Thus the diffusion rate increases with α which increases for small μ (thin box) and small β (small booms).

Using the specific value $b = 2a$ ($\mu = 2$) find the shear stresses at the constraint for extreme values $\beta = 0$ and ∞. You may care to note that the shears at the constraint predicted by Argyris and Dunne[44] are $\sigma_{zy} = -2\sigma_{zx} = -T/3a^2t$.

P 4.12

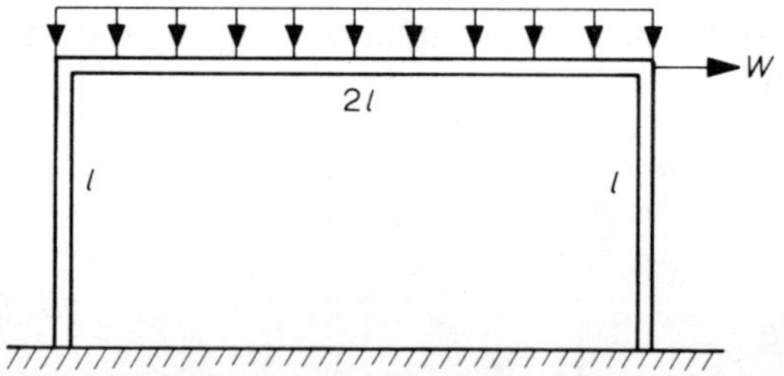

Fig. P. 4.12

The distributed load on the horizontal beam in the portal frame shown has an intensity W/l per unit length. All three beam components have the same full plastic moment capability M_p. Find an upper bound on the collapse load by assuming possible hinges at the beam ends and in the middle of the horizontal beam. Find the moment distribution in the predicted collapse mode and so obtain a lower bound also.

*P 4.13**

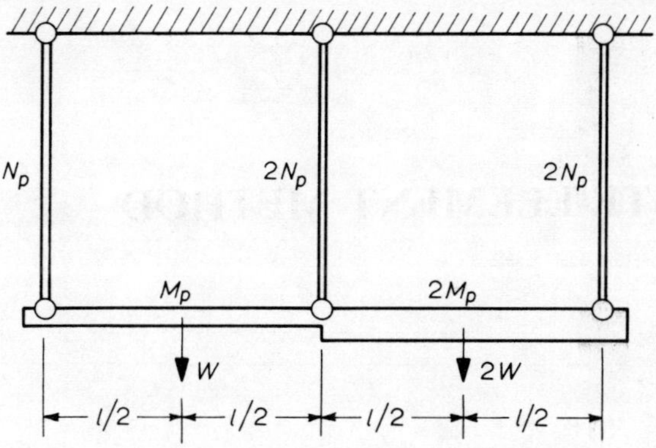

Fig. P. 4.13

A horizontal beam is suspended from three pin-ended rods as shown. The fully plastic loads of the rods are N_p and $2N_p$, and the beam plastic moments are M_p and $2M_p$, where $M_p = \frac{1}{3}N_p l$. Find the collapse load W.

5

THE FINITE-ELEMENT METHOD

5.1 Introduction

In all examples of the Rayleigh–Ritz technique in the previous chapter, we approximated unknown displacement or stress fields as a complete series, each of which was continuous over the whole structure. This ruse, allied to the PVD or PVF, then relieved us of the task of solving differential equations; instead we merely had to evaluate volume and surface integrals which can be done even though the structural geometry may vary through V and may even be discontinuous. We chose, at the risk of boredom, necessarily simple structures so that the basic principles were not lost in a welter of algebra. Even so, many of the integrals were tiresome, and had we examined more two- and three-dimensional problems, the integrals would have been even worse. It is obvious that in practice many real structures, even if they are assemblies of bars and beams, are an order of magnitude more complex than our simple examples. Even after the integrals have been evaluated, the number of unknowns to be solved is likely to be large. This in itself should be a straightforward computing task provided that the equations are not badly conditioned, but the real problem is firstly in organizing the integrals and unknowns; and secondly – and much more importantly – the choice of a global series is unlikely to be appropriate. The character of structures such as aircraft, buildings, and ships is not continuous. Instead they consist of many substructures which themselves may be complex. Even in a simple plate or shell the stress field is very unlikely to be smooth with gentle variations everywhere. We must be able to choose approximations which can cope with rapid variations in both stress and geometry in a certain region and yet be economical in regions where variations are perhaps more gentle. To do this we must modify the *nature* as opposed to the *principle* of the Rayleigh–Ritz method so that the computer can be readily programmed to generate the description of the structure, list the discrete unknowns, and assemble their governing equations – in addition to simply solving them. This is the motivation behind the finite-element method which since the 1960s has revolutionized structural mechanics and freed it from the essentially geometric straightjacket.

The finite-element method was probably first proposed as such by Courant in 1943 who applied it to the torsion problem. The proposal was largely ignored by engineers, possibly because it was published in a mathematical journal[47], and also, who would care to use a method which invariably led to solving a large set of simultaneous equations? (This lack of interest was at the same time displayed in a method proposed by J. M. Bennet for solving redundant frameworks on the new-fangled computer, and by G. Kron for solving similar network problems.) The pioneers of the modern finite-element method must include Argyris[14] and Clough *et al.*[49], who foresaw the impact of the computer and proposed methods for utilizing its capabilities – aimed initially at the aircraft industry which at the time was probably one of the few to be able to afford such luxuries. The aircraft designer of course has always had a vested interest in precise analysis and saving weight to the last percentage.

Mathematically the method can be described simply as a piecewise Rayleigh–Ritz procedure using either the PVD or the PVF. The idea is that we choose to visualize a structure covered by 'elements' which are sufficiently small that the variation of displacements or stresses over them may be approximated by some relatively crude *shape* or *interpolation function.* The anticipated stress field in the aircraft wing of Fig. 5.1 could for example be approximated as a series of

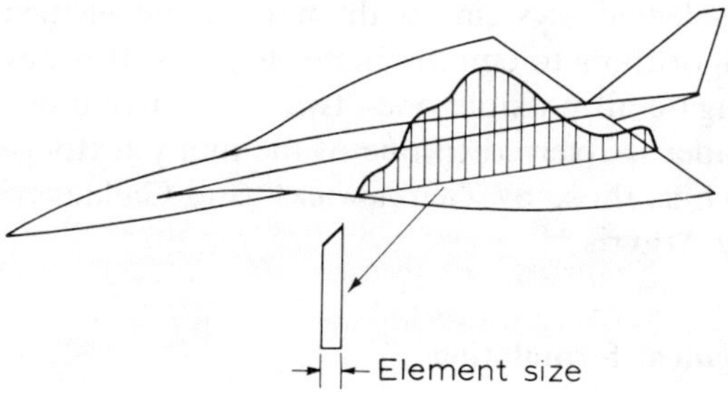

Fig. 5.1 Assumed element variation

linear shapes over small elements. Having specified the *shapes,* it merely remains to find the *magnitudes* of these local approximations. These magnitudes constitute our new discrete unknowns – hence the description 'piecewise Rayleigh–Ritz'. The field variation will usually be continuous in some sense, but we leave these aspects for a while.

Recalling that the PVD or PVF are simply integrals, we formally write the virtual work for the whole structure as

$$\int_V \boldsymbol{\sigma}^t \boldsymbol{\varepsilon} \, dv = \sum_{g=1,2,\ldots} \int_{V_g} \boldsymbol{\sigma}^t \boldsymbol{\varepsilon} \, dv$$

and are able to perform these integrals since we have now specified the shape of the field over the element. (The finite-element method can also be thought of as a method of numerical integration.) Although the absolute values of displacement or stress depend on the magnitude of these unknown amplitudes, the element integrals themselves depend only on the shape of the assumed field and on the shape of the element itself which may be one-, two-, or three-dimensional. The integrals can therefore be programmed once only for a particular type of element. We can leave to the computer the repetitive task of evaluating these for all elements. Sometimes this can be done in closed form for certain types of element: often it has to be done numerically.

The finite-element idealization can be used with either the PVD or the PVF, but it is the former which has virtually monopolized numerical solutions in structural mechanics. The reasons are not difficult to appreciate. It is straightforward to assume a likely form of displacement field $\mathbf{u}$ and then to differentiate it, $\boldsymbol{\varepsilon} = \mathbf{D}^t\mathbf{u}$, to find the strains. It is not as straightforward to assume a stress field which satisfies equilibrium, and, because we have to integrate $\mathbf{D}\boldsymbol{\sigma} + \mathbf{p}_v = \mathbf{0}$, as we have seen earlier we have to cope with both a particular integral and a complementary function. We will therefore devote this chapter to the displacement method and leave until Chapter 6 the use of the PVF and 'equilibrium' elements, together with mixed elements and further generalizations. The straightforward nature in principle of the finite element method will be demonstrated using fairly simple elements. In practice there are almost as many special-purpose elements as there are practitioners using them. Such has been the advance in computational mechanics that there are now a large number of commercial finite-element systems with many finite elements, and solution procedures and algorithms to employ these elements. It is beyond the scope of this text to investigate all possible one-, two- and three-dimensional elements. The interested reader is referred to one of the many textbooks specifically on finite elements such as those by Zienkiewicz[48] and Gallagher[51] and a series of technical notes by Argyris.[52]

5.2 The Displacement formulation

The finite-element method can be thought of as two distinct processes. Firstly, we approximate and discretize the displacement field at an element level and apply the PVD to obtain the internal and external virtual work for an element in terms of local forces and displacements. Secondly, we sum these expressions for all elements over the entire structure and eventually accumulate a global relationship between forces and displacements – $\mathbf{R} = \mathbf{Kr}$. We have in fact already done both parts of this process separately in Chapters 2 and 4. The accumulation of a global stiffness matrix in Chapter 2 was for frameworks in which the elements' behaviour was described exactly. In Chapter 4 we formulated approximate solutions in which the displacement field was not exact. Now we bring them both together, and accumulate the effects of many elements in each one of which the description is approximate. As we have already analysed beam elements, these will form a convenient starting point to fix ideas without having to introduce novel element properties. We will carry through this approach to two-dimensional elements and then find that we have been smarter than we knew.

A beam element has the virtue that it is one-dimensional and yet still exhibits quite a sophisticated displacement field along its length. Remembering that the PVD obliges us to satisfy all kinematic compatibility requirements, consider what is necessary to define the displacement field in a beam element. The displacements should be continuous throughout a beam structure when all elements are assembled, so the displacement and rotation of one end of an element should match those of its neighbour. This can be achieved by simply using these quantities as the discrete unknowns to be solved, and then defining the beam's internal displacement field in terms of them. By assuming that contiguous elements have the same unknown at their interface, we will be able to achieve displacement and slope continuity throughout. The ends of a beam element will henceforth be referred to as '*nodes*' and the *nodal displacements* of the gth element, $\boldsymbol{\rho}_g$, will be the two displacements and rotations first used in Fig. 2.18 and repeated here in Fig. 5.2. Now a cubic polynomial in z is completely defined by the four coefficients of z°, z^1, z^2, and z^3, so these can be related to the four geometrical unknowns in Fig. 5.2. It will be much more convenient from now on

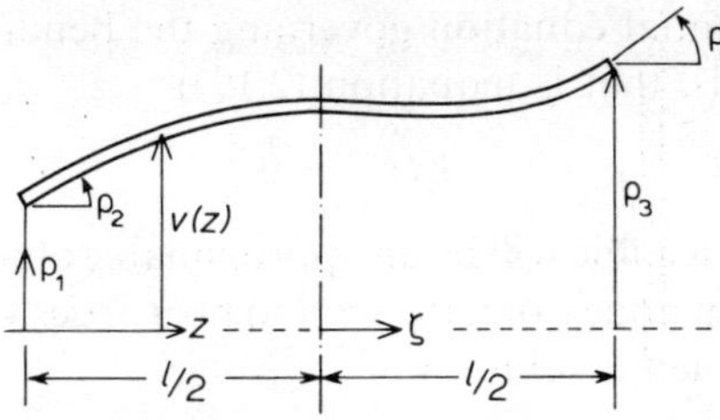

Fig. 5.2 Degrees of freedom for a beam element

to use non-dimensional variables when describing the shape of a field over an element. All necessary virtual-work integrals can then be performed for any general element with non-dimensional limits, leaving the actual size of a specific element to be inserted after the integration. Here, in Fig. 5.2 we use the ordinate $\zeta = z/(l/2) = 2z/l$, measured from the centre so that $\zeta = \pm 1$ at the two nodes. The deflection of the beam may then be written

$$v(\zeta) = \omega_1\rho_1 + \omega_2\rho_2 + \omega_3\rho_3 + \omega_4\rho_4$$

or

$$v = \boldsymbol{\omega}\boldsymbol{\rho}_g \tag{5.1}$$

where

$$\boldsymbol{\omega} = [\omega_1 \quad \omega_2 \quad \omega_3 \quad \omega_4]; \qquad \boldsymbol{\rho}_g = \{\rho_1 \quad \rho_2 \quad \rho_3 \quad \rho_4\}$$

The interpolation functions $\boldsymbol{\omega}$ must clearly have unit value at the relevant node and zero values elsewhere. Thus $\omega_1(\zeta)$ must have a value of unity at $\zeta = -1$, must be zero at $\zeta = +1$, and the slopes must be zero at both $\zeta = \pm 1$. The forms of all four functions $\boldsymbol{\omega}$ are shown in Fig. 5.3. These are known as *Hermitian polynomials*[50] and the reader is invited to verify that ω and $\partial\omega/\partial z$ have the values stated at $\zeta = \pm 1$ (noting that $\partial/\partial z = (2/l)\partial/\partial\zeta$). The cubic polynomial was chosen because it is an obvious choice of curve to fit prescribed values of deflection and slope at the two end nodes. As it happens, this shape is an exact

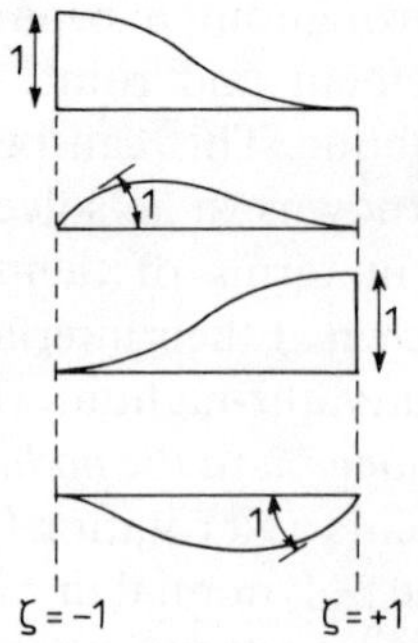

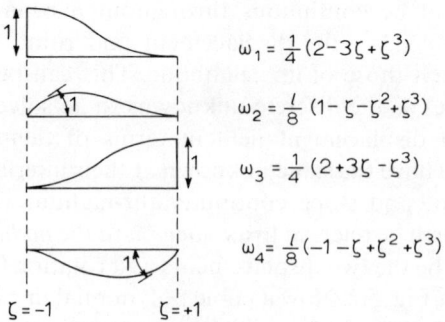

Fig. 5.3 Hermitian interpolation functions

solution to the differential equation governing the bending of a uniform beam loaded only at its ends, that is (equation (2.19)),

$$EIv'''' = 0$$

But the assumption of a cubic will be an approximate solution if, for example, the beam is loaded between nodes, or if the section properties vary in some way, when the deflection $v(z)$ is then a solution of

$$(EIv'')'' = p(z)$$

The next step is to enforce compatibility, $\boldsymbol{\varepsilon} = \mathbf{D}^{\mathrm{t}}\mathbf{u}$, and assuming simple beam theory we have just the single strain component

$$\varepsilon_{zz} = -y\frac{\partial^2 v}{\partial z^2} = \frac{-4y}{l^2}\frac{\partial^2 y}{\partial \zeta^2}$$

and using (5.1),

$$\varepsilon_{zz} = \boldsymbol{\alpha}\boldsymbol{\rho}_g \tag{5.2}$$

where

$$\boldsymbol{\alpha} = \frac{-y}{l^2}[6\zeta \quad l(-1+3\zeta) \quad -6\zeta \quad l(1+3\zeta] \tag{5.3}$$

Having the strains, and taking $\sigma_{zz} = E\varepsilon_{zz}$, the element virtual work becomes

$$\int_{V_g} \sigma\bar{\varepsilon}\,\mathrm{d}v = \int_{V_g} \boldsymbol{\rho}_g\boldsymbol{\alpha}^{\mathrm{t}}E\boldsymbol{\alpha}\bar{\boldsymbol{\rho}}_g\,\mathrm{d}v = \boldsymbol{\rho}_g^{\mathrm{t}}\mathbf{k}_g\bar{\boldsymbol{\rho}}_g$$

where

$$\mathbf{k}_g = \int_{V_g} \boldsymbol{\alpha}^{\mathrm{t}}E\,\boldsymbol{\alpha}\,\mathrm{d}v \tag{5.4}$$

This symmetrical element stiffness matrix $\mathbf{k}_g$ has emerged quite naturally from the work integral and it will be better from now on if we realize that it is nothing more than a characteristic element matrix which we happen to label 'stiffness'. The danger of using actual physical concepts of stiffness arises when we try to interpret (5.4) as a work product (forces) × (displacements), or

$$\boldsymbol{\rho}_g^t \mathbf{k}_g \bar{\boldsymbol{\rho}}_g = \mathbf{P}_g^t \bar{\boldsymbol{\rho}}_g; \qquad \text{thus} \quad \mathbf{P}_g = \mathbf{k}_g \boldsymbol{\rho}_g$$

The element 'forces' $\mathbf{P}_g$ must correspond to $\boldsymbol{\rho}_g$, so in this case they will be nodal shear forces and moments, which we have met before. There is therefore no problem imagining such concentrated nodal forces $\mathbf{P}_g$ in a beam element, but they are only *measures* of concentrated forces which will be physically, meaningless in future elements. There are not a few engineers who puzzle about concentrated nodal forces causing undesirable high stress concentrations in their plates!

The stiffness integral (5.4) can be evaluated on using (5.3), and we find

$$\mathbf{k}_g = \frac{2EI}{l^3}\begin{bmatrix} 6 & 3l & -6 & 3l \\ & 2l^2 & -3l & l^2 \\ \text{symm.} & & 6 & -3l \\ & & & 2l^2 \end{bmatrix} \tag{5.5}$$

agreeing with (2.27) which was also obtained using virtual-work arguments.

If the beam element is loaded by $p(z)$ per unit length, then the external work right-hand side of the PVD becomes

$$\int_{S_g} \mathbf{p}_s^t \bar{\mathbf{u}}\,\mathrm{d}s = \int_{S_g} p\bar{v}\,\mathrm{d}s = \int_0^l p\boldsymbol{\omega}\,\mathrm{d}z\bar{\boldsymbol{\rho}}_g \qquad \text{using (5.1) again.}$$

Because $\boldsymbol{\omega}(\zeta)$ has been defined, this integral can also be evaluated once we know the loading $p(z)$. It is convenient once again to use 'forces' to summarize the work integral; thus

$$\int_0^l p\boldsymbol{\omega}\,\mathrm{d}z\bar{\boldsymbol{\rho}}_g \qquad \text{must be of the form} \quad \mathbf{P}_g^t \bar{\boldsymbol{\rho}}_g$$

whence

$$\mathbf{P}_g = \int_0^l \boldsymbol{\omega}^t p\,\mathrm{d}z \tag{5.6}$$

These measures of applied forces were labelled in section 2.7 as *kinematically equivalent* forces since they are derived solely by evaluating the work done by $p(z)$ over the assumed displacement field $\mathbf{u} = \boldsymbol{\omega}\boldsymbol{\rho}$. Equations 2.29 (a) and (b) gave their values for concentrated and distributed loading.

The beam element has now to be connected to its neighbours and the total virtual work summed. We formally implement this by the usual Boolean connectivity matrix $\mathbf{a}_g$ which relates local labels $\boldsymbol{\rho}_g$ to global labels $\mathbf{r}$:

$$\boldsymbol{\rho}_g = \mathbf{a}_g \mathbf{r} \tag{2.10}$$

So

$$\int_V \boldsymbol{\sigma}^t \bar{\boldsymbol{\varepsilon}}\,\mathrm{d}v = \int_S \mathbf{p}_s^t \bar{\mathbf{u}}\,\mathrm{d}s$$

becomes

$$\sum_{g=1,2,\ldots} \int_{V_g} \boldsymbol{\sigma}^t \bar{\boldsymbol{\varepsilon}}\, dv = \sum_{g=1,2,\ldots} \int_{S_g} \mathbf{p}_s^t \bar{\mathbf{u}}\, ds$$

or

$$\sum_g \boldsymbol{\rho}_g^t \mathbf{k}_g \bar{\boldsymbol{\rho}}_g = \sum_g \mathbf{P}_g^t \bar{\boldsymbol{\rho}}_g$$

i.e.,

$$\left[\mathbf{r}^t \sum_g \mathbf{a}_g^t \mathbf{k}_g \mathbf{a}_g - \sum_g \mathbf{P}_g^t \mathbf{a}_g\right] \bar{\mathbf{r}} = 0$$

whence $\mathbf{r}^t\mathbf{K} - \mathbf{R}^t = \mathbf{0}$ as before, and

$$\mathbf{Kr} = \mathbf{R} \tag{5.7}$$

where

$$\mathbf{K} = \sum \mathbf{a}_g^t \mathbf{k}_g \mathbf{a}_g \tag{5.8}$$

and

$$\mathbf{R} = \sum \mathbf{a}_g^t \mathbf{P}_g \tag{5.9}$$

All the above nine lines of algebra, assembling the global equations, are identical to the previous framework assemblies of Chapter 2, and the details remain the same. The summations (5.8) and 5.9) are not performed as such with Boolean arrays $\mathbf{a}_g$. Once we have identified local displacements ρ_i and ρ_j with global listings r_I and r_J, etc., then (5.8) merely states that k_{ij} is inserted at K_{IJ}. Likewise all nodal forces P_I, arising from elements which have a common displacement r_I, are summed as R_I. The assembly and solution of $\mathbf{r}$ is now easily demonstrated on the uniform beam in Fig. 5.4 which was previously solved (Fig. 4.11) using a global Rayleigh–Ritz approximation. The beam is loaded by a uniformly distributed load w per unit length, so the exact solution to $EIv'''' = w$ must be a quartic, and our cubic element solution will therefore be approximate. Three elements have been used, $g = 1, 2, 3$, all of length l, and the global displacements $\mathbf{r} = \{r_1 \quad r_2 \quad r_3 \quad r_4\}$ are shown in the figure. The element stiffness is given by equation (5.5) and the kinematically equivalent loads from (2.29(b)) are

$$\mathbf{P}_g = \{-wl/2 \quad -wl^2/12 \quad -wl/2 \quad wl^2/12\}$$

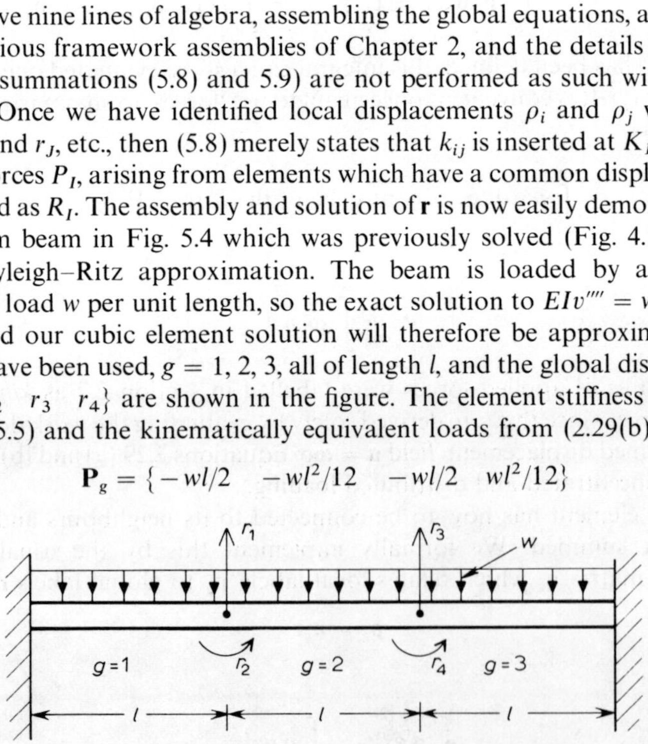

Fig. 5.4 A three-element solution

The kinematic connectivity relationships, and the boundary conditions are as follows:

$$\begin{aligned} g=1:&\quad \rho_1=0,\quad \rho_2=0,\quad \rho_3=r_1,\quad \rho_4=r_2\\ g=2:&\quad \rho_1=r_1,\quad \rho_2=r_2,\quad \rho_3=r_3,\quad \rho_4=r_4\\ g=3:&\quad \rho_1=r_3,\quad \rho_2=r_4,\quad \rho_3=0,\quad \rho_4=0 \end{aligned} \tag{5.10}$$

The assembled global stiffness matrix is therefore:

$$\mathbf{K}=\left[\begin{array}{c|c|c|c} 6+6 & -3l+3l & -6 & 3l \\ =12 & =0 & =-6 & =3l \\ \hline & 2l^2+2l^2 & -3l & l^2 \\ & =4l^2 & =-3l & =l^2 \\ \hline & & 6+6 & -3l+3l \\ \text{symm.} & & =12 & =0 \\ \hline & & & 2l^2+2l^2 \\ & & & =4l^2 \end{array}\right]$$

and the assembled load matrix is:

$$\mathbf{R}=\left\{\begin{array}{c|c|c|c} \dfrac{-wl}{2}-\dfrac{wl}{2} & \dfrac{wl^2}{12}-\dfrac{wl^2}{12} & \dfrac{-wl}{2}-\dfrac{wl}{2} & \dfrac{wl^2}{12}-\dfrac{wl^2}{12} \\ =-wl & =0 & =-wl & =0 \end{array}\right\}$$

The routine nature of these assemblies is clearly suited to computers. Had the elements been of different rigidities or lengths, the summation would still have been straightforward. Solving $\mathbf{Kr}=\mathbf{R}$, we find

$$\mathbf{r}=\{r_1\quad r_2\quad r_3\quad r_4\}=\frac{wl^3}{6EI}\{-l\quad -1\quad -l\quad 1\} \tag{5.11}$$

It is routine to return to the element level to find the displacements and stresses; for example, the displacement at the centre of element $g=2$ is obtained using (5.1), (5.10), and (5.11) as

$$v=\boldsymbol{\omega}(0)\boldsymbol{\rho}_2=\left[\frac{1}{2}\quad\frac{l}{8}\quad\frac{1}{2}\quad\frac{-l}{8}\right]\{-l\quad -1\quad -l\quad 1\}\frac{wl^3}{6EI}=-0.2083\frac{wl^4}{EI}$$

The exact solution is $-0.2109\,wl^4/EI$. The displacements along the whole beam are shown in Fig. 5.5 and are indistinguishable from the exact solution. The bending moments are not so accurate, being the second derivative of an approximate solution, that is, using (5.1),

$$M=-EI\frac{\partial^2 v}{\partial z^2}=\frac{-EI}{l^2}\frac{\partial^2\boldsymbol{\omega}}{\partial\zeta^2}\boldsymbol{\rho}_g=\frac{-EI}{l^2}[6\zeta\quad -l+3\zeta l\quad -6\zeta\quad l+3\zeta l]$$

This linear variation is seen to be a fair approximation to the exact parabola, being both greater and smaller. In fact it can be shown that the finite-element displacement solution is a least-squares fit between the exact stresses and the

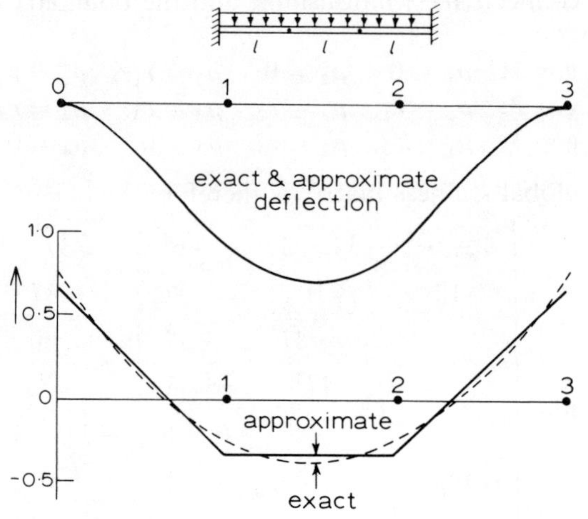

Fig. 5.5 Displacements and moments for distributed load

approximate stresses. In this example the good agreement is somewhat fortuitous since it so happens that the curvature $v''(z)$ is continuous at the nodes even though this was no part of the original assumptions. In fact, if we change the loading to a single concentrated load W at the middle of the beam, the kinematically equivalent loads (2.29(a)) become

$$\mathbf{R} = W\left\{ -\frac{1}{2} \quad -\frac{l}{8} \quad -\frac{1}{2} \quad \frac{l}{8} \right\}$$

and this time the solution in Fig. 5.6 is seen to be less accurate. The displacements are still a good approximation; the maximum displacement at the single load is

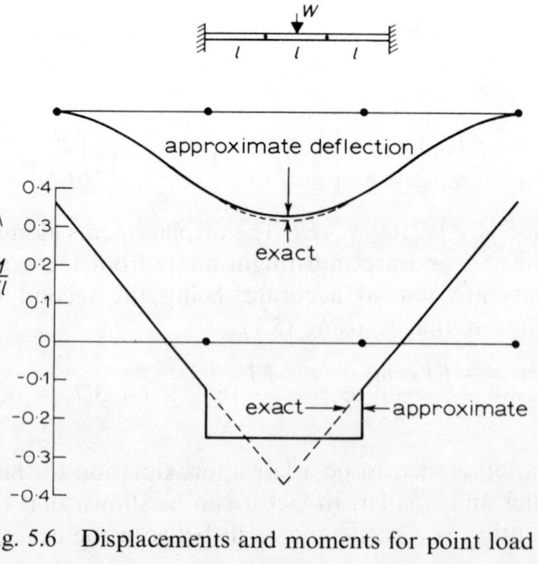

Fig. 5.6 Displacements and moments for point load

just apparent as a lower bound. The bending moment is now discontinuous at the nodes, although it is clearly not possible for the linear approximations to represent the exact variation any better than this. (A two-element solution would obviously be exact in this example.) This then is one of the important features of a finite-element displacement solution – the stresses, unlike displacements, may be discontinuous at the element interfaces. The question then arises whether we should attempt to do anything about this. There are several ways of overcoming this feature without resorting to changing the nature of the element itself. One way is simply to increase the number of elements, in the region of unacceptable discontinuities, until they are small. This is the most respectable ploy since it uses the discontinuities as a diagnostic warning that the idealization may be too coarse. Unfortunately, in the real commercial world the customer may not be able to afford a second, more refined solution. It is quite common therefore for the commercial finite-element packages to have a post-processor which contains a stress-smoothing algorithm to iron out the discontinuities. Neither the efficiency nor the morality of doing this concerns us here.

Another way is to choose higher-order interpolation functions which use second derivatives as nodal freedoms and so ensure continuity of bending moment. At least this would work if EI was constant as well: if EI was discontinuous then it would ensure discontinuous bending moments as in Problem 4.2. However, using continuity of an order higher than simple kinematics requires is not popular. Such elements tend to be overstiff in a solution which is too stiff anyway, and a little flexibility is a good thing. A final alternative is of course to use a force or mixed method which uses nodal bending moments as unknowns (Chapter 6). However, this question of the continuity of displacements and the stresses is a little more profound, as we see when we turn to two dimensions.

5.3 Two-dimensional elements

The simplest form of two-dimensional structure is the flat plate in a state of plane stress ($\sigma_{xx}, \sigma_{yy}, \sigma_{xy}; \sigma_{xz} = \sigma_{yz} = \sigma_{zz} = 0$). Fortunately, it is also very common: many parts of quite complex structures are thin flat plates or assemblies of plates. These plates are best used if they are in a state of plane (membrane) stress; indeed it could be a measure of bad design if the plate is subjected to bending also. We will look at a simple rectangular element first, and then by discerning its limitations, extend our modelling to other sorts.

Consider the rectangular element of Fig. 5.7. We wish to express the displacements $\mathbf{u} = \{u \quad v\}$ throughout the element in terms of the displacements $\boldsymbol{\rho} = \{\rho_1 \quad \rho_2 \quad \ldots \quad \rho_8\}$ at the four corner nodes. Taking just one component, say u, we wish our expansion to have the four labelled values at the corner nodes, so we should be able to use a four-term polynomial

$$u = a_1 + a_2 x + a_3 y + a_4 xy \tag{5.12}$$

On substituting successively

$$u = \rho_1, \rho_3, \rho_5, \quad \text{and} \quad \rho_7$$

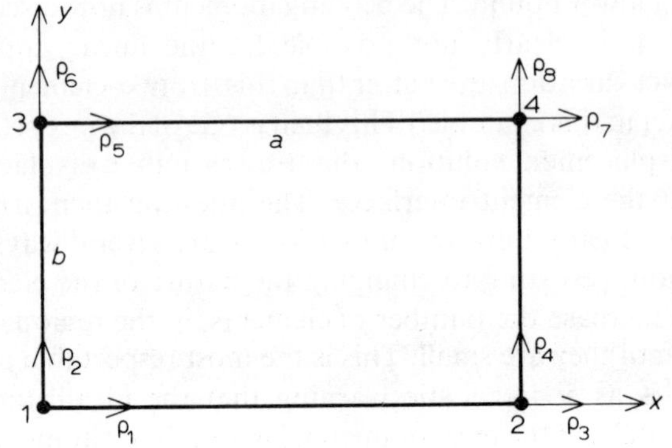

Fig. 5.7 A rectangular element

at

$$(x, y) = (0, 0), (a, 0), (0, b), (a, b)$$

we solve for the above coefficients as

$$a_1 = \rho_1; \qquad a_2 = (\rho_3 - \rho_1)/a;$$
$$a_3 = (\rho_5 - \rho_1)/b; \qquad a_4 = (\rho_7 - \rho_3 - \rho_5 + \rho_1)/ab$$

On substituting back into 5.12),

$$u = \left(1 - \frac{x}{a}\right)\left(1 - \frac{y}{b}\right)\rho_1 + \left(\frac{x}{a}\right)\left(1 - \frac{y}{b}\right)\rho_3 + \left(\frac{y}{b}\right)\left(1 - \frac{x}{a}\right)\rho_5 + \left(\frac{x}{a}\right)\left(\frac{y}{b}\right)\rho_7$$

On looking closely at this expression we now see a more methodical way of generating our shape functions which avoids having to invert expressions like (5.12). Firstly, we should again use non-dimensional x and y where possible; and secondly, the shape functions for a specific node can be deduced directly in orthogonal coordinates when we realize that they must be zero on all sides except the two meeting at that relevant node. So we move the origin and choose $\zeta_1 = 2x/a, \zeta_2 = 2y/b$ as in Fig. 5.8. The equations of the four sides are shown, so it is now possible to write

$$u(\zeta_1, \zeta_2) = \omega_1\rho_1 + \omega_2\rho_3 + \omega_3\rho_5 + \omega_4\rho_7$$

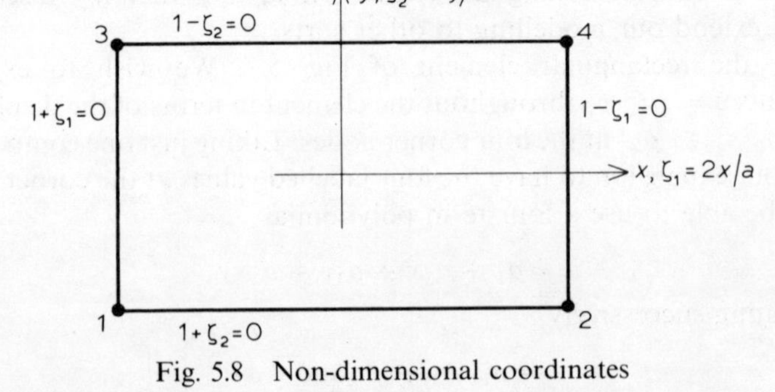

Fig. 5.8 Non-dimensional coordinates

where the four interpolation functions $\omega(\zeta_1, \zeta_2)$ associated with the numbered nodes are

$$\omega_1 = \tfrac{1}{4}(1-\zeta_1)(1-\zeta_2)$$
$$\omega_2 = \tfrac{1}{4}(1+\zeta_1)(1-\zeta_2)$$
$$\omega_3 = \tfrac{1}{4}(1-\zeta_1)(1+\zeta_2)$$
$$\omega_4 = \tfrac{1}{4}(1+\zeta_1)(1+\zeta_2)$$

These (Lagrange) interpolation functions will do for both u and v so we may write

$$\mathbf{u} = \begin{bmatrix} u \\ v \end{bmatrix} = \begin{bmatrix} \omega_1 & 0 & \omega_2 & 0 & \omega_3 & 0 & \omega_4 & 0 \\ 0 & \omega_1 & 0 & \omega_2 & 0 & \omega_3 & 0 & \omega_4 \end{bmatrix} \mathbf{p}_g$$

or

$$\mathbf{u} = \boldsymbol{\omega}\mathbf{p}_g \tag{5.13}$$

This simple technique can be readily extended to other shapes and higher-order fields. The next three steps are familiar: apply compatibility $\boldsymbol{\varepsilon} = \mathbf{D}^t\mathbf{u}$, state the stress–strain law $\boldsymbol{\sigma} = \boldsymbol{\kappa}\boldsymbol{\varepsilon}$, and use the PVD to find the element stiffness $\mathbf{k}_g$. In so doing we will construct the individual steps of the process, but will not carry them out algebraically with the aim of producing a final explicit algebraic expression for $\mathbf{k}_g$. We must recognize that finite elements contain much geometrical information and it is often pointless to assemble a final answer in terms of nodal coordinates, material properties, etc. It is easier to understand, and to programme, the individual steps (some of which may have to be numerical anyway) and then perform the steps in sequence.

The first step is

$$\boldsymbol{\varepsilon} = \mathbf{D}^t\mathbf{u} = \mathbf{D}^t\boldsymbol{\omega}\boldsymbol{\rho}_g$$

thus

$$\boldsymbol{\varepsilon} = \boldsymbol{\alpha}\mathbf{p}_g$$

where

$$\boldsymbol{\alpha} = \mathbf{D}^t\boldsymbol{\omega} \tag{5.14}$$

In this example

$$\boldsymbol{\alpha} = \begin{bmatrix} \dfrac{\partial}{\partial x} & 0 \\ 0 & \dfrac{\partial}{\partial y} \\ \dfrac{\partial}{\partial y} & \dfrac{\partial}{\partial x} \end{bmatrix} \boldsymbol{\omega} = \begin{bmatrix} \dfrac{1}{a}\dfrac{\partial}{\partial \zeta_1} & 0 \\ 0 & \dfrac{1}{b}\dfrac{\partial}{\partial \zeta_2} \\ \dfrac{1}{b}\dfrac{\partial}{\partial \zeta_2} & \dfrac{1}{a}\dfrac{\partial}{\partial \zeta_1} \end{bmatrix} \boldsymbol{\omega}$$

$$
=\frac{1}{4}\begin{bmatrix} -\frac{1}{a}(1-\zeta_2) & 0 & \frac{1}{a}(1-\zeta_2) & 0 & -\frac{1}{a}(1+\zeta_2) & 0 & \frac{1}{a}(1+\zeta_2) & 0 \\ 0 & -\frac{1}{b}(1-\zeta_1) & 0 & -\frac{1}{b}(1+\zeta_1) & 0 & \frac{1}{b}(1-\zeta_1) & 0 & \frac{1}{b}(1+\zeta_1) \\ -\frac{1}{b}(1-\zeta_1) & -\frac{1}{a}(1-\zeta_2) & -\frac{1}{b}(1+\zeta_1) & \frac{1}{a}(1-\zeta_2) & \frac{1}{b}(1-\zeta_1) & -\frac{1}{a}(1-\zeta_2) & \frac{1}{b}(1+\zeta_1) & \frac{1}{a}(1+\zeta_2) \end{bmatrix} \quad (5.15)
$$

Because the element has a plane stress field it is necessary to put $\sigma_{zz} = \sigma_{zx} = \sigma_{zy} = 0$ in (2.1), and after inverting, the isotropic material stiffness is

$$
\boldsymbol{\kappa} = \frac{E}{(1-v^2)}\begin{bmatrix} 1 & v & 0 \\ v & 1 & 0 \\ 0 & 0 & \dfrac{1-v}{2} \end{bmatrix} \quad (5.16)
$$

The internal virtual work

$$
\int_{V_g} \boldsymbol{\sigma}^{\mathrm{t}}\bar{\boldsymbol{\varepsilon}}\,\mathrm{d}v = \int_{V_g} \boldsymbol{\varepsilon}^{\mathrm{t}}\boldsymbol{\kappa}\bar{\boldsymbol{\varepsilon}}\,\mathrm{d}v = \boldsymbol{\rho}_g^{\mathrm{t}}\int_{V_g} \boldsymbol{\alpha}^{\mathrm{t}}\boldsymbol{\kappa}\boldsymbol{\alpha}\,\mathrm{d}v\bar{\boldsymbol{\rho}}_g
$$

and the element stiffness

$$
\mathbf{k}_g = \int_{V_g} \boldsymbol{\alpha}^{\mathrm{t}}\boldsymbol{\kappa}\boldsymbol{\alpha}\,\mathrm{d}v \quad (5.17)
$$

is the same as in the beam (5.4) except that now $\boldsymbol{\kappa}$ is a matrix in view of the three stress components. The external virtual work in the presence of body forces $\mathbf{p}_v$ and surface forces $\mathbf{p}_s$ (if an element edge S_g has surface forces) is equal to

$$
\int_{V_g} \mathbf{p}_v^{\mathrm{t}}\bar{\mathbf{u}}\,\mathrm{d}v + \int_{S_g} \mathbf{p}_s^{\mathrm{t}}\bar{\mathbf{u}}\,\mathrm{d}s = \int_{V_g} \mathbf{p}_v^{\mathrm{t}}\boldsymbol{\omega}\,\mathrm{d}v\bar{\boldsymbol{\rho}}_g + \int_{S_g} \mathbf{p}_s^{\mathrm{t}}\boldsymbol{\omega}\,\mathrm{d}s\bar{\boldsymbol{\rho}}_g
$$
$$
= \left[\int_{V_g} \mathbf{p}_v^{\mathrm{t}}\boldsymbol{\omega}\,\mathrm{d}v + \int_{S_g} \mathbf{p}_s^{\mathrm{t}}\boldsymbol{\omega}\,\mathrm{d}s\right]\bar{\boldsymbol{\rho}}_g
$$

The kinematically equivalent forces are therefore

$$
\mathbf{P}_g = \int_{V_g} \boldsymbol{\omega}^{\mathrm{t}}\mathbf{p}_v\,\mathrm{d}v + \int_{S_g} \boldsymbol{\omega}^{\mathrm{t}}\mathbf{p}_s\,\mathrm{d}s \quad (5.18)
$$

The assembly of $\mathbf{k}_g$ and $\mathbf{P}_g$ into $\mathbf{K}$ and $\mathbf{R}$ then follows as usual.

This is a very simple element with several shortcomings, but the stiffness integral $\mathbf{k}_g$ can actually be evaluated in closed form if desired. For other elements we look at, it will be necessary to evaluate $\mathbf{k}_g$ numerically by Gaussian quadrature in which it is necessary only to evaluate $\alpha(\zeta_1, \zeta_2)$ at a few selected (Gauss) points before finding the product $\boldsymbol{\alpha}^{\mathrm{t}}\boldsymbol{\kappa}\boldsymbol{\alpha}$ and then performing a quick summation. This procedure has become routine in commercial systems, even for simple elements (see Appendix D).

Two things should be satisfied when modelling a finite element, if we expect the answers to converge to the correct values as the mesh is refined and the element sizes reduced. Thus the strain field should firstly contain uniform components which will be the asymptotic nature of very small elements. Equation (5.15) shows that these are present: for example, the strains

$$-(1-\zeta_2)\rho_1/a + (1-\zeta_2)\rho_3/a \qquad \text{produce} \quad (\rho_3 - \rho_1)/a$$

Also, although our approximate formulation may not satisfy equilibrium exactly at a point, it is desirable that equilibrium *in the mean* is satisfied, so a single element should be in equilibrium with the resultant (if any) of the body and surface forces acting upon it. Now the overall equilibrium equations of no resultant force or no resultant moment acting upon an element arise from the coefficients of the constant virtual displacement and virtual rotation terms, so the assumed displacement field should contain these *rigid-body movements.* A small rigid-body movement, including rotation θ, is represented by $\{u, v\} = \{a + \theta y, b - \theta x\}$. On putting $u = a$, ie: $\rho_i = a$ for all nodes, we have $a = \sum \omega_i a$ or $\sum \omega_i = 1$. The $\{\theta y, -\theta x\}$ terms produce the requirement that $\sum \zeta_{1i}\omega_i = \zeta_1$ and $\sum \zeta_{2i}\omega_i = \zeta_2$. The reader may care to check that these conditions are met by the interpolation functions ω_i in equation (5.13). There remain several deficiencies in this simple element.

Firstly, the displacement field is not complete: there are constant terms and linear terms but only xy of second order. The absence of x^2 and y^2 means that the element is not isotropic, so if we change variables from $\{x, y\}$ to $\{x', y'\}$ inclined to $\{x, y\}$, knowing that the latter are a linear combination of the former, the expression for **u** will not look the same in $\{x', y'\}$. Thus the element may be sensitive or insensitive to certain directional strain fields. It is not easy to choose element displacement fields which are complete polynomials as can be seen when we examine the so-called Pascal triangle:

$$\begin{array}{ccccccccc} & & & & x^\circ & & & & \\ & & & x^1 & & y^1 & & & \\ & & x^2 & & xy & & y^2 & & \\ & x^3 & & x^2y & & xy^2 & & y^3 & \\ x^4 & & x^3y & & x^2y^2 & & xy^3 & & y^4 \end{array}$$

The four-noded rectangle uses four terms for both u and v and employs the first two rows plus xy from row 3. We would have to go to terms of order six, that is the seventh row, to achieve a multiple of four. This means a displacement field with 28 unknowns so we would need 28 nodes to define such a field! The insertion of interior nodes will liberate us from the multiple of four, but again no easy symmetric choice is available. In four-sided figures it is therefore necessary to forego completeness. On the other hand, multiples of three are readily available for *triangular* elements, but if we also make the number of nodes per side match the degree of the displacement function – so that a unique displacement field is matched between elements – then this reduces our choice to a meagre three- or six-noded triangle.

Also, a rectangular shape is very restricted and suitable only for regular rectangular plates; even then it is not easy to use a fine mesh locally without committing ourselves to the same degree of fineness everywhere. The only feasible way to introduce flexibility into the mesh would be to employ transition elements with fewer nodes on one face than the opposite. A triangular element is much more adaptable in this respect, or a four-sided element with curved sides – we will develop these alternatives.

A linear displacement field produces restricted variations in stress and strain so that small elements are likely to be needed to avoid large discontinuities in stress between elements. One way of increasing the order of the displacement field would be to use the previous Lagrange interpolation and *nine* nodes at the intersections of the lines $\zeta_1 = 0$, $1 + \zeta_1 = 0$, $1 - \zeta_1 = 0$, $\zeta_2 = 0$, $1 + \zeta_2 = 0$, $1 - \zeta_2 = 0$. An internal node at $\zeta_1 = \zeta_2 = 0$ is not very popular as it complicates the data generation and processing and also disproportionately increases the bandwidth of **K**. A universally popular element is therefore the eight-noded, four-sided rectangle shown in Fig. 5.9. Noting the equations of the new dotted lines, the interpolation functions for the numbered nodes will be:

$$
\begin{aligned}
\omega_1 &= -\tfrac{1}{4}(1-\zeta_1)(1-\zeta_2)(1+\zeta_1+\zeta_2) & \omega_2 &= \tfrac{1}{2}(1-\zeta_1)(1+\zeta_1)(1-\zeta_2)\\
\omega_3 &= -\tfrac{1}{4}(1+\zeta_1)(1-\zeta_2)(1-\zeta_1+\zeta_2) & \omega_4 &= \tfrac{1}{2}(1-\zeta_1)(1-\zeta_2)(1+\zeta_2)\\
\omega_5 &= \tfrac{1}{2}(1+\zeta_1)(1-\zeta_2)(1+\zeta_2) & \omega_6 &= -\tfrac{1}{4}(1-\zeta_1)(1+\zeta_2)(1+\zeta_1-\zeta_2)\\
\omega_7 &= \tfrac{1}{2}(1-\zeta_1)(1+\zeta_1)(1+\zeta_2) & \omega_8 &= -\tfrac{1}{4}(1+\zeta_1)(1+\zeta_2)(1-\zeta_1-\zeta_2)
\end{aligned}
$$

$$\boldsymbol{\alpha} = \mathbf{D}^{\mathrm{t}}\boldsymbol{\omega} \quad \text{then follows}$$

Before looking at some examples using these elements, it is worth while looking exactly at what we have done in making the element approximation in two dimensions and – equally important – what we do when we sum the contributions from all elements.

5.4 The finite-element approximation and relaxed continuity

We summarize the story so far, before conducting a postmortem. Firstly, state the PVD as a summation:

$$\sum_{g=1,2,\ldots} \int_{V_g} \boldsymbol{\sigma}^{\mathrm{t}}\bar{\boldsymbol{\varepsilon}}\,\mathrm{d}v = \sum_{g=1,2,\ldots} \left(\int_{V_g} \mathbf{p}_v^{\mathrm{t}}\bar{\mathbf{u}}\,\mathrm{d}v + \int_{S_g} \mathbf{p}_s^{\mathrm{t}}\bar{\mathbf{u}}\,\mathrm{d}s \right)$$

Then assume

$$\mathbf{u} = \boldsymbol{\omega}\boldsymbol{\rho}_g$$

enforce

$$\boldsymbol{\varepsilon} = \mathbf{D}^{\mathrm{t}}\mathbf{u} = \mathbf{D}^{\mathrm{t}}\boldsymbol{\omega}\boldsymbol{\rho}_g = \boldsymbol{\alpha}\boldsymbol{\rho}_g$$

and put

$$\boldsymbol{\sigma} = \boldsymbol{\kappa}\boldsymbol{\varepsilon}$$

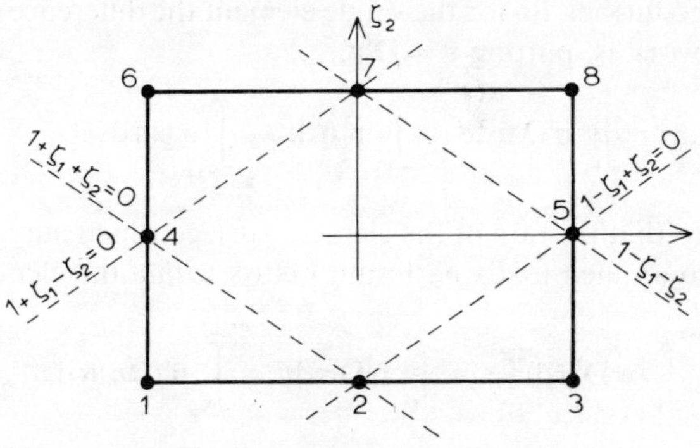

Fig. 5.9 An eight-noded rectangle

Evaluate element integrals:

$$\mathbf{k}_g = \int_{V_g} \boldsymbol{\alpha}^t \boldsymbol{\kappa} \boldsymbol{\alpha}\, dv; \quad \mathbf{P}_g = \int_{V_g} \boldsymbol{\omega}^t \mathbf{p}_v\, dv + \int_{S_g} \boldsymbol{\omega}^t \mathbf{p}_s\, ds$$

Connect all elements:

$$\boldsymbol{\rho}_g = \mathbf{a}_g \mathbf{r}$$

Carry out the summation:

$$\left[\mathbf{r}^t \sum_g \mathbf{a}_g^t \mathbf{k}_g \mathbf{a}_g\right] \bar{\mathbf{r}} = \left[\sum_g \mathbf{P}_g^t \mathbf{a}_g\right] \bar{\mathbf{r}}, \qquad \text{or} \qquad \mathbf{Kr} = \mathbf{R}$$

When discussing the consequences of approximate solutions using the Rayleigh–Ritz and global expansions in the PVD, we started off with the PVD for the whole structure, replaced $\bar{\boldsymbol{\varepsilon}}$ by $\mathbf{D}^t\bar{\mathbf{u}}$ and used the Gauss identity to convert the term $\boldsymbol{\sigma}^t\mathbf{D}^t\mathbf{u}$, so obtaining the Galerkin or weighted residual form

$$\int_V \bar{\mathbf{u}}^t(\mathbf{D}\boldsymbol{\sigma} + \mathbf{p}_v)\, dv - \int_{S_p} \bar{\mathbf{u}}^t[(\mathbf{D}n)\boldsymbol{\sigma} - \mathbf{p}_s]\, ds = 0 \tag{1.72}$$

But we must not be tempted to replace the first term of this expression by a sum of all element volume integrals and the second term by a summation around the entire surface S_p, because the Gauss theorem cannot be applied to functions which have discontinuous derivatives. Suppose then that our finite-element model uses interpolation functions $\boldsymbol{\omega}$ which ensures that displacements are continuous at element interfaces, between nodes and not just at them. Such elements are known as kinematically *conformal* and not all elements are so. But we do not insist that stresses and strains are continuous, in other words the gradients $\mathbf{D}^t\mathbf{u}$ of displacements $\mathbf{u}$ may be discontinuous. It is not possible therefore to apply the Gauss identity across the whole structure embracing all elements at once. However, we may apply it across a *single* element and convert the element volume integral of $\boldsymbol{\sigma}^t\mathbf{D}^t\bar{\mathbf{u}}$ in V_g to a form containing $\bar{\mathbf{u}}^t(\mathbf{D}n)\boldsymbol{\sigma}$ on the surface S_g – and no further. So for the single element the difference in internal and external virtual work is, putting $\bar{\boldsymbol{\varepsilon}} = \mathbf{D}^t\bar{\mathbf{u}}$,

$$\int_{V_g} \boldsymbol{\sigma}^t\mathbf{D}^t\bar{\mathbf{u}}\,\mathrm{d}v - \int_{V_g} \mathbf{p}_v^t\bar{\mathbf{u}}\,\mathrm{d}v - \int_{S_{gp}} \mathbf{p}_s^t\bar{\mathbf{u}}\,\mathrm{d}s$$

where S_{gp} refers to that portion of the element surface which may be a structural surface subject to applied loads $\mathbf{p}_s$. Using Gauss *within* the element,

$$\int_{V_g} \boldsymbol{\sigma}^t\mathbf{D}^t\bar{\mathbf{u}}\,\mathrm{d}v = -\int_{V_g} \bar{\mathbf{u}}^t\mathbf{D}\boldsymbol{\sigma}\,\mathrm{d}v + \int_{S_g} \bar{\mathbf{u}}^t(\mathbf{D}n)\boldsymbol{\sigma}\,\mathrm{d}s$$

the previous difference becomes

$$-\int_{V_g} \bar{\mathbf{u}}^t[\mathbf{D}\boldsymbol{\sigma} + \mathbf{p}_v]\,\mathrm{d}v + \int_{S_g} \bar{\mathbf{u}}^t[(\mathbf{D}n)\boldsymbol{\sigma}]\,\mathrm{d}s - \int_{S_{gp}} \mathbf{p}_s^t\bar{\mathbf{u}}\,\mathrm{d}s$$

By putting the total sum of all these expressions, over all elements, to zero for a large number of virtual displacements $\bar{\mathbf{u}} = \boldsymbol{\omega}\bar{\boldsymbol{\rho}}_g$, we know we are trying to force the residual $\mathbf{D}\boldsymbol{\sigma} + \mathbf{p}_v$ to zero within each element; that is we satisfy equilibium in the mean over V_g. Turning now to the surface terms, when we sum all elements' contributions we may bring together two elements' surface integrals with the same portion of common interface, labelled with the normal n in Fig. 5.10. This

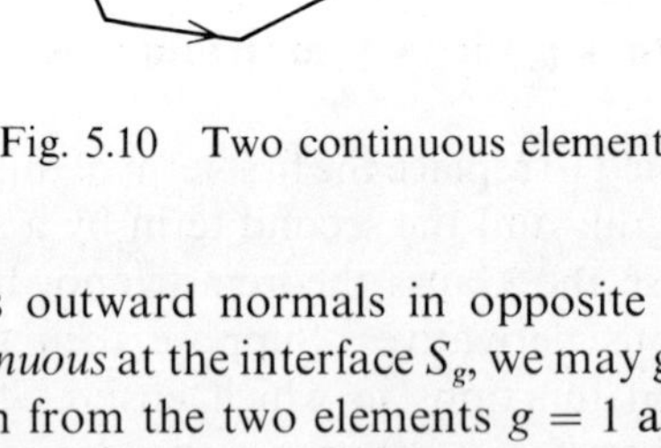

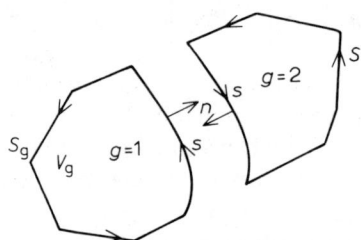

Fig. 5.10 Two continuous elements

common interface has outward normals in opposite senses, so *providing the displacements are continuous* at the interface S_g, we may group together the above weighted residual form from the two elements $g = 1$ and $g = 2$, as

$$\int_{S_g} \bar{\mathbf{u}}^t[(\mathbf{D}n)\boldsymbol{\sigma}_1 - (\mathbf{D}n)\boldsymbol{\sigma}_2]\,\mathrm{d}s - \int_{S_{gp}} \mathbf{p}_s^t\bar{\mathbf{u}}\,\mathrm{d}s$$

So putting this integral to zero over a whole series of displacements $\bar{\mathbf{u}} = \boldsymbol{\omega}\bar{\boldsymbol{\rho}}_g$ on S_g will be attempting to satisfy in the mean

$$(\mathbf{D}n)\boldsymbol{\sigma}_1 - (\mathbf{D}n)\boldsymbol{\sigma}_2 - \mathbf{p}_s = \mathbf{0}$$

If there is no external loading on any part of this interface, then the tractions – or resolved components of stress – $(\mathbf{D}n)\boldsymbol{\sigma}$ will be continuous in the mean, even though they may be discontinuous at any point. We now see why the displacements should be continuous but the stresses may not be. In other words, the finite-element version of the PVD is capable of handling relaxed continuity conditions on stress but not on displacement. The above equation is also meaningful only if $\boldsymbol{\sigma}$ is finite, so we expect that if $\mathbf{u}$ is continuous, the stresses $\boldsymbol{\sigma} = \boldsymbol{\kappa}\mathbf{D}^t\mathbf{u}$ must be finite. Thus the assumed displacement field can have discontinuous gradients but they must be finite. The concept of discontinuous tractions across an element surface should be seen as a consistent feature of our approximation, for it is logical to satisfy inter-element equilibrium only 'in the mean' when that is precisely what we are doing inside the element.

Two examples will illustrate these features using our simple rectangular elements.

Firstly, we see in Fig. 5.11 a finite-element solution of the diffusion problem previously solved in Chapter 3 (Fig. 3.40) using the PVF. Eight-noded elements have been used, connected to three-noded bars on the edges. Only the centre

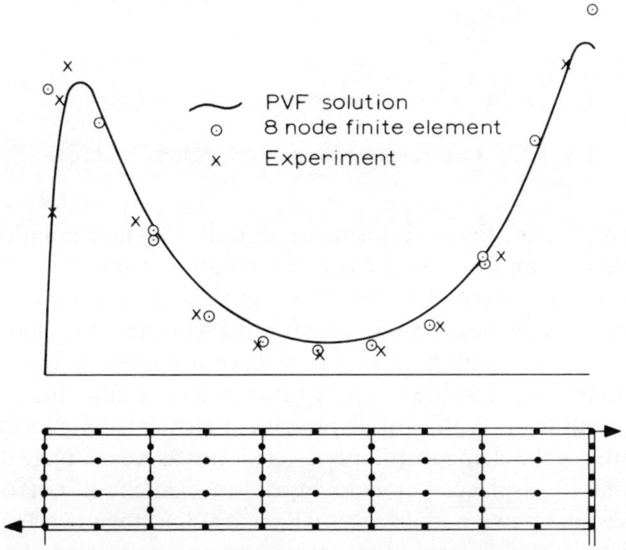

Fig. 5.11 Shear stresses in a diffusion panel

element shears are shown since these are the stresses most difficult to model near the free end. The finite-element solutions have completely missed the rapid fall to zero but elsewhere are more accurate than the PVF solution.

The second example, in Fig. 5.12, shows a uniform cantilever beam subjected to tip forces; the whole beam being divided into three elements. Only the maximum bending stress σ_{xx} on the upper surface is shown, and three cases (a), (b), and (c) show the effect of increasing the aspect ratio (length : depth) of the elements, using both the four-node and eight-node models. The eight-node element with

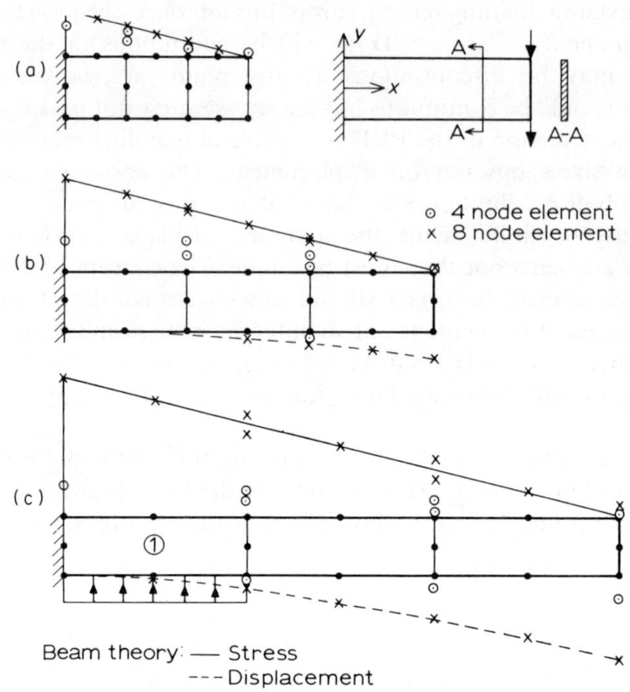

Fig. 5.12 Cantilever beam and poor element choice

quadratic displacement fields performs quite well. The interpolation functions are not complete, so for example the bending strain ε_{xx} varies linearly with x but quadratically with y; nevertheless, the stress agrees well with simple beam theory even in the short stocky beam where constraint effects near the clamped end are small for this loading. The biggest stress discontinuities in these eight-node predictions occur with the high aspect-ratio elements and this characteristic reduction in accuracy is typical of displacement elements. Numerically we find the conditioning of the element stiffness deteriorates as we elongate it; physically this means that a displacement field inside an element is not defined with sufficient precision by two nodal values which are close together. However, if we take the average of the discontinuous stresses at any node, then we see that the mean value is virtually exact; and it is common for finite-element programmes simply to take the arithmetic mean of stresses at the same node point.

Turning to the four-node element solutions, we see that they are a numerical disaster in all but the short stocky beam, and they illustrate the dangers of using a displacement field which cannot accurately simulate the real one. This linear displacement field is inadequate when confronted with a slender beam because the bending deformations will be much greater than the shear deformations. If a four-node element is called upon to simulate a pure bending mode like Fig. 5.13(a) then it can only respond in the symmetrical pattern shown in Fig. 5.13(b). If we feed this latter displacement pattern into the strains $\boldsymbol{\varepsilon} = \boldsymbol{\alpha\rho}$ (5.15) we find

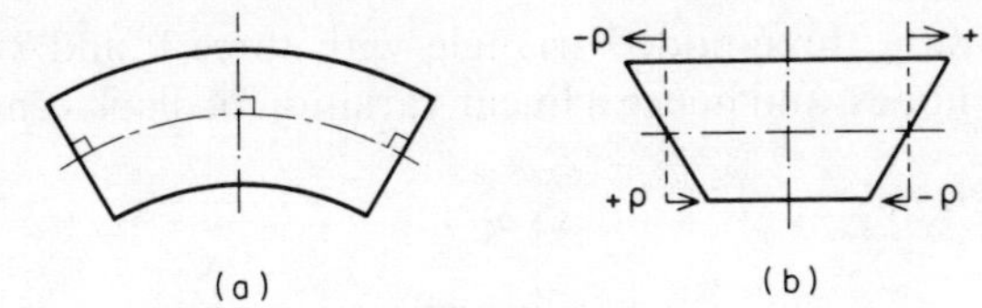

Fig. 5.13

that the maximum shear strains are of the same order as the bending strains whereas they should be zero. The effect of this constraining action is to make the element inordinately stiff when flexible bending action should dominate – hence the displacements and stresses in Fig. 5.12c which are only about a quarter of their true values. Now the bending stresses σ_{xx}, although far too small, do vary linearly through zero on the centre-line of the beam, and hence do add up to a pure couple, but the couple is far too small to equilibrate the moment due to the tip loads. Similarly, the shear stress σ_{yx} through the depth will not add up to the applied shear. But we know that the finite element model will satisfy equilibrium in the mean, so the question arises 'Where has the applied load gone?' The answer lies in the residual stresses on faces which should have been stress-free. The compressive stress σ_{yy} is shown on the lower face of element 1 in Fig. 5.12c and there are similar nonzero residuals on 'stress-free' faces. It is these errors which are keeping the element in equilibrium.

The rectangular element serves our purpose of displaying the virtues and shortcomings of the finite-element idealization, but it is pretty incompetent at fitting into awkward shapes. Even in a rectangular region, if there is a local area of stress concentration, then a cluster of small rectangles will result in a mesh whose local fineness propogates in two orthogonal directions everywhere. We therefore need a more geometrically versatile element such as a triangle or a curvilinear quadrilateral.

5.5 Triangular elements

The triangular element is much more adaptable than the rectangle, and it allows the user to tailor the element mesh in an infinite number of ways to suit any structural geometry or likely distribution of stress concentrations. A large number of small elements can be crowded into a region of expected high stress gradients, and uniformly stressed regions can be left with a small number of large elements – this geometrical or topographical freedom is not available to the same extent with four-sided quadrilaterals examined in the next section, although it must be admitted that a quadrilateral is computationally more efficient than two component triangles with the same total degrees of freedom. The very simple triangular element with *constant strain* has a long and honourable history[49] and is still popular today for solving inelastic problems where the stress–strain relationship is not constant and the structure has to be loaded incrementally, updating the current stiffness at each small increment. The updating of an element stiffness is much less expensive if the stress field does not vary through it.

Figure 5.14 shows a three-noded triangle with three u and v displacement components at the nodes, and hence a linear variation of displacements across it.

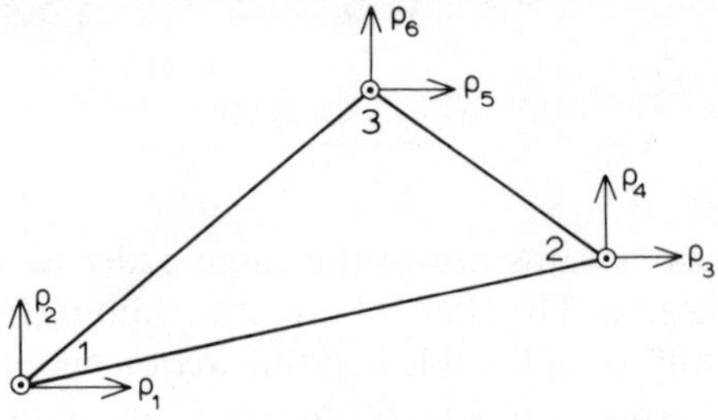

Fig. 5.14 Three-noded triangle

A choice of suitable non-dimensional coordinates ζ is not immediately obvious so we could resort to the method of equation (5.12) and write

$$u = a_1 + a_2 x + a_3 y \tag{5.19}$$

On substituting $u = \rho_i$ at (x_i, y_i) for $i = 1, 2, 3$, we generate the three equations for a_i as

$$\begin{bmatrix} 1 & x_1 & y_1 \\ 1 & x_2 & y_2 \\ 1 & x_3 & y_3 \end{bmatrix} \begin{bmatrix} a_1 \\ a_2 \\ a_3 \end{bmatrix} = \begin{bmatrix} \rho_1 \\ \rho_1 \\ \rho_3 \end{bmatrix} \tag{5.20}$$

Knowing the coordinates (x_i, y_i), these equations can be inverted. Notice that if two vertices are too close then two of the rows in the above matrix will become nearly identical and a singular solution is obtained. This confirms our assertion that equilateral triangles are desirable and high aspect ratios to be avoided. An alternative to inversion is to move node 1 to the origin and align side 1–2 along the x-axis. This simplifies the solution of (5.20) and the resulting element stiffness matrix can always be transformed by rotating it through an angle θ and using (2.34), $\mathbf{k}_g = \mathbf{T}^t\mathbf{k}\mathbf{T}$. So if we put $x_1 = y_1 = y_2 = 0$ in (5.20) and solve for a_1, a_2, a_3, and substitute back into (5.19), we find

$$\begin{aligned} u &= \left[1 - \frac{x}{x_2} - \left(1 - \frac{x_3}{x_2}\right)\frac{y}{y_3}\right]\rho_1 + \left[\frac{x}{x_2} - \left(\frac{x_3}{x_2}\right)\frac{y}{y_3}\right]\rho_3 + \left[\frac{y}{y_3}\right]\rho_5 \\ &= \boldsymbol{\omega}_1\{\rho_1 \quad \rho_3 \quad \rho_5\} \end{aligned}$$

Thus

$$\mathbf{u} = \begin{bmatrix} u \\ v \end{bmatrix} = \begin{bmatrix} \boldsymbol{\omega}_1 & \mathbf{0} \\ \mathbf{0} & \boldsymbol{\omega}_1 \end{bmatrix} \boldsymbol{\rho} = \boldsymbol{\omega}\boldsymbol{\rho}$$

The matrix $\boldsymbol{\alpha} = \mathbf{D}^t\boldsymbol{\omega}$ follows, and consists only of constants for this constant-strain triangle, so (5.17) is simply

$$\mathbf{k}_g = \boldsymbol{\alpha}^t \boldsymbol{\kappa} \boldsymbol{\alpha} A t$$

where A and t are the area and thickness of the triangular element.

Now the form of $\boldsymbol{\omega}$, even after aligning the triangle with the x-axis, is inelegant and lacks symmetry. More importantly, it is not easy to generate Lagrange-type interpolation functions for higher-order fields using a distribution of nodes equally spaced on the three sides of the triangle. We can, however, restore symmetry by introducing homogeneous or area coordinates ζ_1, ζ_2, and ζ_3 shown in Fig. 5.15. The

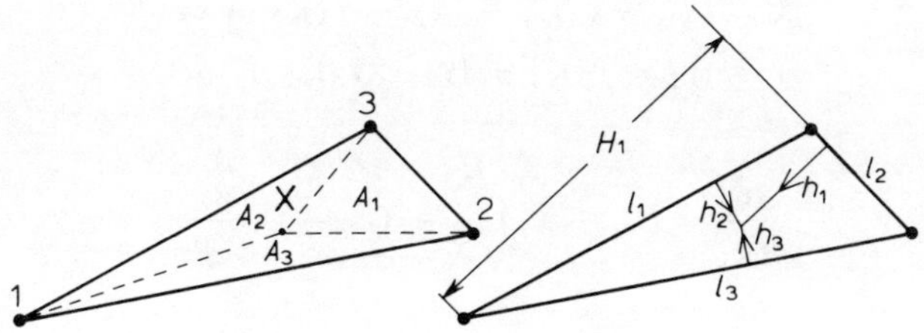

Fig. 5.15 Area coordinates

position of the point X inside the triangle can be identified by $\zeta_1 = A_1/A = h_1/H_1$, and similarly for ζ_2 and ζ_3. These three coordinates are not independent since clearly $\zeta_1 + \zeta_2 + \zeta_3 = 1$, but they do satisfy the requirement of dimensionless coordinates that $\zeta_1 = 1$ at node 1 and zero at nodes 2 and 3, etc. It is now straightforward to generate interpolation functions for any symmetrical array of nodes. Consider for example a six-noded triangle (Fig. 5.16) which will have a quadratic variation of

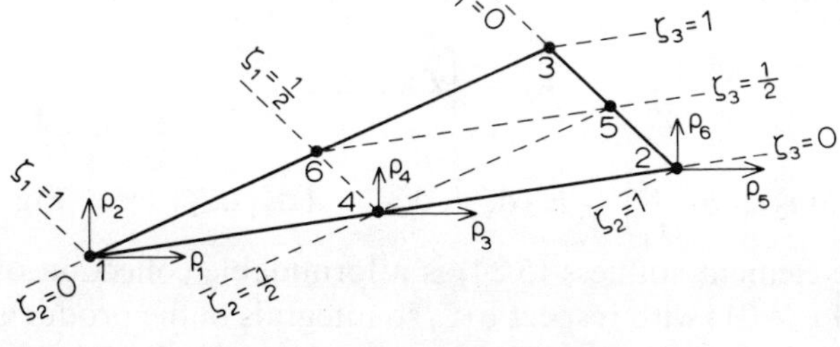

Fig. 5.16 Six-noded triangle

displacements across it. The interpolation functions which are unity at one node and zero at all others are therefore simply

$$\begin{aligned} \omega_1 &= 2(\zeta_1 - \tfrac{1}{2})\zeta_1; \quad & \omega_2 &= 2(\zeta_2 - \tfrac{1}{2})\zeta_2; \quad & \omega_3 &= 2(\zeta_3 - \tfrac{1}{2})\zeta_3 \\ \omega_4 &= 4\zeta_1\zeta_2; & \omega_5 &= 4\zeta_2\zeta_3; & \omega_6 &= 4\zeta_1\zeta_3 \end{aligned} \tag{5.21}$$

So the displacements $\mathbf{u} = \{u, v\}$ in terms of $\boldsymbol{\rho} = \{\rho_1 \quad \rho_2 \quad \ldots \quad \rho_{12}\}$ are:

$$\mathbf{u} = \boldsymbol{\omega\rho} = \begin{bmatrix} \omega_1 & 0 & \omega_4 & 0 & \omega_2 & 0 & \omega_5 & 0 & \omega_3 & 0 & \omega_6 & 0 \\ 0 & \omega_1 & 0 & \omega_4 & 0 & \omega_2 & 0 & \omega_5 & 0 & \omega_3 & 0 & \omega_6 \end{bmatrix} \boldsymbol{\rho}$$

To form $\boldsymbol{\alpha} = \mathbf{D}^t\boldsymbol{\omega}$ we need to find $\partial\omega_1/\partial x$, etc., so we must transform operators from $\{x, y\}$ to ζ_i. Now x and y are linear functions of ζ_i in this problem and ζ_i have unit values at the three nodes, so we must be able to write

$$
\begin{aligned}
x &= \zeta_1 x_1 + \zeta_2 x_2 + \zeta_3 x_3 \\
y &= \zeta_1 y_1 + \zeta_2 y_2 + \zeta_3 y_3
\end{aligned}
\tag{5.22}
$$

and recalling that $\zeta_1 + \zeta_2 + \zeta_3 = 1$,

$$
\begin{aligned}
x &= (x_1 - x_3)\zeta_1 + (x_2 - x_3)\zeta_2 + x_3 \\
y &= (y_1 - y_3)\zeta_1 + (y_2 - y_3)\zeta_2 + y_3
\end{aligned}
$$

Then

$$\frac{\partial}{\partial \zeta_1} = (x_1 - x_3)\frac{\partial}{\partial x} + (y_1 - y_3)\frac{\partial}{\partial y}$$

and

$$\frac{\partial}{\partial \zeta_2}(x_2 - x_3)\frac{\partial}{\partial x} + (y_2 - y_3)\frac{\partial}{\partial y}$$

so

$$
\begin{bmatrix} \dfrac{\partial}{\partial x} \\ \dfrac{\partial}{\partial y} \end{bmatrix} = \frac{1}{A}\begin{bmatrix} y_2 - y_3 & y_3 - y_1 \\ x_3 - x_2 & x_1 - x_3 \end{bmatrix}\begin{bmatrix} \dfrac{\partial}{\partial \zeta_1} \\ \dfrac{\partial}{\partial \zeta_2} \end{bmatrix}
\tag{5.23}
$$

where $A = (y_2 - y_3)(x_1 - x_3) - (y_1 - y_3)(x_2 - x_3)$ is the area of the triangle. The differentiations $\boldsymbol{\alpha} = \mathbf{D}^t\boldsymbol{\omega}$ can now be performed using (5.23) on (5.21), and the stiffness matrix

$$\mathbf{k}_g = \int \boldsymbol{\alpha}^t \boldsymbol{\kappa} \boldsymbol{\alpha} t \, \mathrm{d}A \tag{5.24}$$

where $\mathrm{d}A = (\mathrm{d}h_2)\left(\mathrm{d}h_1 \dfrac{l_1}{H_1}\right) = H_2 \, \mathrm{d}\zeta_2 l_1 \, \mathrm{d}\zeta_1 = A \, \mathrm{d}\zeta_1 \, \mathrm{d}\zeta_2$ (see Fig. 5.15). The integral in the element stiffness (5.24) is a formidable collection of products of derivatives of ω_i (5.21) with respect to ζ_j so integrals of the products $\zeta_i \zeta_j$ must be evaluated over the triangle. As it happens, this can be done in closed form

$$\int_A \zeta_i^a \zeta_j^b \, \mathrm{d}A = 2A \frac{a!\, b!}{(2 + a + b)!}$$

The above idea can be extended to higher-order elements using higher-order Lagrange interpolation functions in the manner of (5.21) so simplicity of the integrals looks attractive. Moreover the displacement field will contain all possible products and we have shown that completeness is a desirable attribute. Having in mind the geometrical versatility of triangles, therefore, it seems surprising that higher-order triangles are not commonly used. One reason is the presence of unpopular internal nodes, but another reason is that the rectangular element can be adapted or 'mapped' into a very versatile four-sided shape having curved sides; this makes it very competitive. The trick is to employ the same interpolation functions for the *geometry* that we have hitherto used for the displacements, and so describe any four-sided shape in terms of discrete coordinates. Both the displacement field (u, v) and the geometry (x, y) are then discretized. These 'isoparametric' elements[55] are universally popular and in fact we have already inadvertently used the germ if this idea in the triangular element by employing (5.22) which describes exactly the position of any point in the element in terms of the vertices' coordinates. The isoparametric mapping simply carries this idea further and describes the geometry approximately in democratically the same fashion as the displacements. (There are other options in which the interpolation functions for the geometry may be of different order to the displacements.)

5.6 Isoparametric curved quad

It will help to imagine an eight-noded curvilinear quadrilateral element to illustrate the isoparametric approach (see Fig. 5.17), since we do not then have to

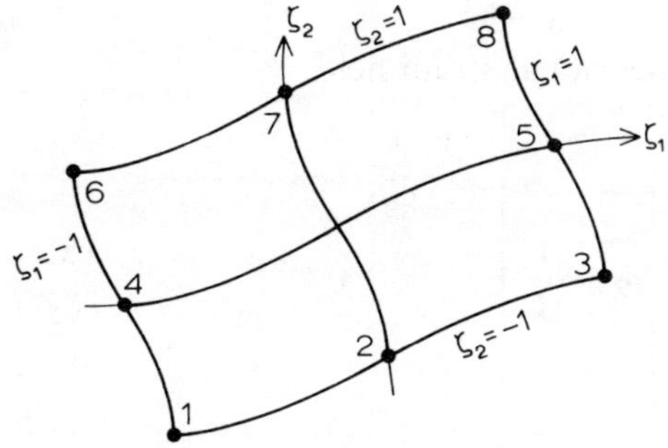

Fig. 5.17 Curvilinear coordinates

concern ourselves with the interpolation details which have already been described for the rectangular form (see Fig. 5.9). Using these again we may simply put

$$x = \omega_1 x_1 + \omega_2 x_2 + \omega_3 x_3 + \ldots + \omega_8 x_8$$

or

$$\begin{bmatrix} x \\ y \end{bmatrix} = \boldsymbol{\omega}\mathbf{x}, \qquad \text{where } \mathbf{x} = \{x_1 \quad y_1 \quad x_2 \quad y_2 \quad \ldots \quad x_8 \quad y_8\} \tag{5.25}$$

To obtain the strain $\boldsymbol{\varepsilon} = \mathbf{D}^t\mathbf{u}$ we have to form $\partial/\partial x$ and $\partial/\partial y$ using differential calculus from

$$\begin{bmatrix} \dfrac{\partial}{\partial \zeta_1} \\ \dfrac{\partial}{\partial \zeta_2} \end{bmatrix} = \begin{bmatrix} \dfrac{\partial x}{\partial \zeta_1} & \dfrac{\partial y}{\partial \zeta_1} \\ \dfrac{\partial x}{\partial \zeta_2} & \dfrac{\partial y}{\partial \zeta_2} \end{bmatrix} \begin{bmatrix} \dfrac{\partial}{\partial x} \\ \dfrac{\partial}{\partial y} \end{bmatrix} = \mathbf{J} \begin{bmatrix} \dfrac{\partial}{\partial x} \\ \dfrac{\partial}{\partial y} \end{bmatrix} \tag{5.26}$$

The array **J** is known as the Jacobian, whose elements we can form from (5.25), so

$$\begin{bmatrix} \dfrac{\partial x}{\partial \zeta_1} \\ \dfrac{\partial y}{\partial \zeta_1} \end{bmatrix} = \frac{\partial}{\partial \zeta_1}\begin{bmatrix} x \\ y \end{bmatrix} = \frac{\partial}{\partial \zeta_1}[\boldsymbol{\omega} \;\; \mathbf{x}] = \left[\frac{\partial \boldsymbol{\omega}}{\partial \zeta_1}\right]\mathbf{x}$$

Therefore

$$\mathbf{J} = \begin{bmatrix} \dfrac{\partial x}{\partial \zeta_1} & \dfrac{\partial x}{\partial \zeta_2} \\ \dfrac{\partial y}{\partial \zeta_1} & \dfrac{\partial y}{\partial \zeta_2} \end{bmatrix} = \left[\frac{\partial \boldsymbol{\omega}}{\partial \zeta_1}\mathbf{x} \quad \frac{\partial \boldsymbol{\omega}}{\partial \zeta_2}\mathbf{x}\right] \tag{5.27}$$

Having **J** and assuming it to be nonsingular, we must invert (5.26) to form

$$\begin{bmatrix} \dfrac{\partial}{\partial x} \\ \dfrac{\partial}{\partial y} \end{bmatrix} = \mathbf{J}^{-1}\begin{bmatrix} \dfrac{\partial}{\partial \zeta_1} \\ \dfrac{\partial}{\partial \zeta_2} \end{bmatrix} = \frac{1}{|J|}\begin{bmatrix} \dfrac{\partial y}{\partial \zeta_2} & -\dfrac{\partial y}{\partial \zeta_1} \\ -\dfrac{\partial x}{\partial \zeta_2} & \dfrac{\partial x}{\partial \zeta_1} \end{bmatrix}\begin{bmatrix} \dfrac{\partial}{\partial \zeta_1} \\ \dfrac{\partial}{\partial \zeta_2} \end{bmatrix}$$

and so for this two-dimensional strain field,

$$\boldsymbol{\alpha} = \mathbf{D}^{\mathrm{t}}\boldsymbol{\omega} = \begin{bmatrix} \dfrac{\partial}{\partial x} & 0 \\ 0 & \dfrac{\partial}{\partial y} \\ \dfrac{\partial}{\partial y} & \dfrac{\partial}{\partial x} \end{bmatrix}\boldsymbol{\omega} = \frac{1}{|\mathbf{J}|}\begin{bmatrix} \dfrac{\partial y}{\partial \zeta_2}\dfrac{\partial}{\partial \zeta_1} - \dfrac{\partial y}{\partial \zeta_1}\dfrac{\partial}{\partial \zeta_2} & 0 \\ 0 & -\dfrac{\partial x}{\partial \zeta_2}\dfrac{\partial}{\partial \zeta_1} + \dfrac{\partial x}{\partial \zeta_1}\dfrac{\partial}{\partial \zeta_2} \\ -\dfrac{\partial x}{\partial \zeta_2}\dfrac{\partial}{\partial \zeta_1} + \dfrac{\partial x}{\partial \zeta_1}\dfrac{\partial}{\partial \zeta_2} & \dfrac{\partial y}{\partial \zeta_2}\dfrac{\partial}{\partial \zeta_1} - \dfrac{\partial y}{\partial \zeta_1}\dfrac{\partial}{\partial \zeta_2} \end{bmatrix}\boldsymbol{\omega}$$

Substituting for $\boldsymbol{\omega}$ gives us $\boldsymbol{\alpha}$, but for reasons of space we do not evaluate all the terms here. The stiffness matrix then follows, and here we use the fact that $\mathrm{d}A = |\mathbf{J}|\,\mathrm{d}\zeta_1\,\mathrm{d}\zeta_2$ so

$$\mathbf{k}_g = \int_{-1}^{1}\int_{-1}^{1} \boldsymbol{\alpha}^{\mathrm{t}}\boldsymbol{\kappa}\boldsymbol{\omega}|\mathbf{J}|\,\mathrm{d}\zeta_1\,\mathrm{d}\zeta_2 \tag{5.28}$$

The length of the product in this integral, and the nature of $\boldsymbol{\alpha}$ which involves **x**, make it profitless even to attempt an explicit expression, and so the stiffness integral is performed numerically evaluating $\boldsymbol{\alpha}$ and $|\mathbf{J}|$ at the Gauss points, as detailed in Appendix D.

The use of isoparametric mapping can be extended to three-dimensional 'brick elements' or tetrahedra using 'volume' instead of area coordinates: the details can be found in many finite-element texts.[48, 51] The isoparametric formulation is a convenient way of describing curvilinear displacements and geometry, but it also has the added bonus that linear displacement fields in the variable $\boldsymbol{\zeta}$ map into similar fields in the real world of x and y. This is important when we recall that linear and constant displacement fields are essential to satisfy overall element equilibrium and to ensure favourable convergence to a constant strain as the mesh size is reduced. This attribute can be seen when we consider the linear field

$$u = a + bx + cy$$

This field produces a displacement $\rho_i = a + bx_i + cy_i$ at any point (x_i, y_i).

But

$$u = \boldsymbol{\omega\rho} = \sum \omega_i \rho_i = \sum \omega_i (a + bx_i + cy_i)$$
$$= a \sum \omega_i + b \sum \omega_i x_i + c \sum \omega_i y_i$$

And because of the mapping $x = \sum \omega_i x_i$ and $y = \sum \omega_i y_i$, then the field $u = a + bx + cy$ is recovered if the condition is satisfied that

$$\sum_i \omega_i = 1$$

This has already been demonstrated for the rectangular element in section 5.3. In fact it should be true for all isoparametric elements since if we contract the element to a point and let all the nodes (x_i, y_i) move to a single point (x_0, y_0) we have

$$x_0 = \sum \omega_i x_0 \qquad \text{or} \qquad \sum \omega_i = 1$$

One word of caution is necessary. The curvilinear mesh chosen must not be too ambitious. There are dangers in choosing a very distorted mesh and hoping it will map numerically onto a two- or three-dimensional rectilinear mesh. Severe distortion can cause the Jacobian to go singular. Unreasonable distortion may take the form of highly angular shapes or simply a distribution of nodal points which differ too much from the regularly spaced nodes in the ζ-plane. For example, even if the side of a quadrilateral was straight, we could divide the length l into two unequal segments βl and $(1-\beta)l$ so that we attempted to map $\mathbf{x} = \{-\beta l \quad 0 \quad (1-\beta)l\}$ into $\boldsymbol{\zeta} = \{-1 \quad 0 \quad 1\}$ by $\mathbf{x} = \boldsymbol{\omega}\mathbf{x}$.

Then

$$\mathrm{x} = [-\tfrac{1}{2}(1-\zeta)\zeta \quad \tfrac{1}{2}(1-\zeta)(1+\zeta) \quad \tfrac{1}{2}(1+\zeta)\zeta]\mathbf{x}$$

so

$$\mathrm{x} = \tfrac{1}{2}(1-\zeta)\zeta\beta l + \tfrac{1}{2}(1+\zeta)\zeta(1-\beta)l$$

Now form

$$\frac{\partial x}{\partial \zeta} = \frac{l}{2}[1 + (2 - 4\beta)\zeta]$$

This derivative vanishes at the nodes $\zeta = \pm 1$ for $\beta = \frac{1}{4}$ or $\frac{3}{4}$, and infinite strains would be the consequence. Actually this singular behaviour can be used to represent approximately a crack-tip stress field along the edge of an element.[53] The modelling of such a field is necessary in determining the stress-intensity or energy-release rate in fracture-mechanics predictions.

5.7 Plate-bending elements

In Chapter 2 we successfully used the PVD to formulate the plate-bending equations simply by extending the beam-bending assumptions into two dimensions. Having solved beams approximately by the finite-element method it might seem routine to extend the model into two dimensions and formulate a plate-bending element. This does turn out to be true but there are frustrating difficulties. Consider first the simple rectangular element of Fig. 5.18 and suppose

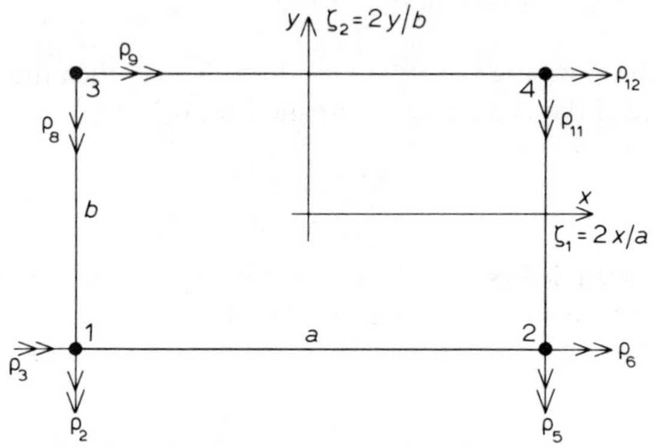

Fig. 5.18 Four-noded plate bending element

we wish to construct a four-noded element. We have shown that equilibrium will be satisfied in the mean over the sides of an element provided there are continuous virtual displacements over which the inter-element forces do work. In this case these inter-element 'forces' are the shear, the moment, and the twisting moment; thus we require continuity of deflection and slope along the edges. If we can uniquely define the edge displacement and slope in terms of their nodal values then continuity is assured. It seems sensible then to try cubic interpolation functions. The normal displacements at nodes 1, 2, 3, and 4 are labelled ρ_1, ρ_4, ρ_7, and ρ_{10}, and the eight remaining degrees of freedom are the rotations as shown in Fig. 5.18. For example, rotations $\partial w/\partial x = \rho_2$ and $\partial w/\partial y = \rho_3$ at node 1 are indicated by double-arrow vectors. We now employ the four beam functions $\omega_1(\zeta)$, $\omega_2(\zeta)$, $\omega_3(\zeta)$, and $\omega_4(\zeta)$, first introduced in Fig. 5.3, and representing unit values of either deflection or slope at a node. Thus

$$w(\zeta_1, \zeta_2) = \omega_1(\zeta_1)\omega_1(\zeta_2)\rho_1 + \omega_2(\zeta_1)\omega_1(\zeta_2)\rho_2 + \omega_2(\zeta_2)\omega_1(\zeta_1)\rho_3 + \ldots \quad (5.29)$$

where $\omega_1(\zeta)$ and $\omega_2(\zeta)$ are the Hermitian polynomials for unit displacement and slope shown again in Fig. 5.19. These nodes do the job of interpolation functions perfectly in one respect: the node shape associated with (say) the deflection ρ_1 has zero slope on all four edges and zero displacement on ζ_1 and $\zeta_2 = +1$. Likewise, the shape associated with rotation $\partial w/\partial x = \rho_2$ has zero displacement on all four edges and zero slope on ζ_1 and $\zeta_2 = +1$. Therefore when node 1 is connected to adjacent elements it will ensure continuous slope and deflection along contiguous

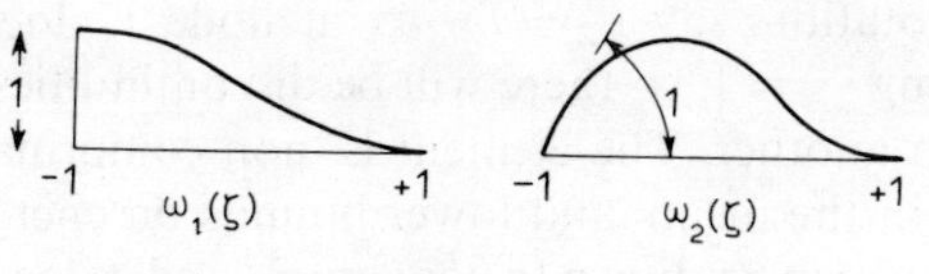

Fig. 5.19 Two Hermitian polynomials

sides $\zeta_1 = -1$ and $\zeta_2 = -1$. The values of ρ_1, ρ_2, and ρ_3 will not affect the deflections and slopes on the other edges $\zeta_1, \zeta_2 = +1$, so continuity there will be ensured by nodal displacements at 2, 3, and 4. But there is one flaw. Because the slope of $\omega_1(\zeta)$ is zero at $\zeta = -1$, then either $\partial\omega_1/\partial\zeta_1$ or $\partial\omega_1/\partial\zeta_2$ are zero in all three modes associated with node 1, consequently $\partial^2\omega/\partial\zeta_1\,\partial\zeta_2$ is zero at the node and there is no way that continuity of just ρ_1, ρ_2, and ρ_3 at node 1 can ensure continuous twist, which we showed, in equation (2.36), controls the shear strain $\varepsilon_{xy} = -2z\,\partial^2 w/\partial x\,\partial y$ in a plate. This deficiency in shear strain unfortunately goes further when we form $\partial^2\omega/\partial\zeta_1\,\partial\zeta_2$ from (5.29) and find that there is no constant term. This inability to represent a constant shear strain will surely affect the convergence of such an element as mesh size is reduced. One way to overcome this is to lower the order of the interpolation to linear shapes $\omega_5(\zeta)$ and $\omega_6(\zeta)$ shown in Fig. 5.20. Of course such shapes are unsatisfactory in controlling both deflections and slopes since the derivative of ω_5 is not zero at $\zeta = +1$. Nevertheless a popular element with quite good convergence properties adopts this trick for the rotation modes ρ_2 and ρ_3, but uses true Hermitians for the deflection mode ρ_1. The complete interpolation functions $\boldsymbol{\omega}$ now produces

$$\begin{aligned} w &= \omega_1(\zeta_1)\omega_1(\zeta_2)\rho_1 + \omega_2(\zeta_1)\omega_5(\zeta_2)\rho_2 + \omega_5(\zeta_1)\omega_2(\zeta_2)\rho_3 + \omega_3(\zeta_1)\omega_1(\zeta_2)\rho_4 \\ &\quad + \omega_4(\zeta_1)\omega_5(\zeta_2)\rho_5 + \omega_6(\zeta_1)\omega_2(\zeta_2)\rho_6 + \omega_1(\zeta_1)\omega_3(\zeta_2)\rho_7 + \omega_2(\zeta_1)\omega_6(\zeta_2)\rho_8 \\ &\quad + \omega_5(\zeta_1)\omega_4(\zeta_2)\rho_9 + \omega_3(\zeta_1)\omega_3(\zeta_2)\rho_{10} + \omega_4(\zeta_1)\omega_6(\zeta_2)\rho_{11} + \omega_6(\zeta_1)\omega_4(\zeta_2)\rho_{12} \\ &= \boldsymbol{\omega\rho} \end{aligned} \tag{5.30}$$

The matrix $\boldsymbol{\alpha}$ now follows from (2.36) as

$$\boldsymbol{\alpha} = -z\begin{bmatrix} \dfrac{\partial^2}{\partial x^2} \\ \dfrac{\partial^2}{\partial y^2} \\ 2\dfrac{\partial^2}{\partial x\,\partial y} \end{bmatrix}\boldsymbol{\omega} = -4z\begin{bmatrix} a^2\dfrac{\partial^2}{\partial\zeta_1^2} \\ b^2\dfrac{\partial^2}{\partial\zeta_2^2} \\ 2ab\dfrac{\partial^2}{\partial\zeta_1\partial\zeta_2} \end{bmatrix}\boldsymbol{\omega} \tag{5.31}$$

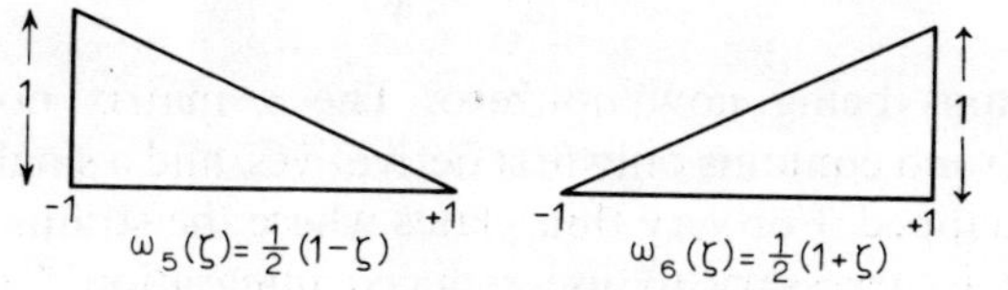

Fig. 5.20 Linear interpolation

Now of course a rotation, say $\rho_2 = \partial w/\partial x$ at node 1, does produce nonzero rotations $\partial w/\partial y$ along $\zeta_2 = 1$, so there will be discontinuities in slope when this element is joined to another. The element is 'non-conformal' and the remarks about 'equilibrium in the mean' and lower bounds on energy no longer apply. However, this element can be shown to converge[54] and it does seem that the extra kinematic flexibility, introduced by allowing incompatible rotations or 'kinks', does compensate for the overstiff PVD solutions. If we use this element for quadrilaterals by using a lower order linear mapping $\{x \quad y\} = \boldsymbol{\omega}\mathbf{x}$, the formulation becomes cumbersome, owing to the fact that the Jacobian transformation (5.26) has to be performed twice, and inverted, before we can recover $\boldsymbol{\alpha}$ in (5.31). For this reason, another popular approach[56] which also resolves the problem of kinematic continuity, is to drop the zero shear assumption of Engineers' bending theory which says that the (clockwise) rotations of the normal about the x and y directions are

$$\theta_x = \frac{\partial w}{\partial y} \qquad \text{and} \qquad \theta_y = -\frac{\partial w}{\partial x}$$

but instead we expand *separately* θ_x, θ_y, and w in terms of their nodal values $\boldsymbol{\rho}'_x$, $\boldsymbol{\rho}'_y$ and $\boldsymbol{\rho}_w$.

That is,

$$\theta_x = \boldsymbol{\omega}\boldsymbol{\rho}'_x; \qquad \theta_y = \boldsymbol{\omega}\boldsymbol{\rho}'_y; \qquad w = \boldsymbol{\omega}\boldsymbol{\rho}_w \tag{5.32}$$

We still preserve the other beam-like assumptions that normals through the plate thickness remain straight and that strains ε_{zz} can be ignored. The displacement field is now

$$u = z\theta_y; \qquad v = -z\theta_x; \qquad w$$

So using $\boldsymbol{\varepsilon} = \mathbf{D}^t\mathbf{u}$,

$$\varepsilon_{xx} = \frac{\partial u}{\partial x} = z\frac{\partial \theta_y}{\partial x}$$

$$\varepsilon_{yy} = \frac{\partial v}{\partial y} = -z\frac{\partial \theta_x}{\partial y}$$

$$\varepsilon_{xy} = \frac{\partial u}{\partial y} + \frac{\partial v}{\partial x} = z\left(\frac{\partial \theta_y}{\partial y} - \frac{\partial \theta_x}{\partial x}\right)$$

$$\varepsilon_{zx} = \frac{\partial w}{\partial x} + \frac{\partial u}{\partial z} = \frac{\partial w}{\partial x} + \theta_y$$

$$\varepsilon_{zy} = \frac{\partial w}{\partial y} + \frac{\partial v}{\partial z} = \frac{\partial w}{\partial y} - \theta_x$$

the last two strains being now nonzero. The $\boldsymbol{\alpha}$ matrix now follows after substituting (5.32) and contains only first derivatives, and a single Jacobian if the quadrilateral is mapped. For very thin plates where the strains ε_{zx} and ε_{zy} must approach zero it is necessary to use reduced integration[59] to avoid the ill-conditioning that arises as w and θ become no longer independent.

The difficulties in achieving a plate-bending element with satisfactory continuity and having constant strain capability have been demonstrated. These problems have prompted the search for a different but simple solution using nodal moments as unknowns, and so ensuring bending stress continuity directly. This alternative is examined in the problems at the end of the next chapter.

5.8 Gross deformations

Having posed the problem of gross deformations in Chapter 2 and developed, via the incremental PVD (2.56), formulae suitable for solving strut buckling and arch snap-through, it is tempting simply to discretize these existing equations into finite elements and produce an answer. However, we will go back to first principles since many of the points are quite general in nature.

The incremental PVD is valid for gross deformations but it does contain actual forces and stresses $\boldsymbol{\sigma}$ as well as the incremental strains $\delta\boldsymbol{\varepsilon}$ and displacements $\delta\mathbf{u}$. Although we may have analysed exactly the simple initial buckling of a strut, and the complex post-buckling of a shallow arch, it is generally impossible to evaluate $\boldsymbol{\sigma}$ in explicit form in terms of the large displacements $\mathbf{r}$; in which case we have no other recourse but to load the structure in successive increments $\delta\mathbf{R}$, starting from zero, and solving for the consequent increments $\delta\mathbf{r}$ and $\delta\boldsymbol{\sigma}$. During this incremental load it is safe to take the current values of stress to be constant; the current values having been *accumulated* from all previous load increments, thus

$$\mathbf{R} = \sum \delta\mathbf{R} \qquad \text{and} \qquad \boldsymbol{\sigma} = \sum \delta\boldsymbol{\sigma}$$

Consider the problem in Fig. 5.21 of a string stretched between two fixed supports and then loaded by a single transverse force R. The displacement r must be a manifest change in geometry which cannot be ignored, for until a rotation θ is

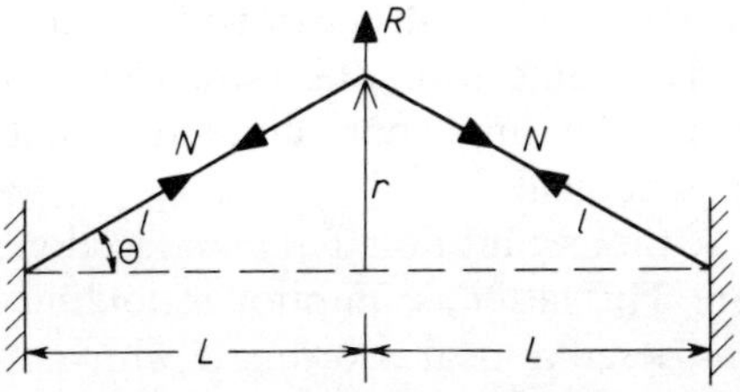

Fig. 5.21 Nonlinear deformations of a plucked string

developed, the string tension N will have no component opposing R. This problem happens to be so simple that we can estimate the strain in the string explicitly in terms of r,

$$\varepsilon = (l - L)/L = (1 + r^2/L^2)^{1/2} - 1,$$

but this so rarely happens that we ignore this special feature. We will continue to assume that material failure inhibits the allowable strain to a few per cent, so the strains are small and we may assume Hooke's law again. To apply incremental loading we note that equilibrium requirements are

$$R = 2N \sin\theta$$

therefore

$$\delta R = 2\,\delta N \sin\theta + 2N \cos\theta\,\delta\theta \tag{5.33}$$

The first term is the usual change in internal stress produced by an incremental strain $\delta\varepsilon$ and displacement δr. The second term is the additional resistance to load caused by *existing* stresses N being realigned by $\delta\theta$. Both these terms can be converted to functions of δr.
Thus

$$\tan\theta = r/L; \qquad \text{therefore} \quad \delta\theta = \cos^2\theta\,\delta r/L$$

Also

$$\delta N = AE\delta\varepsilon, \qquad \text{where} \quad \delta\varepsilon = \delta r \sin\theta/1$$

Therefore substituting into (5.33),

$$\delta R = [(2\sin^2\theta\cos\theta)\,AE/L + (2N\cos\theta\cos^2\theta)/L]\delta r$$

or

$$\delta R = (K_E + K_G)\delta_r = K_T\,\delta r \tag{5.34}$$

The symbols K_G, K_E, and K_T are explained as follows. K_T is naturally a 'tangent stiffness' since it relates the current incremental load to incremental displacement, that is $K_T = \partial R/\partial r$. This tangent stiffness is composed of two components, the first term K_E being the usual 'elastic' stiffness, but the second term K_G, called the *geometric stiffness*, arises from the realignment of the internal stresses. In principle all structural finite elements have both elastic stiffness and geometric stiffness, the latter usually evaluated for some normalized stress level (N here). We now look at some examples – scorning the simple bar – and return firstly to the beam, but this time using the above arguments.

We continue to make the assumption that plane sections remain plane and that rotations are moderate. This latter assumption is not limiting since the beam axis (z) can always be considered a local coordinate which may have to be updated during a loading programme. Assuming $v'(z)$ is small enough for its cosine to be unity, we ignore the departures of y from its original orientation and write once again the axial displacement as

$$w(z) = w_0(z) - yv'(z)$$

where $w_0(z)$ and $v(z)$ are the displacements of the centroid axis (see Fig. 5.22). The nonlinear compatibility relationship (2.61) for the total strain is still

$$\varepsilon_{zz} = \frac{\partial w}{\partial z} + \frac{1}{2}\left(\frac{\partial v}{\partial z}\right)^2$$

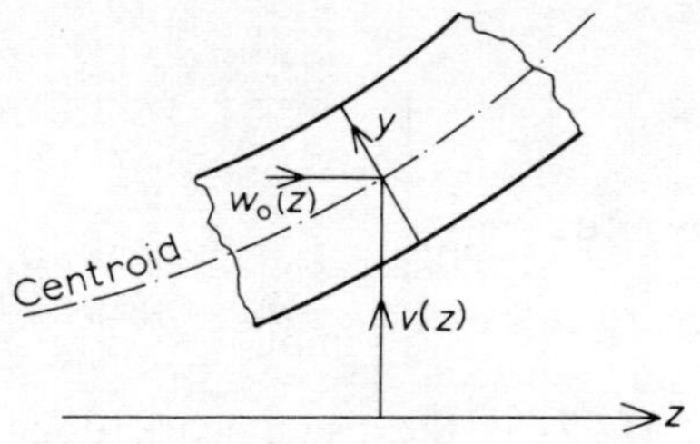

Fig. 5.22 Beam theory again

Substituting for $w(z)$,

$$\varepsilon_{zz} = w_0' - yv'' + \tfrac{1}{2}(v')^2 \tag{5.35}$$

and

$$\delta\varepsilon_{zz} = \delta w_0{}' - y\,\delta v'' + v'\,dv' \tag{5.36}$$

The internal incremental virtual work, integrating over the cross-sectional area A, is then

$$\int_A \sigma_{zz}\,\delta\bar{\varepsilon}_{zz}\,\mathrm{d}A = \int_A E[w_0' - yv'' + \tfrac{1}{2}(v')^2]\,[\delta w_0' - y\,\delta v'' + v'\,\delta v']\,\mathrm{d}A$$

The coefficient of the virtual displacement $\delta w_0'$ simply gives the equation of axial equilibrium

$$\textstyle\int E w_0'\,\mathrm{d}A = \int \sigma\,\mathrm{d}A = N$$

where N and σ are taken to be tensile to be consistent, and we assume that $(v')^2$ is small compared with w_0'. The rest of the integral – noting that $\int y\,\mathrm{d}A = 0$ – becomes

$$A\sigma v'\,\delta v' + EIv''\,\delta v'' + \tfrac{1}{2}AE(v')^2 v'\,\delta v'$$

Discarding the last term compared with the others – for moderately small rotations $v'(z)$ – we are left with an integral over the length which is much as before (equation (2.67)) when we add the loading:

$$\int_0^l (EIv''\,\delta v'' + Nv'\delta v')\,\mathrm{d}z = \int_0^l p(z)\,\delta v\,\mathrm{d}z \tag{5.37}$$

Using the beam interpolation function $v = \boldsymbol{\omega}\boldsymbol{\rho}_g$ of Fig. 5.3, the first term is again

$$\int_0^l EIv''\,\delta v''\,\mathrm{d}z = \boldsymbol{\rho}_g^{\mathrm{t}}\mathbf{k}_{\mathrm{E}}\,\delta\boldsymbol{\rho}_g$$

with the elastic stiffness $\mathbf{k}_{\mathrm{E}}$ as before (equation (5.5)). The second term containing the internal stress N is the geometric stiffness $\mathbf{k}_{\mathrm{G}}$, and on using $v = \boldsymbol{\omega}\boldsymbol{\rho}_g$ again,

$$\int_0^l v'\,\delta v'\,\mathrm{d}z = \boldsymbol{\rho}_g \int_0^l \boldsymbol{\omega}'^{\mathrm{t}}\boldsymbol{\omega}'\,\mathrm{d}z\,\delta\boldsymbol{\rho}_g = \boldsymbol{\rho}_g k_{\mathrm{G}}\,\delta\boldsymbol{\rho}_g$$

where

$$\mathbf{k}_G = \int_0^l \boldsymbol{\omega}'^t\boldsymbol{\omega}'\,dz = \frac{1}{30l}\begin{bmatrix} 36 & 3l & -36 & 3l \\ & 4l^2 & -3l & -l^2 \\ & & 36 & -3l \\ \text{symm.} & & & 4l^2 \end{bmatrix} \tag{5.38}$$

The right-hand side of (5.37) is the usual work done by the kinematically equivalent forces $\mathbf{P}_g^t\delta\boldsymbol{\rho}_g$, so that the equation of equilibrium for a single element would be

$$(\mathbf{k}_E + N\mathbf{k}_G)\boldsymbol{\rho}_g = \mathbf{P}_g \qquad \text{or} \qquad \mathbf{k}_T\boldsymbol{\rho}_g = \mathbf{P}_g$$

If the beam is divided into many elements, $\boldsymbol{\rho}_g = \mathbf{a}_g\mathbf{r}$, then summing the above,

$$(\mathbf{K}_E + N\mathbf{K}_G)\mathbf{r} = \mathbf{K}_T\mathbf{r} = \mathbf{R} \tag{5.39}$$

and $\mathbf{K}_T = \sum \mathbf{a}_g^t\mathbf{k}_T\mathbf{a}_g$ as usual. This equation still assumes that the displacements $\mathbf{r}$ are moderate. If they are not then (5.39) must be written in incremental form $\mathbf{K}_T\,\delta\mathbf{r} = \delta\mathbf{R}$, in which $\mathbf{K}_T$ is the current value. If the displacements are then accumulated and continuously updated ($\mathbf{r} = \sum \delta\mathbf{r}$) so must be the stresses and the current coordinates of the nodes if the geometrical changes are significant. This means that $\mathbf{a}_g$ would cease to be a simple Boolean selection as the element changes its orientation, so instead we evaluate, say, $\mathbf{k}_T'$ and $\mathbf{P}_g'$ for the current *local* axes and then replace $\mathbf{k}_T = \mathbf{T}\mathbf{k}_T'\mathbf{T}$ and $\mathbf{P}_g = \mathbf{T}\mathbf{P}_g'$ (equation (2.34)) before assembling $\mathbf{K}_T$ in the usual way and solving $\delta\mathbf{r} = \mathbf{K}_T^{-1}\delta\mathbf{R}$. The increments $\delta\mathbf{R}$ should be chosen small enough for the tangent stiffness $\mathbf{K}_T$ to be sensibly constant during the displacement $\delta\mathbf{r}$; even so this can cause the summed solution to drift away from the exact values (see Fig. 5.23). It is possible to correct this error by iteration during the increment to ensure that no errors remain at the end of the incremental step. A simple device is to take the average of tangent stiffness at the beginning ($\mathbf{K}_{T0}$) and at the end ($\mathbf{K}_{T1}$) putting $\mathbf{K}_T = \frac{1}{2}(\mathbf{K}_{T0} + \mathbf{K}_{T1})$. The initial value of $\mathbf{K}_{T1}$ is set to $\mathbf{K}_{T0}$ and then updated as the new geometry $\mathbf{x}_1 = \mathbf{x}_0 + \delta\mathbf{r}$ is established. This procedure converges to the exact position if the $\mathbf{R} \sim \mathbf{r}$ curves are quadratic – this is a fair approximation to any nonlinear behaviour, so quite large

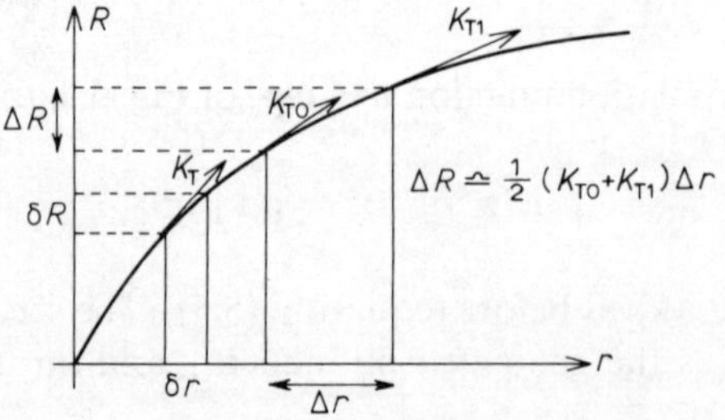

Fig. 5.23 Incremental summation

steps ($\delta\mathbf{R}$, $\delta\mathbf{r}$) can then be tolerated, and this is useful if the structure is about to buckle, that is tending to a singular matrix.

Another popular alternative is the Newton–Raphson method, in which the first displacements $\delta\mathbf{r} = \mathbf{K}_T^{-1}\,\delta\mathbf{R}$ are used to update $\mathbf{K}_T$ and to find the consequent shortfall from $\delta\mathbf{R}$. These residual forces are then applied with the corrected $\mathbf{K}_T$, and so on. A cheaper form is to stick with the original inverse $\mathbf{K}_T^{-1}$ and accept the greater number of iterations then needed for residual $\delta\mathbf{R}$ to become negligible. Both these methods are at risk if $\mathbf{K}_{T1}$ is singular.

Initial buckling problems do not have to be incremented since $\mathbf{r}$ is moderate and we may solve (5.39) directly:

$$\mathbf{r} = \mathbf{K}_T^{-1}\mathbf{R} \tag{5.40}$$

If the initial stresses are negative–say $N = -P$–then it is possible for $\mathbf{K}_T$ to become singular, and then initial buckling is predicted by $|\mathbf{K}_T| = 0$, or

$$|\mathbf{K}_E - P\mathbf{K}_G| = 0 \tag{5.41}$$

Equation (5.40) can be used to estimate stresses before buckling, and (5.41) is the usual eigenvalue problem for finding critical values of P_{crit} at buckling. For example, suppose we have a uniform strut of length $2l$ clamped at both ends; and we divide it into two equal elements. The central deflection r_1 and rotation r_2 are the only nonzero displacements. The two element stiffnesses may be summed as usual:

$$\mathbf{K}_E = \frac{2EI}{l^3}\begin{bmatrix} 6+6 & -3l+3l \\ -3l+3l & 2l^2+2l^2 \end{bmatrix} = \frac{8EI}{l^3}\begin{bmatrix} 3 & 0 \\ 0 & l^2 \end{bmatrix}$$

$$\mathbf{K}_G = \frac{1}{30l}\begin{bmatrix} 36+36 & -3l+3l \\ -3l+3l & 4l^2+4l^2 \end{bmatrix} = \frac{8}{30l}\begin{bmatrix} 9 & 0 \\ 0 & l^2 \end{bmatrix}$$

Solving $|\mathbf{K}_E - P\mathbf{K}_G| = 0$, we obtain two independent solutions

$$\left(\frac{3EI}{l^3} - \frac{9P}{30l}\right) r_1 = 0, \quad P_{crit} = \frac{10EI}{l^2}$$

and

$$\left(\frac{EI}{l^3} - \frac{P}{30l}\right) r_2 = 0, \quad P_{crit} = \frac{30EI}{l^2}$$

The first value with $r_2 = 0$ is the symmetrical buckling load and compares well with the exact solution of $\pi^2 EI/l^2 = 9.88EI/l^2$. The second antisymmetrical mode ($r_1 = 0$) compares badly with the exact solution of $20.2EI/l^2$ – it is fortunate that it is the larger non-critical value. In fact it is usual for the lower-order modes to be the most accurate in both buckling and vibration analysis, and in this example it confirms that the deflected shape of a single element, spanning a half of the strut, is not accurately represented by a cubic. This is in contrast to the same beam loaded only by lateral loads where a cubic was found to be an accurate representation. (Fig. 5.5).

The extension of the geometrical stiffness concept to two dimensions is fairly obvious. Instead of a single nonlinear compatibility relation we start off with three:

$$\varepsilon_{xx} = \frac{\partial u}{\partial x} + \frac{1}{2}\left(\frac{\partial w}{\partial x}\right)^2$$

$$\varepsilon_{yy} = \frac{\partial v}{\partial y} + \frac{1}{2}\left(\frac{\partial w}{\partial y}\right)^2$$

$$\varepsilon_{xy} = \frac{\partial u}{\partial y} + \frac{\partial v}{\partial x} + \frac{\partial w}{\partial x}\frac{\partial w}{\partial y}$$

The geometric stiffness arises again from the second terms in the strain, coupled with the existing in-plane forces (per unit length) N_{xx}, N_{yy}, and N_{xy}. Their contribution to the incremental virtual work, taking increments of the above, is

$$\int_V \boldsymbol{\sigma}^t \delta\boldsymbol{\varepsilon}\, \mathrm{d}v = \int\left[N_{xx}\frac{\partial w}{\partial x}\delta\left(\frac{\partial w}{\partial x}\right) + N_{yy}\frac{\partial w}{\partial y}\delta\left(\frac{\partial w}{\partial y}\right) + N_{xy}\left(\frac{\partial w}{\partial x}\delta\left(\frac{\partial w}{\partial y}\right) + \frac{\partial w}{\partial y}\delta\left(\frac{\partial w}{\partial x}\right)\right)\right]\mathrm{d}A$$

and the integrand can be summarized as

$$\begin{bmatrix} \dfrac{\partial w}{\partial x} & \dfrac{\partial w}{\partial y} \end{bmatrix} \begin{bmatrix} N_{xx} & N_{xy} \\ N_{xy} & N_{yy} \end{bmatrix} \begin{bmatrix} \delta\left(\dfrac{\partial w}{\partial x}\right) \\ \delta\left(\dfrac{\partial w}{\partial y}\right) \end{bmatrix}$$

If $w(x, y)$ is interpolated as usual,

$$w = \boldsymbol{\omega}\boldsymbol{\rho}; \qquad \frac{\partial w}{\partial x} = \frac{\partial \boldsymbol{\omega}}{\partial x}\boldsymbol{\rho}; \qquad \frac{\partial w}{\partial y} = \frac{\partial \boldsymbol{\omega}}{\partial y}\boldsymbol{\rho};$$

then the work is $\boldsymbol{\rho}^t \mathbf{k}_G \delta\boldsymbol{\rho}$ where

$$\mathbf{k}_G = \int \begin{bmatrix} \dfrac{\partial \boldsymbol{\omega}}{\partial x} \\ \dfrac{\partial \boldsymbol{\omega}}{\partial y} \end{bmatrix}^t \begin{bmatrix} N_{xx} & N_{xy} \\ N_{xy} & N_{yy} \end{bmatrix} \begin{bmatrix} \dfrac{\partial \boldsymbol{\omega}}{\partial x} \\ \dfrac{\partial \boldsymbol{\omega}}{\partial y} \end{bmatrix} \mathrm{d}A \tag{5.42}$$

This is a symmetric matrix of the same form as the one-dimensional beam geometric stiffness (5.38).

5.9 Dynamics

The discretized equations of motion are usully obtained by formulating Lagrange's equations in terms of kinetic and strain energy, but this is unnecessary. We simply have to reverse the sign of the inertia 'forces' and treat the structure as if it were in equilibrium. Denoting the acceleration as $\partial^2 \mathbf{u}/\partial t^2 = \ddot{\mathbf{u}}$, we put the body force $\mathbf{p}_v = -\rho\ddot{\mathbf{u}}$, where the scalar ρ is the density. The PVD for free vibration then becomes

$$\int_V \boldsymbol{\sigma}^t \bar{\boldsymbol{\varepsilon}}\, dv = -\int_V (\rho\ddot{\mathbf{u}})^t \bar{\mathbf{u}}\, dv$$

Putting $\mathbf{u} = \boldsymbol{\omega}\boldsymbol{\rho}$ inside V_g and summing,

$$\sum_g \boldsymbol{\rho}_g^t \mathbf{k}_g \bar{\boldsymbol{\rho}}_g = -\sum_g \ddot{\boldsymbol{\rho}}_g^t \int_{V_g} \boldsymbol{\omega}^t \rho \boldsymbol{\omega}\, dv \bar{\boldsymbol{\rho}}_g$$

We denote the new element integral as an *element mass matrix*

$$\mathbf{m}_g = \int_{V_g} \boldsymbol{\omega}^t \rho \boldsymbol{\omega}\, dv \tag{5.43}$$

This matrix is of similar form to the stiffness matrix, and identical routines can be used to evaluate it. The elements are now assembled in the usual way by denoting $\boldsymbol{\rho}_g = \mathbf{a}_g \mathbf{r}$, so we have

$$\mathbf{r}^t \sum_g \mathbf{a}_g^t \mathbf{k}_g \mathbf{a}_g \bar{\mathbf{r}} = -\ddot{\mathbf{r}}^t \sum_g \mathbf{a}_g^t \mathbf{m}_g \mathbf{a}_g \bar{\mathbf{r}}$$

or

$$\mathbf{K}\mathbf{r} = -\mathbf{M}\ddot{\mathbf{r}}$$

If applied forces $\mathbf{R}(t)$ are also present then

$$\mathbf{M}\ddot{\mathbf{r}} + \mathbf{K}\mathbf{r} = \mathbf{R}(t) \tag{5.44}$$

The assembly of the global mass matrix

$$\mathbf{M} = \sum_g \mathbf{a}_g^t \mathbf{m}_g \mathbf{a}_g \tag{5.45}$$

is seen to be identical to the stiffness assembly. The solution of the equations of motion (5.44) depends on the nature of the problem, but essentially any of the techniques used to solve one degree of freedom are extendable to the many degrees of freedom that may occur in (5.44). For example, if the structure is periodically excited by a set of forces $\mathbf{R}(t) = \mathbf{R}\sin\Omega t$, all in phase at the same frequency, we may assume $\mathbf{r}(t) = \mathbf{r}\sin\Omega t$ and then (5.44) simplifies to

$$[\mathbf{K} - \Omega^2\mathbf{M}]\mathbf{r} = \mathbf{R} \tag{5.46}$$

The 'dynamic stiffness matrix' can then be inverted to solve for the amplitudes of motion $\mathbf{r}$. The new term $-\Omega^2\mathbf{M}$ which modifies the stiffness is completely analogous to the similarly modifying geometric stiffness of the previous section, and an analogous loss of stiffness can occur if the determinant

$$|\mathbf{K} - \Omega^2\mathbf{M}| = 0 \tag{5.47}$$

The amplitudes $\mathbf{r}$ will diverge if Ω tends to any of the eigenvalues given by this equation, that is the structure *resonates* at this particular *resonant frequency*.

These eigenvalues are also known as the 'natural' frequencies since they are also the only frequencies at which the structure can vibrate freely. This can be seen if we put $\mathbf{R} = \mathbf{0}$ in (5.46), whereupon the displacements $\mathbf{r}$ must be zero unless the aforementioned determinant is zero.

For many structures it is sufficient simply to know the natural frequencies and to avoid any excitation at them. (This is easier said than done; for example, a typical helicopter may have natural frequencies every 3 or 4 Hz from 16 Hz upwards, so keeping these away from cyclic rotor loading can be fraught.) The dynamic problem differs from buckling in one respect: whereas the latter involves finding only the smallest eigenvalue, it is necessary to avoid many resonant frequencies. Merely escaping the lowest natural frequency does not ensure salvation from the higher ones.

The number of degrees of freedom needed to obtain an accurate distribution of stresses may run into thousands, but to retain all these in a dynamic analysis would be costly and unnecessary. The very high frequencies would be quickly damped even if they were excited at all. It is possible to use eigenvalue algorithms which evaluate the lower eigenvalues first, but even so it is an effort to operate with the full stiffness and mass matrices. Also, if it is necessary to examine the transient response to non-periodic loading it may be necessary to integrate numerically the equations of motion directly. This technique [57] is beyond the scope of this text, but the incremental algorithms used will often employ a time step that must be small compared with the smallest natural period of the structure. There are therefore good reasons for reducing the degrees of freedom in order to eliminate the higher natural frequencies prior to calculating the dynamic response.

One way of simplification is to identify all those displacements ($\mathbf{r}_1$) of a structure with significant mass and then ignore the mass of the rest ($\mathbf{r}_2$). Equation (5.46) then becomes of the form

$$\left[\begin{array}{c|c} \mathbf{K}_{11} & \mathbf{K}_{12} \\ \hline \mathbf{K}_{21} & \mathbf{K}_{22} \end{array}\right]\left[\begin{array}{c} \mathbf{r}_1 \\ \hline \mathbf{r}_2 \end{array}\right] - \Omega^2\left[\begin{array}{c|c} \mathbf{M}_{11} & \mathbf{0} \\ \hline \mathbf{0} & \mathbf{0} \end{array}\right]\left[\begin{array}{c} \mathbf{r}_1 \\ \hline \mathbf{r}_2 \end{array}\right] = \mathbf{0} \tag{5.48}$$

The last equation can be solved for $\mathbf{r}_2 = -\mathbf{K}_{22}^{-1}\mathbf{K}_{21}\mathbf{r}_1$ and on back-substituting,

$$[\mathbf{K}_c - \Omega^2\mathbf{M}_{11}]\mathbf{r}_1 = 0$$

where the condensed stiffness $\mathbf{K}_c = \mathbf{K}_{11} - \mathbf{K}_{12}\mathbf{K}_{22}^{-1}\mathbf{K}_{21}$. It is possible that all of the elements of a structure have masses which cannot be ignored, but that the set of displacements $\mathbf{r}_1$ have more associated mass than the others. The above simplification may then be too drastic, and an approximation has to be made in order to include the mass associated with $\mathbf{r}_2$. The Guyan reduction[58] condenses the stiffness matrix and the mass matrix without removing the missing elements in (5.48), that is, the submatrices $M_{12} = M_{21}$, and M_{22}. The previous reduction which had these masses as zero, is now assumed to be an approximation only, but we still put $\mathbf{r}_2 = -\mathbf{K}_{22}^{-1}\mathbf{K}_{21}\mathbf{r}_1$ and write the reduction as a transformation

$$\mathbf{r} = \begin{bmatrix} \mathbf{r}_1 \\ \mathbf{r}_2 \end{bmatrix} = \begin{bmatrix} \mathbf{I} \\ -\mathbf{K}_{22}^{-1}\mathbf{K}_{21} \end{bmatrix}\mathbf{r}_1 = \mathbf{T}\mathbf{r}_1$$

where $\mathbf{I}$ is the unit diagonal matrix of order $\mathbf{r}_1$. Viewing this as the usual coordinate transformation we would expect that $\mathbf{K}_c = \mathbf{T}^t\mathbf{K}\mathbf{T}$, which in fact it is. The Guyan approximation simply makes the assertion that the mass matrix transforms to the reduced $\mathbf{M}_c = \mathbf{T}^t\mathbf{K}\mathbf{T}$ also. This ensures that the inertia forces $\mathbf{M}\ddot{\mathbf{r}}$ still do the same virtual work over displacements $\bar{\mathbf{r}}$ that $\mathbf{M}_c\ddot{\mathbf{r}}_1$ do over $\bar{\mathbf{r}}_1$. This approximation raises the frequencies of the fundamental modes as we might expect, but the distortion of the complete frequency spectrum is tolerable if no large masses are present in $\mathbf{M}_{22}$.

5.10 Summary

The finite-element displacement method proceeds as follows.

1. Assume a displacement field $\mathbf{u} = \boldsymbol{\omega}\boldsymbol{\rho}_g$ in which the interpolation functions $\boldsymbol{\omega}$ are unity at the relevant node and zero at others, and have desired continuity between nodes.
2. $\boldsymbol{\varepsilon} = \mathbf{D}^t\mathbf{u} = \boldsymbol{\alpha}\boldsymbol{\rho}_g$: therefore find $\boldsymbol{\alpha} = \mathbf{D}^t\boldsymbol{\omega}$.
3. Evaluate element properties:

$$\mathbf{k}_g = \int_{V_g} \boldsymbol{\alpha}^t\boldsymbol{\kappa}\boldsymbol{\alpha}\,\mathrm{d}v; \qquad \mathbf{P}_g = \int_{V_g} \boldsymbol{\omega}^t\mathbf{p}_v\,\mathrm{d}v + \int_{S_g} \boldsymbol{\omega}^t\mathbf{p}_s\,\mathrm{d}s$$

4. Rotate element to align with global coordinates:

$$\mathbf{k}_g \to \mathbf{T}^t\mathbf{k}_g\mathbf{T}; \qquad \mathbf{P}_g \to \mathbf{T}\mathbf{P}_g$$

5. Assemble $\mathbf{K}$ and $\mathbf{R}$.
   (Symbolically, $\mathbf{K} = \sum \mathbf{a}_g^t\mathbf{k}_g\mathbf{a}_g; \qquad \mathbf{R} = \sum \mathbf{a}_g^t\mathbf{P}_g$)
6. Solve displacements.
   (Symbolically, $\mathbf{r} = \mathbf{K}^{-1}\mathbf{R}$)
7. Recover element displacements and stresses:

$$\boldsymbol{\rho}_g = \mathbf{a}_g\mathbf{r}; \qquad \boldsymbol{\rho}_g \to \mathbf{T}^{-1}\boldsymbol{\rho}_g = \mathbf{T}^t\boldsymbol{\rho}_g; \qquad \boldsymbol{\sigma} = \boldsymbol{\kappa}\boldsymbol{\alpha}\boldsymbol{\rho}_g$$

Isoparametric elements: $\mathbf{x} = \boldsymbol{\omega}\boldsymbol{\rho}$ as well. $\sum \omega_i = 1$.

Gross deformations: $\mathbf{K}_\mathrm{T} = \mathbf{K}_\mathrm{E} + \sigma\mathbf{K}_\mathrm{G}$

Initial buckling: $|\mathbf{K}_\mathrm{E} + \sigma\mathbf{K}_\mathrm{G}| = 0$

Large deformations: $\delta\mathbf{r} = \mathbf{K}_\mathrm{T}^{-1}\,\delta\mathbf{R}; \quad \mathbf{r} = \sum \delta\mathbf{r}; \quad \mathbf{R} = \sum \delta\mathbf{R}$

update $\quad \mathbf{x} \to \mathbf{x} + \delta\mathbf{r}; \quad \theta \to \theta + \delta\theta; \quad \mathbf{T} \to \mathbf{T} + \delta\mathbf{T}$

Dynamics:

$$\mathbf{m}_g = \int_{V_g} \boldsymbol{\omega}^t\rho\boldsymbol{\omega}\,\mathrm{d}v; \qquad \mathbf{M} = \sum \mathbf{a}_g^t\mathbf{m}_g\mathbf{a}_g$$

$$\mathbf{M}\ddot{\mathbf{r}} + \mathbf{K}\mathbf{r} = \mathbf{R}(t); \qquad \text{natural frequencies: } |\mathbf{K} - \Omega^2\mathbf{M}| = 0.$$

Problems

*P 5.1**

A slender tapered bar element with a simple uniaxial stress field is to be used and connected to the edge of plate elements, so the axial load may vary. Choosing three nodes – displacements $\{\rho_1 \quad \rho_2 \quad \rho_3\}$ at the ends and in the middle, construct the stiffness matrix in terms of the length l, and cross-sectional area which tapers linearly from $2A$ at one end to A at the other. Find the displacements for a simple uniform tension and compare with the exact solution.

P 5.2

The strain–stress law, including 'initial' strains $\boldsymbol{\eta}$, has been written (equation (2.2)) as $\boldsymbol{\varepsilon} = \boldsymbol{\phi}\boldsymbol{\sigma} + \boldsymbol{\eta}$, where $\boldsymbol{\eta}$ may be due to any cause other than stress. Premultiplying this by $\boldsymbol{\kappa} = \boldsymbol{\phi}^{-1}$, we obtain stresses as

$$\boldsymbol{\sigma} = \boldsymbol{\kappa}\boldsymbol{\varepsilon} - \boldsymbol{\kappa}\boldsymbol{\eta}$$

Show that if there are initial strains, we merely add to the kinematically equivalent element forces the extra vector

$$\mathbf{P}_g = \int_{V_g} \boldsymbol{\alpha}^t \boldsymbol{\kappa}\boldsymbol{\eta}\, dv$$

A beam element (Fig. 5.2) has a temperature variation given by $T = T_0 + cy$, where T_0 is the temperature at the centroid ($y = 0$) and c is the temperature gradient through the beam's symmetrical section. Both T_0 and c are constant along the element length. Find the thermal loading in terms of c and the coefficient of expansion α.

P 5.3

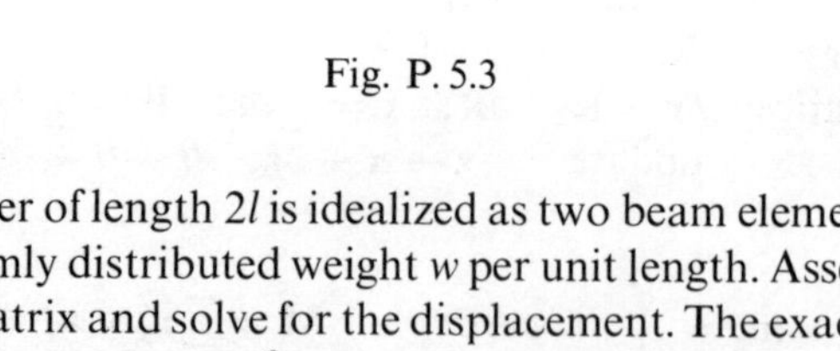

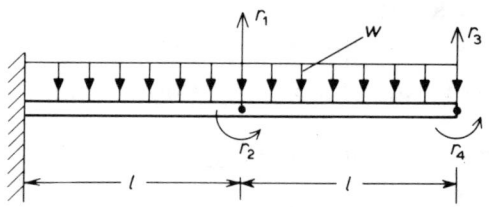

Fig. P. 5.3

A uniform cantilever of length $2l$ is idealized as two beam elements as shown. It is loaded by a uniformly distributed weight w per unit length. Assemble the stiffness matrix and load matrix and solve for the displacement. The exact solution is $r_1 = -0.708wl^4/EI$; $\quad r_2 = -1.167wl^3/EI$; $\quad r_3 = -2wl^4/EI$; $\quad r_4 = -1.333wl^3/EI$

Sketch also the bending moment, and compare with the exact solution.

P 5.4

In using four-sided, two-dimensional plane-stress elements, it is often a problem designing a graded mesh – see Fig. P 5.4. A useful transition element might be the

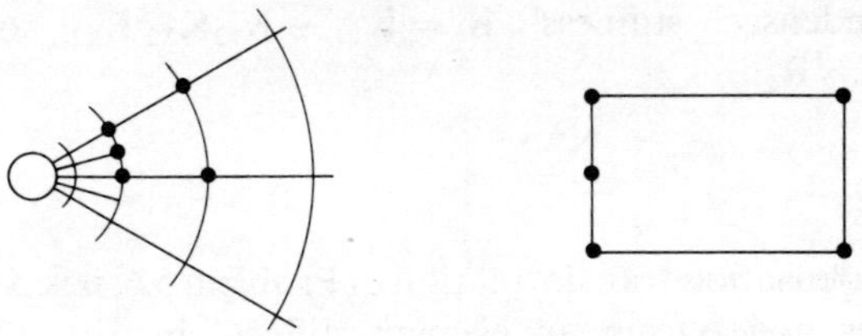

Fig. P. 5.4

five-noded element shown in the figure. Construct the interpolation functions for this element.

P 5.5

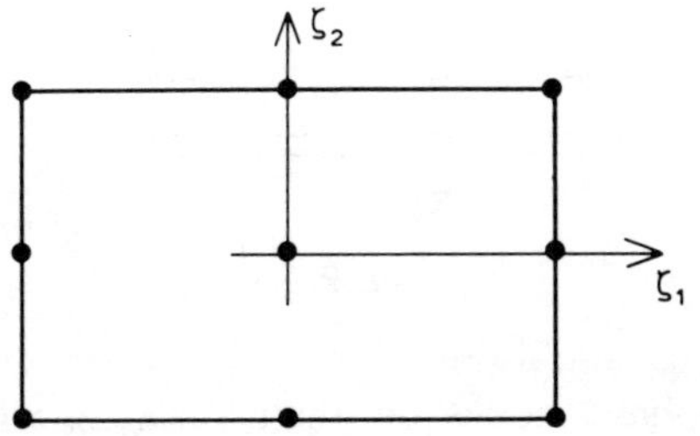

Fig. P. 5.5

Construct the Lagrange-type interpolation functions for the nine-noded plane stress element shown, using the correct products of ζ, $1 - \zeta$, $1 + \zeta$. Compare the nature of the strain field, associated with both centre and corner nodes, with that obtained from the element of Fig. 5.9.

P 5.6

A substructure within a larger structure has displacements $\mathbf{r}$ which can be partitioned into $\mathbf{r} = \{\mathbf{r}_1 \quad \mathbf{r}_2\}$ where the displacements $\mathbf{r}_1$ are confined to the surface of the substructure and $\mathbf{r}_2$ are interior nodal displacements. The applied loads can be similarly partitioned $\mathbf{R} = \{\mathbf{R}_1 \quad \mathbf{R}_2\}$ and the stiffness

$$\mathbf{K} = \begin{bmatrix} \mathbf{K}_{11} & \mathbf{K}_{12} \\ \mathbf{K}_{21} & \mathbf{K}_{22} \end{bmatrix}$$

In connecting this substructure to others it is convenient to retain only the displacements of the interface $\mathbf{r}_1$. Show that the behaviour of the substructure as a 'super-element' can be represented by the condensed form

$$\mathbf{K}_c \mathbf{r}_1 = \mathbf{R}_c$$

where the 'condensed stiffness' $\mathbf{K}_c = \mathbf{K}_{11} - \mathbf{K}_{12}\mathbf{K}_{22}^{-1}\mathbf{K}_{21}$ and the loading $\mathbf{R}_c = \mathbf{R}_1 - \mathbf{K}_{12}\mathbf{K}_{22}^{-1}\mathbf{R}_2$.

P 5.7

By using the condensation transformation of Problem 5.6, it is possible to remove internal or other nodes from an element stiffness matrix? Can you see any disadvantages in constructing the eight-noded stiffness matrix of section 5.3 by condensing the result of Problem 5.5?

P 5.8

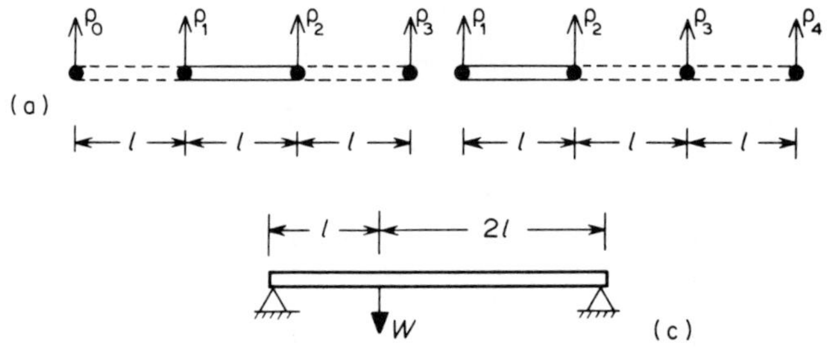

Fig. P. 5.8

Most finite elements have nodes on their surfaces, with internal nodes on sufferance if some extra degree of freedom is sought; but there seems no reason in principle why nodes should not be external to an element. Construct the cubic interpolation functions for the beam element shown, $v = \boldsymbol{\omega\rho}$, where the displacements $\boldsymbol{\rho}$ are at the four nodes shown in the figure, (a) for a general element, and (b) for an edge element. Construct the stiffness matrix, remembering that the virtual-work integral is carried out only *over the element.* Consider the problems in asembling a global stiffness matrix and also the advantage of a higher-order field with fewer unknowns. Solve the problem in Fig. P 5.8(c). Could you construct a special element with three degrees of freedom, $\{\rho_1 \quad \rho_2 \quad \rho_3\}$, to cope with a clamped-end boundary condition at node 1?

*P 5.9**

Using the general equation (5.42), evaluate the 4 × 4 geometric stiffness matrix for a pin-ended bar element under tension N, and aligned with the x-axis. Verify that the columns of this matrix may be generated (as in section 2.3) by imposing a unit displacement at a node and finding the forces needed to maintain it.

P 5.10

Use the geometric stiffness of Problem 5.9 as an approximation to the geometric stiffness (5.38) of a beam with varying curvature. Find the buckling load of a two-element strut with (a) clamped ends, (b) simply supported ends, and compare the buckling load with those predicted using the true geometric stiffness.

P 5.11

In the manner of equation (5.32) express separately the displacement and rotation of a beam element in terms of the displacement and rotation at the end nodes – using a linear interpolation of course. Find the element strains and, ignoring the fact that the shear strains are unlikely to be constant over a beam section, evaluate the element stiffness matrix exactly. Also evaluate the stiffness contribution from the shear strains approximately, using a single Gauss point in the centre of the beam. Using both stiffness matrices find the tip deflection of a single element cantilever due to a tip load and examine the solution as $GA \to \infty$.

*P 5.12**

Find the mass matrix of a uniform beam element, of density ρ and cross-sectional area A, using the displacement field of Fig. 5.2. Idealize a beam of length $2l$, and clamped at both ends, as two elements, and find the lowest natural frequency. (The exact answer is given by $\Omega^2 \rho A l^4/EI = 31.28$.) Replace the true mass matrix with the gross simplification that one half of an element mass should be associated with each end displacement and none at all with the rotation. Find the modified frequency. Did you expect this approximation to yield a *lower* frequency?

P 5.13

A uniform cantilever (ρ, A, EI, and l again) is supported at its tip by a light spring whose stiffness is given by $k = 5EI/l^3$. Construct the stiffness matrix using a single element for the beam, and show that the lowest natural frequency is given by $\Omega^2 \rho A l^4/EI = 32.2$. (The exact value is 31.38.)

6

EQUILIBRIUM ELEMENTS AND OTHERS

6.1 Introduction

The generation of equilibrium elements should be a logical and predictable continuation of all that has so far been done in earlier chapters, although their utility has yet to be proved. After all, the PVF was the logical dual of the PVD and we have already successfully discretized that. Admittedly there were some restrictions in using the PVF which we created in Chapter 3 when we resorted to self-equilibrating stress systems to avoid introducing displacements into the solution. On the positive side, it was found that an approximate PVF solution was an overflexible upper bound so it is tempting to speculate whether we can use equilibrium elements in conjunction with displacement elements and so obtain upper and lower bounds to the correct solution.[60] Another temptation is that a direct solution in terms of stresses will avoid having to differentiate an approximate displacement field and so it should be more accurate for an equivalent number of unknowns. Possible also is a solution in terms of nodal forces which guarantees continuity of tractions between elements even for structures with discontinuous section properties. The problems we met in finding suitable plate-bending elements may yield to a force solution also.

An advantage of the force method, at least for simple frameworks, is the fact that the number of self-equilibrating redundant systems is likely to be much smaller than the unknowns in a displacement solution. This advantage is now considered old hat. The convenience of setting up a displacement finite-element system far outweighs the inconvenience of storing and solving a large number of displacements. This fact of life we will face up to when we have developed a few equilibrium elements.

In formulating equilibrium elements we should approach the problem as the dual of the finite-element displacement approximation. Thus, instead of assuming displacement fields in terms of nodal displacements and then satisfying compatibility, we should assume stress fields which satisfy equilibrium. But do we express a stress field in terms of nodal stresses and interpolation functions? To do so would depart from the spirit of duality, since the dual of a nodal displacement is a nodal force and not a stress. The dual of an assumed stress field, for which $\mathbf{D}\boldsymbol{\sigma} + \mathbf{p}_v = \mathbf{0}$ has been integrated, is actually an assumed *strain* field which has to be integrated to satisfy $\boldsymbol{\varepsilon} = \mathbf{D}^t\mathbf{u}$ and there are only one or two occasions where this has ever been done with advantage.[61] Actually, there is no conflict between prescribing a stress field and then using nodal forces to measure it; after all, the nodal forces in the finite-element displacement formulation are only 'measures' of stress level, so they need only be that here. The beam elements we used were conveniently labelled with physically imaginable shear force and bending moments at element ends, but if we choose to start looking at two- and three-dimensional equilibrium elements then such nodal forces cease to lend themselves to physical visualization. Moreover, the generation of suitable self-equilibrating systems may be fairly straightforward in frameworks but tricks have to be used in a continuum. Nevertheless, the attraction of 'nodal force' measures is overwhelming even in a continuum and we shall have to be able to ensure static compatibility between elements by equating nodal forces at their interface. To bring together nodal stresses at an interface is not so appealing, since there are three stress components in two dimensions, but only continuity of the normal and tangential tractions is required. A valuable bonus of using nodal force measures will be that they have corresponding nodal displacements which we can use to deceive the computer into accepting the element as a displacement model, with advantage. Of course if nodal forces are only abstract measures, the corresponding displacements are even more remote from the real ones.

This preamble about the use and nature of nodal forces and displacements in equilibrium elements is a necessary intellectual safeguard since some of the ideas which were physically obvious for frameworks are positively confusing for a continuum. For instance, it is physically quite clear that we may cut a redundant framework and then restore this violation of displacement compatibility, but it is far from obvious which sort of compatibility we have destroyed in a continuum.

6.2 Plane stress elements

Consider then the problem of generating a stress field satisfying $\mathbf{D}\boldsymbol{\sigma} + \mathbf{p}_v = \mathbf{0}$. The body-force particular integral can presumably be integrated if we know how $\mathbf{p}_v$ varies, so we have to find the self-equilibrating stresses satisfying $\mathbf{D}\boldsymbol{\sigma} = \mathbf{0}$, which in two dimensions reduces to

$$\begin{aligned} \frac{\partial \sigma_{xx}}{\partial x} + \frac{\partial \sigma_{xy}}{\partial y} &= 0 \\ \frac{\partial \sigma_{yy}}{\partial y} + \frac{\partial \sigma_{xy}}{\partial x} &= 0 \end{aligned} \qquad (6.1)$$

One way to generate polynomials which satisfy these equations is to assume $\sigma_{xy}(x, y)$ and integrate both equations to find σ_{xx} and σ_{yy}. (In the special case $\sigma_{xy} =$ constant, then σ_{xx} can be any function of y, and σ_{yy} any function of x.) We can therefore quickly establish a series of equilibrium fields of any order, thus:

Order zero:

$$\sigma_{xx} = \text{constant} \qquad \text{(i)}$$
$$\sigma_{yy} = \text{constant} \qquad \text{(ii)}$$
$$\sigma_{xy} = \text{constant} \qquad \text{(iii)}$$

Order one:

$$
\begin{array}{llll}
\sigma_{xy} = 0, & \sigma_{xx} = y, & \sigma_{yy} = 0 & \text{(iv)} \\
\sigma_{xy} = 0, & \sigma_{xx} = 0, & \sigma_{yy} = x & \text{(v)} \\
\sigma_{xy} = -x, & \sigma_{xx} = 0, & \sigma_{yy} = y & \text{(vi)} \\
\sigma_{xy} = -y, & \sigma_{xx} = x, & \sigma_{yy} = 0 & \text{(vii)}
\end{array}
\qquad (6.2)
$$

Order two:

$$
\begin{array}{llll}
\sigma_{xy} = 0, & \sigma_{xx} = y^2, & \sigma_{yy} = 0 & \text{(viii)} \\
\sigma_{xy} = 0, & \sigma_{xx} = 0, & \sigma_{yy} = x^2 & \text{(ix)} \\
\sigma_{xy} = -x^2, & \sigma_{xx} = 0, & \sigma_{yy} = 2xy & \text{(x)} \\
\sigma_{xy} = -y^2, & \sigma_{xx} = 2xy, & \sigma_{yy} = 0 & \text{(xi)} \\
\sigma_{xy} = -xy, & \sigma_{xx} = \tfrac{1}{2}x^2, & \sigma_{yy} = \tfrac{1}{2}y^2 & \text{(xii)}
\end{array}
$$

and so on.

The number of available stress fields, S, for polynomials of order n is

$$S = \sum_{i=0}^{n} (i + 3)$$

Now we have to order the amplitude of these interpolated fields in terms of nodal forces. Consider the two extreme cases (i) and (x), and suppose we wish to employ a simple rectangular element aligned to the x and y directions. The stresses in the two fields have resultants on the four faces which are summarized in Fig. 6.1. The

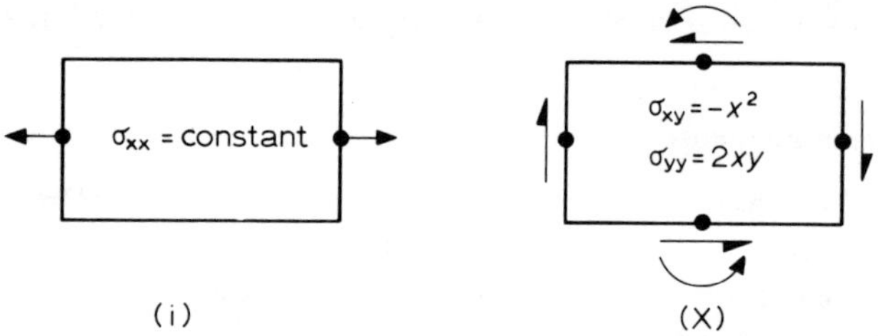

Fig. 6.1 Self-equilibrating fields in a rectangle

nodal forces have been summed as if the chosen nodes were in the centre of each face. This is not necessary: after all, both the type of nodal force and its position merely have to be a unique measure of the tractions $(\mathbf{D}n)\boldsymbol{\sigma}$ on the edge, so that when the nodal forces $\mathbf{P}_g$ on an element's surface are equated to those on the adjacent element, then continuity of $(\mathbf{D}n)\boldsymbol{\sigma}$ is assured. The PVF requires us to satisfy both internal and surface equilibrium, so the order of $\mathbf{P}_g$ must be just right. Too few nodal forces and the stress field will not be uniquely defined by them. Too many and the alloction of edge tractions to them becomes arbitrary. Now the nodal forces $\mathbf{P}_g$ on a single edge do not know whether the stress field is self-equilibrating; indeed, there may be body forces in which case the set of $\mathbf{P}_g$ will have a resultant. For this reason we now find it convenient to distinguish between a set of self-equilibrating forces arising from fields like (6.2), and those arising from the body forces which will have a three-component resultant in two dimensions. Suppose we define the magnitudes of the stress fields in (6.2) by a vector $\mathbf{P}_N$; thus P_{N1} is the magnitude of (i), σ_{xx} = constant, etc. Then by resolving the tractions on an element's edges we will obtain equations of the form

$$\mathbf{P}_g = \mathbf{B}_N \mathbf{P}_N \tag{6.3}$$

The body forces will similarly be distributed to the nodes in some statically equivalent fashion as:

$$\mathbf{P}_g = \mathbf{B}_0 \mathbf{P}_0 \tag{6.4}$$

where the three components of $\mathbf{P}_0$ measure the two resultant force components and the couple. Thus the transformation between $\mathbf{P}_g$ and the others is in general

$$\mathbf{P}_g = \mathbf{B}_N \mathbf{P}_N + \mathbf{B}_0 \mathbf{P}_0 = [\mathbf{B}_N \quad \mathbf{B}_0] \begin{bmatrix} \mathbf{P}_N \\ \mathbf{P}_0 \end{bmatrix} \tag{6.5}$$

The choice of subscript N is not accidental; it referred previously to 'natural' forces being responsible for the *deformations* of an element – a bar in Chapter 2 – and the 'natural' stiffness[62] was a relationship between these natural forces and the deformation modes $\boldsymbol{\rho}_N$. $\mathbf{P}_N$ and $\boldsymbol{\rho}_N$ must *correspond*–that is, the work done equals $\mathbf{P}_N^t \boldsymbol{\rho}_N$ – which means that $\boldsymbol{\rho}_N$ are *generalized* displacements since the forces $\mathbf{P}_N$ refer to patterns of nodal forces. The natural stiffness is nonsingular since it involves no rigid-body modes, and when inverted it gives the flexibility, as we shall see again.

We now have a new problem akin to that of the incomplete displacement fields of Chapter 5. Suppose we select stress fields of order zero, or one, or two in (6.2); then adding resultants $\mathbf{P}_0$ to them produces a set $\mathbf{P}_g$ of order 6, 10, or 15. But if $\mathbf{P}_g$ is to be assembled as a collection of nodal forces on the element edges, there should be an equal number on each side. This means we could have a triangular element, with zero-order stress field, and two nodal forces on each side; but the first possible four-sided complete element would be a fifth-order field with $\mathbf{P}_N$ of order 33 and having therefore 36 nodal freedoms! Suppose then we abandon the quest for completeness, and settle for a system $\mathbf{P}_N$ and $\mathbf{P}_0$ which is a multiple of 4 for a rectangle. If we aim for a complete linear field, then $\mathbf{P}_N$ is of order at least 7 from (6.2), so we require two more modes from the set (6.2) (viii) to (xii) to make $\mathbf{P}_N$ of order 9 and $\mathbf{P}_g$ order 12; that is, three nodal forces per side. These could be distributed as some asymmetric mix of normal and tangential forces, but we choose the system of forces and moments shown in Fig. 6.2, that is,

$$\mathbf{P}_g = \{N_1 \ Q_1 \ M_1/b \ N_2 \ Q_2 \ M_2/a \ N_3 \ Q_3 \ M_3/b \ N_4 \ Q_4 \ M_4/a\} \tag{6.6}$$

Suppose we augment the linear field with the two quadratic $\sigma_{xx} = y^2$ (viii) and $\sigma_{yy} = x^2$ (ix). If the resultant of the stresses σ_{xx} on $x = a$ is put equal to N_1, this will not define the magnitude of this field since N_1 has to be related to the resultant of σ_{xx} = constant (i) also. In other words, both modes (i) and (viii) are statically equivalent to a single force on the edge $x = a$. We cannot add more normal nodal forces to distinguish between the two modes since we would then run out of the necessary tangential shear components. This is a real problem with

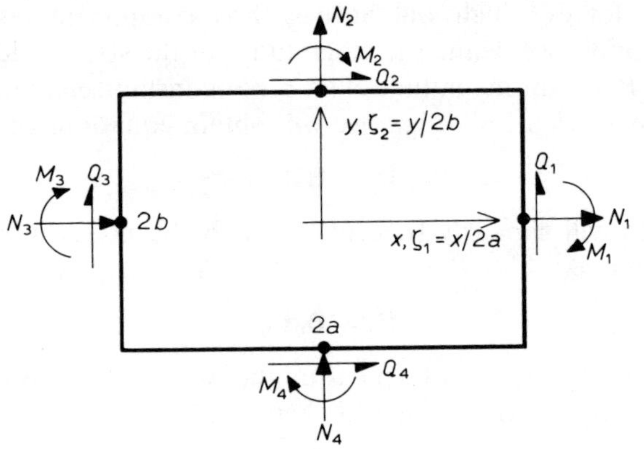

Fig. 6.2 Ninth order equilibrium stress field

high-order equilibrium fields and we have deliberately chosen a non-simple field to illustrate the first of many such problems. It does explain why only the simplest equilibrium elements have been advocated by even their staunchest supporters;[27] and the problems multiply for oblique boundaries. However, in this rectangle it is possible to select modes (x) and (xi) which are uniquely defined in terms of our chosen nodal forces. The complete set of self-equilibrating fields is summarized below for a plate of aspect ratio $\lambda = a/b$ and thickness t. Zero values of stress or force are not quoted.

P_{N1}: $\sigma_{xx} = 1$; $N_1 = -N_3 = 2bt$
P_{N2}: $\sigma_{yy} = 1$; $N_2 = -N_4 = 2at$
P_{N3}: $\sigma_{xy} = 1$; $Q_1 = -Q_3 = 2bt$, $Q_2 = -Q_4 = 2at$
P_{N4}: $\sigma_{xx} = \zeta_1$, $\sigma_{xy} = -\zeta_2/\lambda$; $N_1 = N_3 = -Q_2 = -Q_4 = 2bt$
P_{N5}: $\sigma_{yy} = \zeta_2$, $\sigma_{xy} = -\lambda\zeta_1$; $N_2 = N_4 = -Q_1 = -Q_3 = 2at$
P_{N6}: $\sigma_{xx} = \zeta_2$; $M_1/b = -M_3/b = 2bt/3$
P_{N7}: $\sigma_{yy} = \zeta_1$; $-M_2/a = M_4/a = 2at/3$
P_{N8}: $\sigma_{xx} = \zeta_1\zeta_2$, $\sigma_{xy} = -\zeta_2^2/2\lambda$; $M_1/b = M_3/b = 2bt/3$
  $-Q_1 = Q_3 = bt/3\lambda$, $-Q_2 = Q_4 = bt$
P_{N9}: $\sigma_{xy} = \zeta_1\zeta_2$, $\sigma_{xy} = -\lambda\zeta_1^2/2$; $M_2/a = M_4/a - 2at/3$
  $-Q_1 = Q_3 = at$, $-Q_2 = Q_4 = \lambda at/3$

The stress field can now be summarized in terms of these modes:

$$\boldsymbol{\sigma} = \boldsymbol{\beta}\mathbf{P}_N \tag{6.7}$$

where

$$\mathbf{P}_N = \{P_{N1} \quad P_{N2} \quad P_{N3} \quad P_{N4} \quad P_{N5} \quad P_{N6} \quad P_{N7} \quad P_{N8} \quad P_{N9}\} \tag{6.8}$$

and

$$\boldsymbol{\beta} = \begin{bmatrix} 1 & 0 & 0 & \zeta_1 & 0 & \zeta_2 & 0 & \zeta_1\zeta_2 & 0 \\ 0 & 1 & 0 & 0 & \zeta_2 & 0 & \zeta_1 & 0 & \zeta_1\zeta_2 \\ 0 & 0 & 1 & -\zeta_2/\lambda & -\lambda\zeta_1 & 0 & 0 & -\zeta_2^2/2\lambda & \lambda\zeta_1^2/2 \end{bmatrix} \tag{6.9}$$

This establishes the internal stress field and we should hope that continuous tractions between elements would be secured by matching nodal forces. Unfortunately this is not so. We have already seen how two stress fields can give rise to the same resultant nodal force, in other words, the difference between the two fields is a self-equilibrating edge stress whose magnitude is not controllable. This has happened again in the above field, and if we care to examine the P_{N4} and P_{N5} fields we see that there are linearly varying shears with no resultant on the edges. Although this self-equilibrating error can be expected to be local, if not negligible, it will give rise to discontinuous shear stresses. In other words, the element *is not statically conformal.*

The resultant forces $\mathbf{P}_0$ on the element can be summarized by a force in the x direction, in the y direction, and by a pure couple which we allocate equally to all four nodes. The complete array of natural and resultant forces (6.5) is now $[\mathbf{B}_N \mid \mathbf{B}_0]$ as shown in the following array:

$$[\mathbf{B}_N \mid \mathbf{B}_0] = 2bt \left[\begin{array}{cccccccc|ccc}
1 & & 1 & & & & & & 1 & & \\
 & 1 & & -\lambda & & & -1/6\lambda & -\lambda/2 & & 1 & \\
 & & & & 1/3 & & 1/3 & & & & 1 \\
\lambda & & & \lambda & & & & & & 1 & \\
 & \lambda & -1 & & & & -1/2 & -\lambda^2/6 & 1 & & \\
 & & & & & -\lambda/3 & & -\lambda/3 & & & \lambda \\
-1 & & 1 & & & & & & 1 & & \\
 & -1 & & \lambda & & & 1/6\lambda & \lambda/2 & & 1 & \\
 & & & & -1/3 & & 1/3 & & & & 1 \\
-\lambda & & & -\lambda & & & & & & 1 & \\
 & -\lambda & -1 & & & & 1/2 & \lambda^2/6 & 1 & & \\
 & & & & & \lambda/3 & & -\lambda/3 & & & \lambda
\end{array}\right] \tag{6.10}$$

Having established the stresses (6.7), the next step is the internal virtual work for the element:

$$\int_{V_g} \bar{\boldsymbol{\sigma}}^t \boldsymbol{\varepsilon} \, dv = \bar{\mathbf{P}}_N^t \int_{V_g} \boldsymbol{\beta}^t \boldsymbol{\phi} \boldsymbol{\beta} \, dv \mathbf{P}_N = \bar{\mathbf{P}}_N^t \mathbf{f}_N \mathbf{P}_N \tag{6.11}$$

where, as in Chapter 3,

$$\mathbf{f}_N = \int_{V_g} \boldsymbol{\beta}^t \boldsymbol{\phi} \boldsymbol{\beta} \, dv \tag{3.38}$$

The material flexibility matrix for isotropic plane stress is

$$\boldsymbol{\phi} = \frac{1}{E} \begin{bmatrix} 1 & -\nu & 0 \\ -\nu & 1 & 0 \\ 0 & 0 & 2(1+\nu) \end{bmatrix}$$

On substituting (6.9) for $\boldsymbol{\beta}$ we find the element flexibility

$$\mathbf{f}_N = \frac{4abt}{3E}\begin{bmatrix} 3 & -3\nu & & & & & & & \\ & 3 & & & & & & & \\ & & 6(1+\nu) & & & & & -(1+\nu)/\lambda & -(1+\nu)\lambda \\ & & & 1+2(1+\nu)/\lambda^2 & & & -\nu & & \\ & & & & 1+2\lambda^2(1+\nu) & -\nu & & & \\ & \text{symm.} & & & & 1 & & & \\ & & & & & & 1 & & \\ & & & & & & & \dfrac{1}{3}+\dfrac{3(1+\nu)}{10\lambda^2} & \dfrac{(1-\nu)}{6} \\ & & & & & & & & \dfrac{1}{3}+\dfrac{3(1+\nu)\lambda^2}{10} \end{bmatrix} \tag{6.12}$$

The next stage in a conventional force solution is to sum (6.11); and (3.37) to create the necessary self-equilibrating systems by satisfying overall equilibrium $\mathbf{P}_g = \mathbf{b}_{0g}\mathbf{R} + \mathbf{b}_{xg}\mathbf{X}$. Unfortunately, we have the virtual work in terms of $\mathbf{P}_N$ and not $\mathbf{P}_g$, so we should first invert (6.5) with (6.10) to obtain

$$\begin{bmatrix} \mathbf{P}_N \\ \mathbf{P}_0 \end{bmatrix} = [\mathbf{B}_N \quad \mathbf{B}_0]^{-1}\mathbf{P}_g$$

For high-order elements with complex geometry this inversion is probably unavoidble, but for this current element we can use the PVD to obtain $\mathbf{P}_N$ directly in terms of $\mathbf{P}_g$. Suppose we let the displacements *corresponding* to $\mathbf{P}_g$, $\mathbf{P}_N$, and $\mathbf{P}_0$ be $\boldsymbol{\rho}_g$, $\boldsymbol{\rho}_N$ and $\boldsymbol{\rho}_0$; that is, the generalized displacements over which these named forces do work. The nodal displacements $\boldsymbol{\rho}_g$ are clearly a mixture of displacements and rotations corresponding to $\mathbf{P}_g$ in (6.6); the three rigid-body movements are also obvious. In an analogous fashion to (6.5) we may write

$$\boldsymbol{\rho}_g = [\mathbf{A}_N \quad \mathbf{A}_0]\begin{bmatrix} \mathbf{P}_N \\ \boldsymbol{\rho}_0 \end{bmatrix} \tag{6.13}$$

The PVD then becomes, equating virtual work,

$$\mathbf{P}_N^t\bar{\boldsymbol{\rho}}_N + \mathbf{P}_0^t\bar{\boldsymbol{\rho}}_0 = \mathbf{P}_g^t\bar{\boldsymbol{\rho}}_g = \mathbf{P}_g^t\mathbf{A}_N\bar{\boldsymbol{\rho}}_N + \mathbf{P}_g^t\mathbf{A}_0\bar{\boldsymbol{\rho}}_0$$

As this is true for any $\bar{\boldsymbol{\rho}}_N$ and $\bar{\boldsymbol{\rho}}_0$, we have

$$\mathbf{P}_N = \mathbf{A}_N^t\mathbf{P}_g \tag{6.14}$$

and

$$\mathbf{P}_0 = \mathbf{A}_0\mathbf{P}_g \tag{6.15}$$

Inversion is therefore completely avoided if $\mathbf{A}_N$ can be found. Now $\mathbf{A}_N$ in (6.13) can be generated, as in the unit displacement method, by selecting a natural mode, say $\rho_{Ni} = 1$, corresponding to P_{Ni} one at a time, and noting the $\boldsymbol{\rho}_g$ pattern. To find a particular $\boldsymbol{\rho}_g$ pattern due to $\rho_{Ni} = 1$ we use the fact that $\mathbf{P}_N$ correspond to $\boldsymbol{\rho}_N$ so putting $P_{Ni} = 1$ produces a known pattern $\mathbf{P}_g = \mathbf{B}_N\mathbf{P}_N$ (equation (6.3)) which must do unit work over this $\boldsymbol{\rho}_g$ pattern; but no other component of $\mathbf{B}_N$ must do work over this $\boldsymbol{\rho}_g$. This then fixes the $\boldsymbol{\rho}_g$ pattern. Mathematically, all we are saying is that $\mathbf{B}_N^t\mathbf{A}_N = \mathbf{I} = \mathbf{A}_N^t\mathbf{B}_N$, which follows from the requirement that

$$\begin{bmatrix} \mathbf{A}_N \\ \mathbf{A}_0 \end{bmatrix} [\mathbf{B}_N \quad \mathbf{B}_0] = \mathbf{I}$$

The generated $\boldsymbol{\rho}_g$ patterns are all straightforward except perhaps $\boldsymbol{\rho}_{N3}$ which needs thinking about. We find

$$\begin{array}{l} \rho_{N1} \rightarrow \\ \rho_{N2} \rightarrow \\ \rho_{N3} \rightarrow \\ \rho_{N4} \rightarrow \\ \rho_{N5} \rightarrow \\ \rho_{N6} \rightarrow \\ \rho_{N7} \rightarrow \\ \rho_{N8} \rightarrow \\ \rho_{N9} \rightarrow \end{array} \quad \mathbf{A}_N^t = \frac{1}{2bt} \begin{bmatrix} 1/2 & & & & & & -1/2 & & & & & \\ & & & 1/2\lambda & & & & & & -1/2\lambda & & \\ & 1/4 & 1/2\lambda & & 1/4\lambda & -1/2 & & -1/4 & 1/2\lambda & & -1/4\lambda & 1/2 \\ 1/4 & & & & -1/4 & & 1/4 & & & & -1/4 & \\ & -1/4\lambda & & 1/4\lambda & & & & -1/4\lambda & & 1/4\lambda & & \\ & & 3/2 & & & & & & -3/2 & & & \\ & & & & & -3/2\lambda & & & & & & 3/2\lambda \\ & & 3/2 & & & & & & 3/2 & & & \\ & & & & & -3/2\lambda & & & & & & -3/2\lambda \end{bmatrix}$$

The element virtual work can now be written in the required form

$$\overline{\mathbf{P}}_N^t\mathbf{f}_N\mathbf{P}_N = \overline{\mathbf{P}}_g^t\mathbf{A}_N^t\mathbf{f}_N\mathbf{A}_N\mathbf{P}_g = \overline{\mathbf{P}}_g^t\mathbf{f}_g\mathbf{P}_g \qquad (6.16)$$

This expression is identical to the formulation for frameworks in Chapter 3 (equation (3.38)) and it now awaits the instruction $\mathbf{P}_g = \mathbf{b}_{0g}\mathbf{R} + \mathbf{b}_{xg}\mathbf{X}$ and $\overline{\mathbf{P}}_g = \mathbf{b}_{xg}\overline{\mathbf{X}}$, before repeating the force solution procedure of equation (3.43). But now – the final nail in the coffin of the force method – we must generate the self-equilibrating $\mathbf{P}_g = \mathbf{b}_{xg}\mathbf{X}$ systems which we recall should be as local as possible. For a rectangular mesh it is not too difficult to use collections of four panels as in Chapter 3, but for general quadrilaterals the task is much more complex. The overlapping systems illustrated in Fig. 6.3 will suffice. Figure 6.3 can be turned

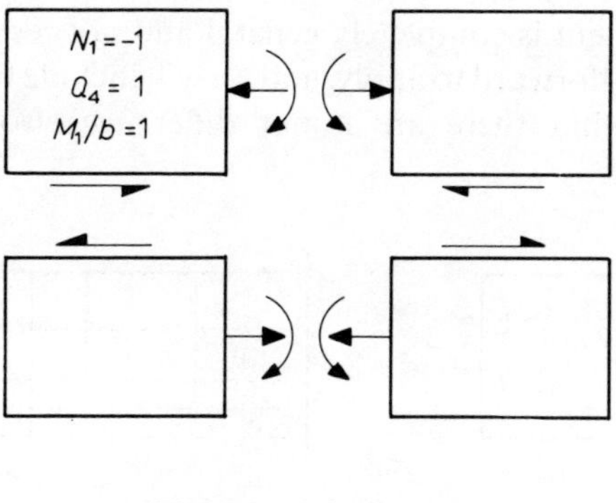

Fig. 6.3 A $\mathbf{b}_{xg}$ system

through a right angle for another type of independent system, and a third system shown in Fig. 6.4 consists of pure moments. In a comparative study of force and displacement methods,[65] a force solution using these systems has been obtained to the diffusion problem shown below in Fig. 6.5, and using a total of 60 $\mathbf{X} \rightarrow \mathbf{b}_{xg}$ systems: the results will be discussed shortly. However, the troublesome generation of both the element matrices $\mathbf{A}_N$ and the system matrices $\mathbf{b}_{xg}$ is only straightforward in this rectangular structure. Moreover, the danger of

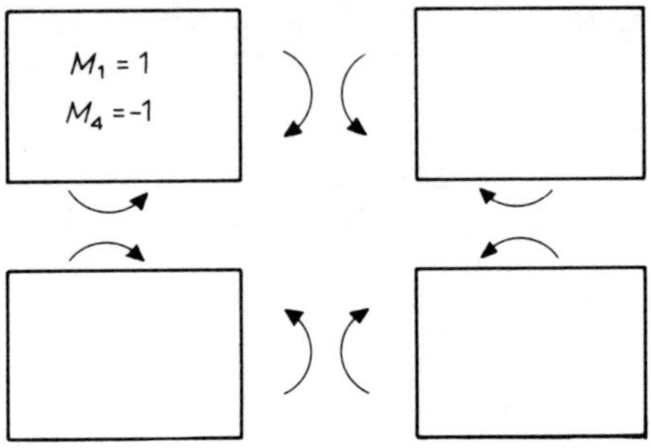

Fig. 6.4 Another $\mathbf{b}_{xg}$ system

programming errors is always much more of a threat than in the routine displacement finite elements. Therefore an alternative way of assembling this equilibrium element will now be shown which simulates a displacement formulation even though we are still using a genuine equilibrium element; it is thus much more systematic and can in fact be incorporated into standard displacement finite-element programmes. In view of the entrenched position of the latter in practice, this is a very attractive feature. Although in the example chosen, the displacement unknowns increase to 218 compared with 60, because of the intricate nature of the normal force formulation, we find that the computing time actually *decreases* by 40 per cent. The need for the matrix $\mathbf{A}_N^t$, orthogonal to $\mathbf{B}_N$, is also avoided.

6.3 The simulated stiffness method[63]

The following argument is completely general and not restricted to rectangular elements. It is straightforward to apply, and we will include body forces and initial strains $\boldsymbol{\eta}$ to show that there are minor differences from the true stiffness formulation.

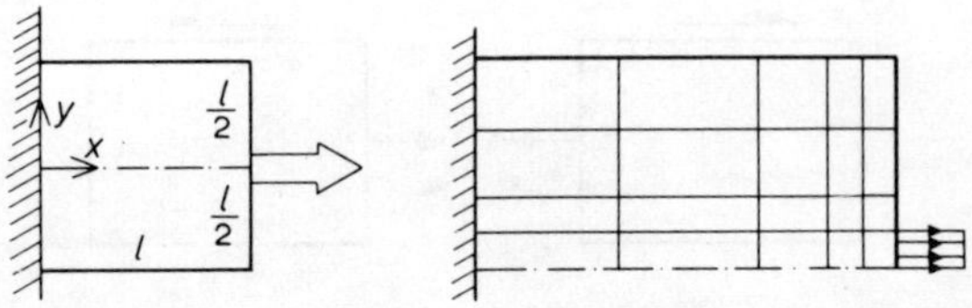

Fig. 6.5 A square plate diffusion problem

We begin with self-equilibrating solutions to the equations of equilibrium, $\mathbf{D\sigma} = \mathbf{0}$, which we write again as $\boldsymbol{\sigma} = \boldsymbol{\beta}\mathbf{P}_N$ where $\mathbf{D}\boldsymbol{\beta} = \mathbf{0}$. If there are body forces we presumably can obtain a particular integral of $\mathbf{D\sigma} + \mathbf{P}_v = \mathbf{0}$, so we write (6.7) as

$$\boldsymbol{\sigma} = \boldsymbol{\beta}\mathbf{P}_N + \boldsymbol{\beta}_0 \tag{6.17}$$

This equation can be inserted into the boundary tractions $(\mathbf{D}n)\boldsymbol{\sigma}$ and a selected and sufficient number of nodal forces $\mathbf{P}_g$ put equal to $\int(\mathbf{D}n)\boldsymbol{\sigma}\,\mathrm{d}s$ over element edges, as before.

$$\text{Then} \qquad \mathbf{P}_g = \mathbf{B}_N\mathbf{P}_N + \mathbf{B}_0\mathbf{P}_0 \qquad \text{again} \tag{6.5}$$

Having introduced the concept of corresponding displacements $\boldsymbol{\rho}_g$, we now use this fact. We recall that the virtual work $\mathbf{P}_g^t\boldsymbol{\rho}_g$ is merely an equivalence; that is,

$$\bar{\mathbf{P}}_g^t\boldsymbol{\rho}_g = \int_{S_g} [(\mathbf{D}n)\bar{\boldsymbol{\sigma}}]^t\mathbf{u}\,\mathrm{d}s \tag{6.18}$$

and the real displacements $\mathbf{u}$ can never be extracted from this expression. Applying the PVF to a single element, and remembering that $\bar{\mathbf{p}}_v = \mathbf{0}$ to eliminate internal displacement variables, we have

$$\int_{V_g} \bar{\boldsymbol{\sigma}}^t\boldsymbol{\varepsilon}\,\mathrm{d}v = \int_{S_g} \bar{\mathbf{p}}_s^t\mathbf{u}\,\mathrm{d}s \tag{6.19}$$

But equilibrium requires that $\bar{\mathbf{p}}_s = (\mathbf{D}n)\bar{\boldsymbol{\sigma}}$, therefore the right-hand side of this equation must be

$$\int_{S_g} \bar{\mathbf{p}}_s^t\mathbf{u}\,\mathrm{d}s = \int_{S_g} [(\mathbf{D}n)\bar{\boldsymbol{\sigma}}]^t\mathbf{u}\,\mathrm{d}s = \bar{\mathbf{P}}_g^t\boldsymbol{\rho}_g$$

Also $\bar{\boldsymbol{\sigma}} = \boldsymbol{\beta}\bar{\mathbf{p}}_N$ in (6.17) if $\bar{\mathbf{p}}_v = 0$, so using $\boldsymbol{\varepsilon} = \boldsymbol{\phi}\boldsymbol{\sigma} + \boldsymbol{\eta}$ equation (6.19) becomes

$$\bar{\mathbf{p}}_N \int_{V_g} [\boldsymbol{\beta}^t(\boldsymbol{\phi}\boldsymbol{\beta}\mathbf{P}_N + \boldsymbol{\phi}\boldsymbol{\beta}_0) + \boldsymbol{\eta}]\,\mathrm{d}v - \bar{\mathbf{P}}_g^t\boldsymbol{\rho}_g = 0$$

or using $\bar{\mathbf{P}}_g = \mathbf{B}_N\bar{\mathbf{P}}_N$,

$$\bar{\mathbf{P}}_N^t\,[\mathbf{f}_N\mathbf{P}_N + \mathbf{f}_0 - \mathbf{B}_N^t\boldsymbol{\rho}_g] = 0$$

where

$$\mathbf{f}_N = \int_{V_g} \boldsymbol{\beta}^t\boldsymbol{\phi}\boldsymbol{\beta}\,\mathrm{d}v \qquad \text{again} \tag{6.11}$$

and

$$\mathbf{f}_0 = \int_{V_g} \boldsymbol{\beta}^t(\boldsymbol{\phi}\boldsymbol{\beta}_0 + \boldsymbol{\eta})\,\mathrm{d}v$$

The above is true for all $\bar{\mathbf{P}}_N$, so

$$\mathbf{P}_N = \mathbf{f}_N^{-1}\mathbf{B}_N^t\boldsymbol{\rho}_g - \mathbf{f}_N^{-1}\mathbf{f}_0 \tag{6.20}$$

and using (6.5),

$$\mathbf{P}_g = [\mathbf{B}_N \mathbf{f}_N^{-1} \mathbf{B}_N^t] \boldsymbol{\rho}_g - \mathbf{B}_N \mathbf{f}_N^{-1} \mathbf{f}_0 + \mathbf{B}_0 \mathbf{P}_0 \tag{6.21}$$

We have therefore created a new element stiffness

$$\mathbf{k}_g = \mathbf{B}_N^t \mathbf{f}_N^{-1} \mathbf{B}_N \tag{6.22}$$

It is interesting to notice that this stiffness is the Argyris form mentioned in earlier chapters, $\mathbf{k}_g = \mathbf{a}_N^t \mathbf{k}_N \mathbf{a}_N$, with the 'natural' stiffness $\mathbf{k}_N = \mathbf{f}_N^{-1}$. However, we have not generated $\mathbf{a}_N$ by kinematic reasoning, but have found $\mathbf{B}_N$ by equilibrium arguments. It now remains to integrate any element surface forces and collect them at our chosen nodes as a global list of statically equivalent surface forces $\mathbf{R}$. The equilibrium equations connecting $\mathbf{P}_g$ and $\mathbf{R}$ must be of the usual form

$$\sum_g \mathbf{a}_g^t \mathbf{P}_g = \mathbf{R} \tag{6.23}$$

since if we apply the PVF

$$\sum_g \bar{\mathbf{P}}_g^t \boldsymbol{\rho}_g = \bar{\mathbf{R}}^t \mathbf{r}$$

and the above produces

$$\sum_g \bar{\mathbf{P}}_g^t \boldsymbol{\rho}_g = [\textstyle\sum \mathbf{a}_g^t \bar{\mathbf{P}}_g]^t \mathbf{r}$$

or

$$\sum_g \bar{\mathbf{P}}_g^t [\boldsymbol{\rho}_g - \mathbf{a}_g \mathbf{r}] = 0 \qquad \text{for all } \mathbf{P}_g$$

Thus $\boldsymbol{\rho}_g = \mathbf{a}_g \mathbf{r}$ for all g elements, and we can therefore enforce global equilibrium simply by assembling the element as if it was a displacement model. Putting (6.21) in to (6.23), the global equations now become

$$\mathbf{K}\mathbf{r} = \mathbf{R} + \sum_g \mathbf{a}_g^t [\mathbf{B}_N^{-1} \mathbf{f}_N^{-1} \mathbf{f}_0 - \mathbf{B}_0 \mathbf{P}_0] \tag{6.24}$$

We have therefore succeeced in simulating a displacement solution except that the loading due to applied forces and initial strains is accumulated in a different way. After solving for $\mathbf{r}$ the element 'displacements' $\boldsymbol{\rho}_g = \mathbf{a}_g \mathbf{r}$ are recovered, hence $\mathbf{P}_N$ from (6.20) and $\boldsymbol{\sigma}$ from (6.17). This method is much preferable for two- and three-dimensional solids, and as we mentioned has been found to be some 40 per cent quicker than the true force method when applied to the following test case (shown in Fig. 6.5) of a square plate diffusing a concentrated load applied to one edge. Only one half of the plate needs to be modelled and the mesh is shown in the figure. For comparison, a finite-element displacement solution was run with double the number of four-noded elements which of course have discontinuous stress fields. The total computing time taken by the simulated stiffness method was almost the same as the displacement solution in spite of the fact that the element flexibility has to be inverted. The solved variation of $\boldsymbol{\sigma}_{xx}$ along the centre-line is shown in Fig. 6.6 and the smooth nature of the equilibrium solution is obvious. The shear stress we recall may not be continuous, and in Fig. 6.7 are depicted – to the same scale – the shears across the plate at $x/l = 5/6$, which is quite close to the applied force. The stress discontinuities are not negligible but are much less common for the equilibrium element.

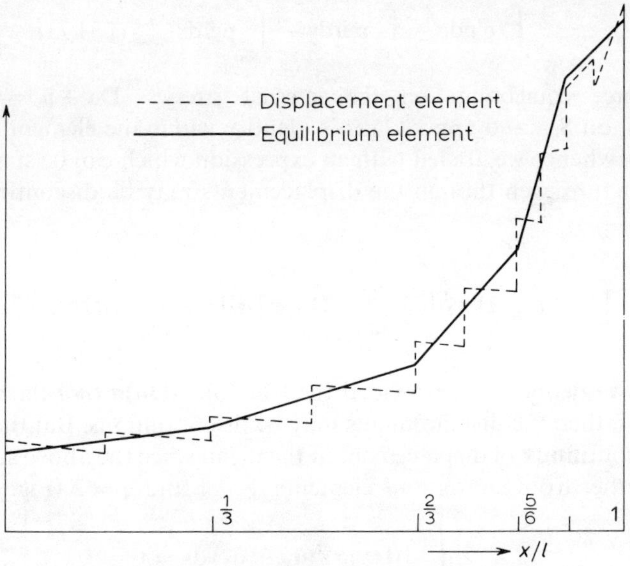

Fig. 6.6 Diffusion of stresses into a square plate

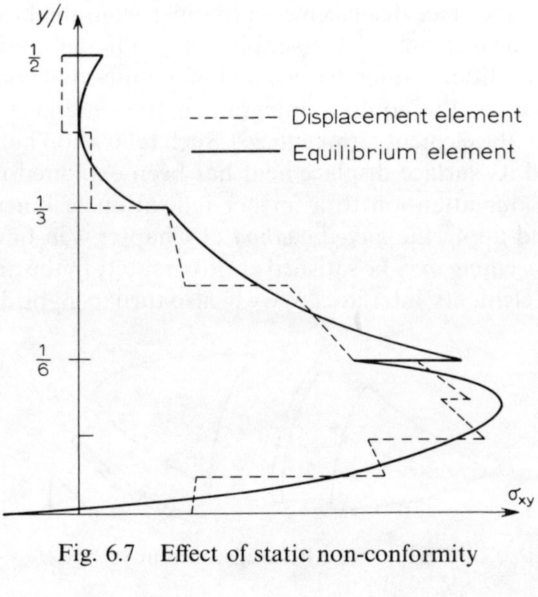

Fig. 6.7 Effect of static non-conformity

6.4 Relaxed continuity and mixed elements

The necessary and relaxed continuity conditions for equilibrium elements can be argued in a precisely similar fashion to those of the displacement formulation in the previous chapter, section 5.4. We start with the PVF for an element, before summing, and examine the difference between internal and external virtual work, that is

$$\int_{Vg} \bar{\boldsymbol{\sigma}}^t \boldsymbol{\varepsilon} \, dv - \int_{Vg} \bar{\mathbf{p}}_v^t \mathbf{u} \, dv - \int_{Sgu} \bar{\mathbf{p}}_s^t \tilde{\mathbf{u}} \, ds \qquad \text{cf. (1.73)}$$

Now enforce equilibrium on the internal stresses, $\mathbf{D}\boldsymbol{\sigma} + \bar{\mathbf{p}}_v = \mathbf{0}$ in V_g and $(\mathbf{D}n)\bar{\boldsymbol{\sigma}} = \bar{\mathbf{p}}_s$ on S_{gu}, and apply Gauss's identity *within* the element to convert the $\mathbf{u}^t\mathbf{D}\bar{\boldsymbol{\sigma}}$ term, whence we are left with an expression which can be summed over the whole structure even though the displacements may be discontinuous.

Thus

$$\sum_g \left[\int_{V_g} \bar{\boldsymbol{\sigma}}^t(\boldsymbol{\varepsilon} - \mathbf{D}^t\mathbf{u}) \, dv + \int_{S_g} [(\mathbf{D}n)\bar{\boldsymbol{\sigma}}]^t \mathbf{u} \, ds - \int_{S_{gu}} [(\mathbf{D}n)\bar{\boldsymbol{\sigma}}]^t \tilde{\mathbf{u}} \, ds \right] = 0 \quad (6.25)$$

This time we deduce that provided the tractions $(\mathbf{D}n)\boldsymbol{\sigma}$ (*not* the stresses $\boldsymbol{\sigma}$) are continuous, then the displacements may be discontinuous. But the PVF will try to satisfy continuity of displacement in the mean since the above summation will bring together from contiguous elements, $g = 1$ and $g = 2$ (Fig. 6.8),

$$\int_{Sg} [(\mathbf{D}n)\bar{\boldsymbol{\sigma}}]^t(\mathbf{u}_1 - \mathbf{u}_2) \, ds = 0$$

Inter-element compatibility in the mean is therefore given no more priority than compatibility in the interior ($\boldsymbol{\varepsilon} \simeq \mathbf{D}^t\mathbf{u}$) in the mean. It is possible to go one stage further and allow interface displacements to differ from the boundary values of the fields within the elements.[13] We simply apply our summation over existing elements and an additional interface element of infinitesimal volume as shown in Fig. 6.8. One half of the contour integral on this interface element is then incorporated into the element term in (6.25). Such relaxation between an element interior field and its surface displacement has been exploited in a few cases.[66]

We now turn our attention to a further relaxation of kinematic and static compatibility and apply the *mixed method* of Chapter 4 in finite-element form also, in which *everything* may be satisfied approximately in the mean, either in the interior or at the elements' interfaces. We will also turn to 'hybrid' formulations in

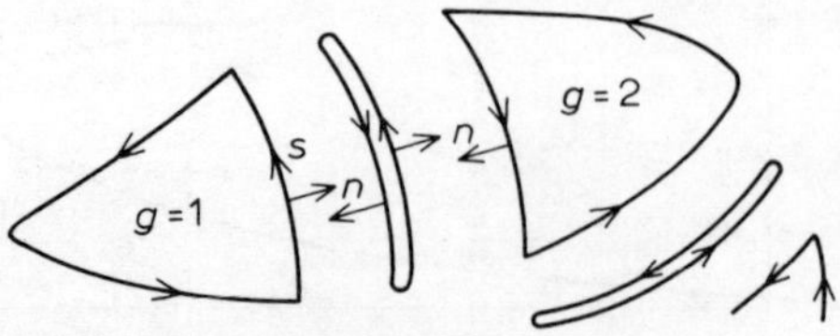

Fig. 6.8 Relaxed continuity

which some of the conditions are satisfied exactly and some in the mean. Clearly there are a large number of such permutations and the usual way to generate these elements is by inventing a new energy theorem with or without Lagrange multipliers. We will be able to dispense with such artifices and simply use the finite-element weighted residual form of the PVF, equation (6.25) above, or of the PVD of section 5.4 which we write again here as:

$$\sum_g \left[\int_{V_g} \bar{\mathbf{u}}^t(\mathbf{D}\boldsymbol{\sigma} + \mathbf{p}_v)\,dv - \int_{S_g} \bar{\mathbf{u}}^t\,(\mathbf{D}n)\boldsymbol{\sigma}\,ds + \int_{S_{gp}} \bar{\mathbf{u}}^t\mathbf{p}_s\,ds \right] = 0 \qquad (6.26)$$

The incentive for using a mixed formulation stems from two causes. Firstly, as we saw in the conventional Rayleigh–Ritz technique, there is a sporting chance that a mixed solution will lie somewhere between the lower and upper bounds of the pure displacement and pure force solutions. Secondly, there were difficulties encountered in both the displacement and equilibrium elements in constructing continuous displacement fields (conformal elements) or the parallel continuous tractions (statically conformal). These difficulties were associated with the desire to choose a complete isotropic polynomial field consistent with a regular number of nodal freedoms on the element edges. By choosing separate expansions for displacements and stresses we have much more freedom in which to manoeuvre.

The term $\tilde{\mathbf{u}}$ will be dropped from (6.25) since nonzero displacements are rarely prescribed between elements. In constructing mixed elements we have to decide whether we wish displacements $\mathbf{u}$ or tractions $(\mathbf{D}n)\boldsymbol{\sigma}$ to be continuous only in the mean across $\mathbf{S}_g$, since then we retain the appropriate weighted residual term in (6.25) or (6.26). As the weight coefficient of the residual has to be continuous, then we must employ either continuous tractions or displacements respectively. If the displacements are discontinuous, we drop the weights $\bar{\mathbf{u}}$ in (6.26); and if the tractions are discontinuous, then the weight $(\mathbf{D}n)\boldsymbol{\sigma}$ goes from (6.25). The two equations are used separately to generate the element's characteristic integrals, the only links being the stress–strain law linking $\boldsymbol{\sigma}$ and $\boldsymbol{\varepsilon}$, and the fact that the virtual weights $\bar{\mathbf{u}}$ and $\bar{\boldsymbol{\sigma}}$ are selected as the same set of interpolation functions which are used in the real variables $\mathbf{u}$ and $\boldsymbol{\sigma}$.

Firstly, the *surface tractions may be discontinuous*, but the stresses continuous within the element so we may use Gauss again to transform back the surface integral over S_g in (6.26), and drop the discontinuous weight $(\mathbf{D}n)\bar{\boldsymbol{\sigma}}$ in (6.25). We are left with the two virtual-work expressions for an element as:

$$\int_{V_g} (\mathbf{p}_v^t\bar{\mathbf{u}} - \boldsymbol{\sigma}^t\mathbf{D}^t\bar{\mathbf{u}})\,dv + \int_{S_{gp}} \mathbf{p}_s^t\,\bar{\mathbf{u}}\,ds \qquad (6.27)$$

and

$$\int_{V_g} \bar{\boldsymbol{\sigma}}^t(\boldsymbol{\varepsilon} - \mathbf{D}^t\mathbf{u})\,dv \qquad (6.28)$$

These compare with (4.33) and (4.32).

Now assume fields for both displacement *and* stresses in terms of nodal measures,

$$\mathbf{u} = \boldsymbol{\omega}\boldsymbol{\rho}_g \qquad \text{and} \qquad \boldsymbol{\sigma} = \boldsymbol{\beta}\mathbf{P}_g$$

Notice that $\boldsymbol{\beta}\mathbf{P}_g$ is not a natural stress field–we are not satisfying $\mathbf{D}\boldsymbol{\sigma} = \mathbf{0}$. The two expressions (6.27) and (6.28) then become, using the stress–strain law $\boldsymbol{\varepsilon} = \boldsymbol{\phi}\boldsymbol{\sigma} + \boldsymbol{\eta}$,

$$\bar{\boldsymbol{\rho}}_g^t(\mathbf{R}_g - \mathbf{Q}_g^t\mathbf{P}_g) \tag{6.29}$$

and

$$\bar{\mathbf{P}}_g^t(\mathbf{f}_g\mathbf{P}_g - \mathbf{Q}_g\boldsymbol{\rho}_g + \mathbf{H}_g) \tag{6.30}$$

where the element matrices are:

$$\mathbf{Q}_g = \int_{V_g} \boldsymbol{\beta}^t\mathbf{D}^t\boldsymbol{\omega}\,\mathrm{d}v \tag{6.31}$$

$$\mathbf{f}_g = \int_{V_g} \boldsymbol{\beta}^t\boldsymbol{\phi}\boldsymbol{\beta}\,\mathrm{d}v \tag{6.32}$$

$$\mathbf{H}_g = \int_{V_g} \boldsymbol{\beta}^t\boldsymbol{\eta}\,\mathrm{d}v \tag{6.33}$$

and

$$\mathbf{R}_g = \int_{V_g} \boldsymbol{\omega}^t\mathbf{p}_v\,\mathrm{d}v + \int_{S_{gp}} \boldsymbol{\omega}^t\mathbf{p}_s\,\mathrm{d}s \tag{6.34}$$

The last equation is the familiar set of kinematically equivalent loads which previously we labelled $\mathbf{P}_g$, but now a different symbol is necessary since nodal forces $\mathbf{P}_g$ and $\mathbf{R}_g$ are not the same. It is permissible to enforce continuity of $\mathbf{P}_g$ even though there are kinematically equivalent forces at the interface. Thus if we prepare a global list of nodal forces

$$\mathbf{P} = \{\mathbf{P}_1 \quad \mathbf{P}_2 \quad \mathbf{P}_3 \quad \ldots \quad \mathbf{P}_g \quad \ldots\}$$

the connectivity matrix $\mathbf{b}_g$, in $\mathbf{P}_g = \mathbf{b}_g\mathbf{P}$, will be a sparse Boolean array. We are obliged to ensure continuity of displacements, so again $\boldsymbol{\rho}_g = \mathbf{a}_g\mathbf{r}$, and summing both (6.29) and (6.30) over all elements we have:

$$\begin{bmatrix} \mathbf{A}^t & \mathbf{0} \\ -\mathbf{F} & \mathbf{A} \end{bmatrix} \begin{bmatrix} \mathbf{P} \\ \mathbf{r} \end{bmatrix} = \begin{bmatrix} \mathbf{R} \\ \mathbf{H} \end{bmatrix} \tag{6.35}$$

where

$$\mathbf{A} = \sum \mathbf{b}_g^t\mathbf{Q}_g\mathbf{a}_g$$
$$\mathbf{F} = \sum \mathbf{b}_g^t\mathbf{f}_g\mathbf{b}_g$$
$$\mathbf{H} = \sum \mathbf{b}_g^t\mathbf{H}_g$$
$$\mathbf{R} = \sum \mathbf{a}_g^t\mathbf{R}_g$$

This mixed formulation has been used widely in constructing plate and shell elements. We recall that in plate bending, the strain and stresses depend on the second derivative of the (normal) displacement, so interpolations of higher order than cubic are needed to better a linear strain field, and if the nodes are confined to the computationally attractive corners of an element, then we have to introduce second derivatives of displacement as nodal freedoms. This will produce an overstiff element, and then there are problems with discontinuous section properties – continuity of strain would then *ensure* a discontinuous stress field. By employing the mixed method above it is possible to have continuity of deflection and slope using cubic interpolation, but to expand separately the bending moment which could be continuous even across discontinuities of structure or geometry. However, the equations (6.35) are not commercially attractive. They are not symmetrical, and many standard algorithms for solving them will not work. It is possible of course to eliminate $\mathbf{P} = \mathbf{F}^{-1}(\mathbf{Ar} - \mathbf{H})$ and solve the reduced set

$$(\mathbf{A}^t\mathbf{F}^{-1}\mathbf{A})\mathbf{r} = \mathbf{R} + \mathbf{A}^t\mathbf{F}^{-1}\mathbf{H}$$

However, the formulation of the 'stiffness' – the congruent transformation $\mathbf{A}^t\mathbf{F}^{-1}\mathbf{A}$ – is expensive. Because these equations are not compatible with most finite-element software, it is common to drop the requirement that $\mathbf{P}_g$ be continuous between elements, and settle for interface equilibrium in the mean. If we do not equate nodal forces across interfaces, then $\mathbf{P}_g$ is a local autonomous element variable, and its coefficient in equation (6.30) must be zero, that is,

$$\mathbf{P}_g = \mathbf{f}_g^{-1}\mathbf{Q}_g\boldsymbol{\rho}_g - \mathbf{f}_g^{-1}\mathbf{H}_g \tag{6.36}$$

Then returning to (6.29) and summing, we have

$$[\textstyle\sum \mathbf{a}_g^t\mathbf{Q}_g^t\mathbf{f}_g^{-1}\mathbf{Q}_g\mathbf{a}_g]\mathbf{r} = \sum \mathbf{a}_g^t(\mathbf{R}_g - \mathbf{Q}_g^t\mathbf{f}_g^{-1}\mathbf{H}_g) \tag{6.37}$$

This set of equations is identical, for assembling and solving, to the conventional displacement equations, with an equivalent element stiffness

$$\mathbf{k}_g = \mathbf{Q}_g^t\mathbf{f}_g^{-1}\mathbf{Q}_g \tag{6.38}$$

and kinematically equivalent loads $\mathbf{R}_g - \mathbf{Q}_g^t\mathbf{f}_g^{-1}\mathbf{H}_g$. This mixed model then simulates a stiffness solution even though the chosen forces and displacements (6.36) are not related by $\mathbf{P}_g = \mathbf{k}_g\boldsymbol{\rho}_g$.

Secondly, we could form the dual to the previous mixed methods *and allow discontinuous displacements* by dropping the surface integral in (6.26) containing $\bar{\mathbf{u}}^t(\mathbf{D}n)\boldsymbol{\sigma}$. Again using Gauss to transform the surface integral in (6.25), the element virtual-work terms are:

$$\int_{V_g} \bar{\mathbf{u}}^t(\mathbf{D}\boldsymbol{\sigma} + \mathbf{p}_v)\,\mathrm{d}v + \int_{S_{gp}} \mathbf{p}_s^t\bar{\mathbf{u}}\,\mathrm{d}s \tag{6.39}$$

and

$$\int_{V_g} (\bar{\boldsymbol{\sigma}}^t\boldsymbol{\varepsilon} + \mathbf{u}^t\mathbf{D}\bar{\boldsymbol{\sigma}})\,\mathrm{d}v \tag{6.40}$$

Upon expanding $\mathbf{u}$ and $\boldsymbol{\sigma}$ as before, we obtain from these two equations identical expressions to (6.29)–(6.34) except that now

$$\mathbf{Q}_g^t = -\int_{V_g} \boldsymbol{\omega}^t\mathbf{D}\boldsymbol{\beta}\,\mathrm{d}v \tag{6.41}$$

If the nodal displacements are continuous – even though the displacements $\mathbf{u} = \boldsymbol{\omega}\boldsymbol{\rho}_g$ may be discontinuous between nodes – we may put $\boldsymbol{\rho}_g = \mathbf{a}_g\mathbf{r}$. We are already committed to continuous tractions in this formulation, so $\mathbf{P}_g = \mathbf{b}_g\mathbf{P}$ again, and we obtain the same set of global equations for $\{\mathbf{P} \quad \mathbf{r}\}$ as before (equation (6.35)). But if we drop the requirement that nodal displacements are continuous, then $\bar{\boldsymbol{\rho}}_g$ becomes a single element variable and we can say

$$\bar{\boldsymbol{\rho}}_g^t(\mathbf{R}_g - \mathbf{Q}_g^t\mathbf{P}_g) = 0$$

But we cannot solve this for $\mathbf{P}_g$ since $\mathbf{Q}_g^t$ cannot be inverted. Consequently we are unable to construct a simulated element stiffness matrix and incorporate this mixed model into a displacement programme. This particular mixed method is not a commercial success.

To be honest, mixed methods are not overwhelmingly popular because industry will always plump for the uncomplicated coding of the pure displacement method. But one problem, for example, that will not easily yield to the displacement formulation is that of the singular material stiffness matrix (incompressible materials), since stresses have to be derived from strains and not vice versa; here a mixed formulation is helpful. However, it has been shown[69] that certain of such mixed formulations are equivalent to using the reduced integration trick referred to in Chapter 5, section 8, in connection with (singular) shear rigidity in thin plates and beams. This is a much more satisfactory way of using a mixed formulation, and it also appeases purists who wish to use the mixed principles when defining convergence criteria.

6.5 A mixed bending element

The generation of element matrices $\mathbf{f}_g$, $\mathbf{Q}_g$, etc., in the mixed formulation does not usually present problems once the variables $\boldsymbol{\rho}_g$ and $\mathbf{P}_g$, and their interpolation functions, have been selected; and there is more freedom in so selecting since neither equilibrium nor compatibility has to be satisfied. We choose as an example a beam bending problem to reveal two further precautions necessary in modelling beams and plates. Firstly, when we referred to continuity of *either* $\mathbf{u}$ or $(\mathbf{D}n)\boldsymbol{\sigma}$, there will be occasions where there are discontinuities in just one component of $\mathbf{u}$ or one of $(\mathbf{D}n)\boldsymbol{\sigma}$, and so combinations of (6.25) and (6.26) must be used. Secondly, the inter-element continuity requirements in bending problems involve derivatives of displacements and stresses as well. This arises from the kinematic connection between $w(z)$ and $v(z)$, viz. $w(z) = -yv'(z)$. A simple example will illustrate these points (see Fig. 6.9).

Define the nodal unknowns as

$$\mathbf{P}_g = \{M_1 \quad M_2\} \qquad \text{and} \qquad \boldsymbol{\rho}_g = \{\rho_1 \quad \rho_2\}$$

Fig. 6.9 Linearly varying moment and deflection

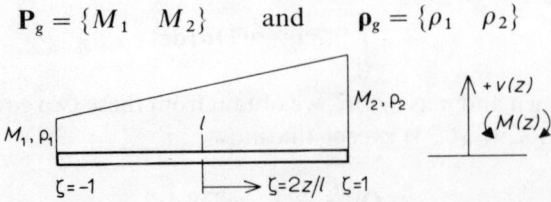

The interpolation functions in

$$M(\zeta) = \boldsymbol{\beta}\mathbf{P}_g \qquad \text{and} \qquad v(\zeta) = \boldsymbol{\omega}\boldsymbol{\rho}_g$$

are both the same, that is

$$\boldsymbol{\omega}(\zeta) = \boldsymbol{\beta}(\zeta) = \tfrac{1}{2}[1+\zeta \quad 1-\zeta] \tag{6.42}$$

The discontinuous displacements are the rotations $\partial v/\partial z$, so the work term which attempts to secure continuity will be $\bar{M}(-v')$, and the corresponding term for shears will be $\bar{v}M'$. The moments and deflections are continuous as required. Reading M for $\boldsymbol{\sigma}$, v for $\mathbf{u}$, and 'curvature $= M/EI$' for $\boldsymbol{\varepsilon} = \boldsymbol{\phi}\boldsymbol{\sigma}$, the element virtual-work expressions become:

from (6.25):

$$\int_0^l \bar{M}\left(\frac{M}{EI} + v''\right)\mathrm{d}z + \Big[\bar{M}(-v')\Big]_0^l$$

from (6.26):

$$\int_0^l \bar{v}(M'' + p)\mathrm{d}z - \Big[\bar{v}M'\Big]_0^l$$

Upon substituting (6.42), the matrix $\mathbf{f}_g$ in $\bar{\mathbf{P}}_g^{\mathrm{t}}\mathbf{f}_g\mathbf{P}_g$ emerges from the first equation as

$$\mathbf{f}_g = \frac{l}{2EI}\int_{-1}^{1}\boldsymbol{\beta}^{\mathrm{t}}\boldsymbol{\beta}\,\mathrm{d}\zeta = \frac{l}{6EI}\begin{bmatrix} 2 & 1 \\ 1 & 2 \end{bmatrix}$$

From the term $\bar{\mathbf{P}}_g^{\mathrm{t}}\mathbf{Q}_g\boldsymbol{\rho}_g$ in the first equation – or equally from $\bar{\boldsymbol{\rho}}_g^{\mathrm{t}}\mathbf{Q}_g^{\mathrm{t}}\mathbf{P}_g$ in the second – we find

$$\mathbf{Q}_g = \frac{2}{l}\Big[\boldsymbol{\omega}^{\mathrm{t}}\boldsymbol{\beta}'\Big]_{-1}^{1} = \frac{4}{l}\begin{bmatrix} 1 & -1 \\ -1 & 1 \end{bmatrix}$$

and

$$\mathbf{R}_g = \frac{l}{2}\int_{-1}^{1}\boldsymbol{\omega}^{\mathrm{t}}p\,\mathrm{d}s$$

Suppose we reconsider the example in Chapter 5, Fig. 5.4, of a clamped beam which we divide into five elements as shown in Fig. 6.10. Thus

$$\mathbf{P} = \{M_0 \quad M_1 \quad M_2 \quad M_3 \quad M_4 \quad M_5\}$$

$$\mathbf{r} = \{r_1 \quad r_2 \quad r_3 \quad r_4\}$$

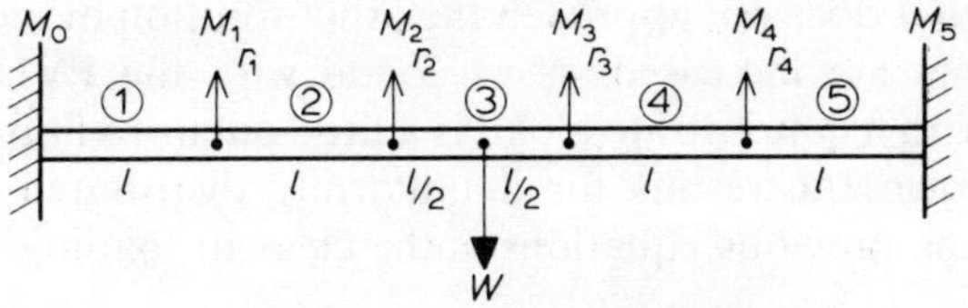

Fig. 6.10 A five-element beam

The loading $\mathbf{R}_g = [0 \quad 0]$ for all elements except $\mathbf{R}_3 = \{-W/2 \quad -W/2\}$. The Boolean connectivity matrices in $\mathbf{P}_g = \mathbf{b}_g\mathbf{P}$ and $\boldsymbol{\rho}_g = \mathbf{a}_g\mathbf{r}$ are clearly

$$\mathbf{a}_1 = \begin{bmatrix} 0 & 0 & 0 & 0 \\ 1 & 0 & 0 & 0 \end{bmatrix}, \quad \mathbf{a}_3 = \begin{bmatrix} 0 & 1 & 0 & 0 \\ 0 & 0 & 1 & 0 \end{bmatrix}$$

$$\mathbf{b}_1 = \begin{bmatrix} 1 & 0 & 0 & 0 & 0 & 0 \\ 0 & 1 & 0 & 0 & 0 & 0 \end{bmatrix} \qquad \text{etc.}$$

We find after assembling:

$$\mathbf{A}^t = \frac{1}{l}\begin{bmatrix} -1 & 2 & -1 & & & \\ & -1 & 2 & -1 & & \\ & & -1 & 2 & -1 & \\ & & & -1 & 2 & -1 \end{bmatrix}$$

$$\mathbf{F} = \frac{l}{6EI}\begin{bmatrix} 2 & 1 & & & & \\ 1 & 4 & 1 & & & \\ & 1 & 4 & 1 & & \\ & & 1 & 4 & 1 & \\ & & & 1 & 4 & 1 \\ & & & & 1 & 2 \end{bmatrix}$$

$$\mathbf{H} = \mathbf{0} \qquad \text{and} \qquad \mathbf{R} = \{0 \quad -W/2 \quad -W/2 \quad 0\}$$

Solving (6.35), we find:

$$\mathbf{P} = \frac{Wl}{10}\{6 \quad 1 \quad -4 \quad -4 \quad 1 \quad 6\}$$

and

$$\mathbf{r} = \frac{-Wl^3}{60EI}\{13 \quad 32 \quad 32 \quad 13\}$$

The moments and deflected shape are shown in Fig. 6.11 together with the exact solution. There cannot be said to be a significant improvement on the previous displacement solution even though the moment distribution is now continuous. The deflections are a poor approximation, although the central deflection happens to be exact with an even number of elements. This central deflection is a direct measure of strain energy and happens to be a lower bound in this particular example, although it does not approach the exact solution monotonically as the degrees of freedom are increased, as we found with the PVD solutions. The central element with a concentrated load is a little unfair to finite elements, so in Fig. 6.12 are shown the results for a uniformly distributed load. The only modification to the previous equations is the element loading

$$\mathbf{R}_g = \{-wl/2 \quad -wl/2\}$$

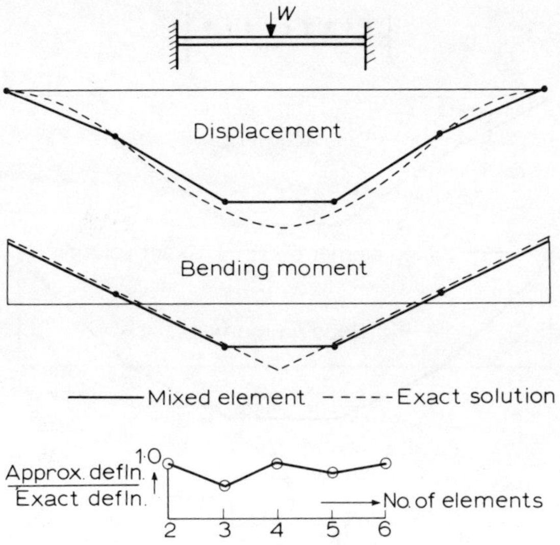

Fig. 6.11 Mixed bending elements

The moments are again more accurate than the deflections. They are bounded *everywhere* by the exact solution as an upper bound. This curious feature could only happen in a mixed formulation and shows that stresses may not always be the mean fit we have come to expect. In contrast, the discontinuous shears are seen to be precisely that. This example is not a very convincing case for mixed methods, and given that a beam has no special problems it has not been worth it. The mixed method is best reserved for the special problems previously mentioned.

6.6 Hybrid elements

The hybrid approach is to use the weighted residual equations (6.25) and (6.26) again but this time to satisfy some equilibrium or some compatibility equations *exactly*; thus some weighted residual terms will disappear. The method is still therefore a mixed formulation and will not lead to lower or upper bounds on energy like the pure PVD or PVF. The various possible permutations have now become very large and it is not the purpose of this text to investigate all of them. There are three sets of equations of equilibrium and compatibility (or continuity) inside the element, between the element's internal fields and its own surface, and between contiguous elements. The choice of these permutations depends very much on the specific structure and the field idealization so there may be considerable variation in the types of equations and their acceptable errors. Without making sweeping generalizations it does seem that the first hybrid[67] has been used more than most, and in the very few commercial systems which employ 'equilibrium' elements, these are often in fact a 'hybrid stress' element employing an assumed stress field which satisfies equilibrium exactly.

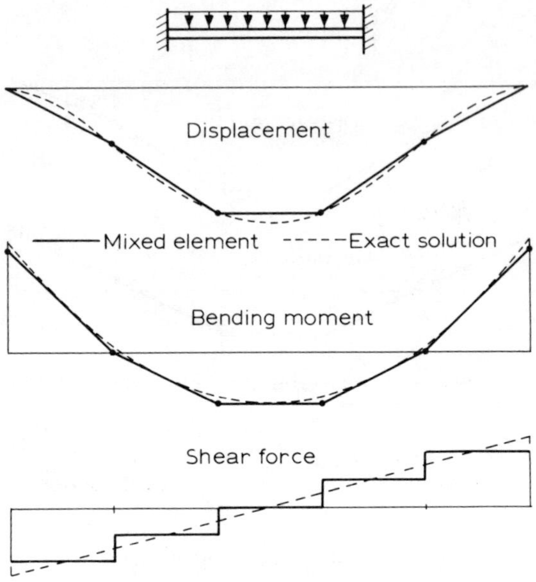

Fig. 6.12 Mixed element solution

If internal equilibrium is satisfied, we write as before, within the element,

$$\boldsymbol{\sigma} = \boldsymbol{\beta}\mathbf{P}_N + \boldsymbol{\beta}_0 \qquad \text{and} \qquad \bar{\boldsymbol{\sigma}} = \boldsymbol{\beta}\mathbf{P}_N \tag{6.17}$$

where $\boldsymbol{\beta}_0$ is the particular integral, and $\mathbf{D}\boldsymbol{\beta} = \mathbf{0}$. This is a restriction on the modelling compared with the mixed formulation but as we have seen is fairly straightforward in two-dimensional plates, in membrane or bending states. It is, however, a formidable task in general shell elements and in three-dimensional solids where the 'stress function' approach[42] has its special difficulties.

We now entirely avoid the previous difficulties encountered with satisfying equilibrium exactly at the element surface S_g in terms of selected nodal force measures. We abandon this idea completely and settle instead for making displacements continuous between elements as in the PVD. We will choose a displacement field quite independently of the internal equilibrium field – with some provisos – and enjoy the luxury of choosing nodal displacements more or less uninhibited. The continuous virtual displacements will enable us to satisfy inter-element equilibrium in the mean, but the tractions $(\mathbf{D}n)\boldsymbol{\sigma}$ will not in general be continuous. Turning to equation (6.26), the equilibrium residual is now zero and we are left with an element virtual work expression,

$$-\int_{S_g} \bar{\mathbf{u}}^t(\mathbf{D}n)\boldsymbol{\sigma}\,\mathrm{d}s + \int_{S_{gp}} \bar{\mathbf{u}}^t\mathbf{p}_s\,\mathrm{d}s \tag{6.43}$$

In the other equation, (6.25), the term in $(\mathbf{D}n)\bar{\boldsymbol{\sigma}}$ must be discarded since it is no longer continuous over S_g. Now we capitalize on the fact that $\mathbf{D}\bar{\boldsymbol{\sigma}} = \mathbf{0}$ in V_g and use the Gauss identity to transform

$$\int_{V_g} (\bar{\boldsymbol{\sigma}}^t\mathbf{D}^t\mathbf{u} + \mathbf{u}^t\mathbf{D}\bar{\boldsymbol{\sigma}})\,\mathrm{d}v \qquad \text{to} \qquad \int_{S_g} \mathbf{u}^t(\mathbf{D}n)\bar{\boldsymbol{\sigma}}\,\mathrm{d}s$$

or

$$\int_{V_g} \bar{\boldsymbol{\sigma}}^t\mathbf{D}^t\mathbf{u}\,\mathrm{d}v = \int_{S_g} \mathbf{u}^t(\mathbf{D}n)\bar{\boldsymbol{\sigma}}\,\mathrm{d}s \qquad \text{since} \quad \mathbf{D}\bar{\boldsymbol{\sigma}} = \mathbf{0}$$

Equation (6.25) now has the element work as

$$\int_{V_g} \bar{\boldsymbol{\sigma}}^t\boldsymbol{\varepsilon}\,\mathrm{d}v - \int_{S_g} \mathbf{u}^t(\mathbf{D}n)\bar{\boldsymbol{\sigma}}\,\mathrm{d}s \tag{6.44}$$

We notice that the displacement field $\mathbf{u}$ has now vanished entirely from the interior of the element and we are free to expand $\mathbf{u}$ *only on the surface*. Indeed, if an element has several geometric surfaces we may expand differently on each provided that piecewise continuity is preserved. In practice, hybrids seem to have employed a continuous displacement field expanded throughout the volume but evaluated only on the surface.

On substituting $\mathbf{u} = \boldsymbol{\omega}\boldsymbol{\rho}_g$ and (6.17) for stress, the two work expressions (6.43) and (6.44) become

$$\bar{\boldsymbol{\rho}}^t\,[\mathbf{R}_g - \mathbf{B}_g^t\mathbf{P}_N] \tag{6.45}$$

and

$$\bar{\mathbf{P}}_N^t\,[\mathbf{f}_N\mathbf{P}_N + \mathbf{f}_0 - \mathbf{B}_g\boldsymbol{\rho}_g] \tag{6.46}$$

where this time the kinematically equivalent loads are

$$\mathbf{R}_g = \int_{S_{gp}} \boldsymbol{\omega}^t\mathbf{p}_s\,\mathrm{d}s - \int_{S_p} \boldsymbol{\omega}^t(\mathbf{D}n)\boldsymbol{\beta}_0\,\mathrm{d}s \tag{6.47}$$

and the integral

$$\mathbf{B}_g^t = \int_{S_g} \boldsymbol{\omega}^t(\mathbf{D}n)\boldsymbol{\beta}\,\mathrm{d}s \tag{6.48}$$

The integrals $\mathbf{f}_N$ and $\mathbf{f}_0$ are the same as in the equilibrium model (6.11) and (6.20). Now that there is no directly enforced equilibrium connection across elements – we are not relating to a global list – then $\bar{\mathbf{P}}_N$ is purely a single element variable and its coefficient in (6.46) must vanish.

Then

$$\mathbf{P}_N = \mathbf{f}_N^{-1}\mathbf{B}_g\boldsymbol{\rho}_g - \mathbf{f}_N^{-1}\mathbf{f}_0$$

and (6.45) becomes

$$\bar{\boldsymbol{\rho}}_g^t\,[\mathbf{R}_g - \mathbf{B}_g^t\mathbf{f}_N^{-1}\mathbf{B}_g\boldsymbol{\rho}_g + \mathbf{B}_g^t\mathbf{f}_N^{-1}\mathbf{f}_0]$$

The element stiffness

$$\mathbf{k}_g = \mathbf{B}_g^t \mathbf{f}_N^{-1} \mathbf{B}_g \tag{6.49}$$

is a familiar form even though $\mathbf{B}_g$ in equation (6.48) is not. On enforcing continuity of inter-element displacements by putting $\boldsymbol{\rho}_g = \mathbf{a}_g \mathbf{r}$, and summing the work over all elements, we obtain the usual global stiffness equation with a right-hand side different from (6.24), that is,

$$\mathbf{K}\mathbf{r} = \sum \mathbf{a}_g^t \left[\mathbf{R}_g + \mathbf{B}_g^t \mathbf{f}_N^{-1} \mathbf{f}_0\right]$$

Comparing this stiffness evaluation with the mixed model, the only difference in $\mathbf{k}_g$ is that the mixed model uses a volume integral

$$\mathbf{Q}_g = \int_{V_g} \boldsymbol{\beta}^t \mathbf{D}^t \boldsymbol{\omega}\, \mathrm{d}v \qquad \text{compared with} \quad \mathbf{B}_g^t = \int_{S_g} \boldsymbol{\omega}^t (\mathbf{D}n) \boldsymbol{\beta}\, \mathrm{d}s$$

The hybrid stress stiffness is therefore easier to model and cheaper to evaluate, although like all mixed and equilibrium elements it is not as straightforward to code either the stiffness or the loading. It is, however, the freedom of the choice of the nodal displacements which makes the hybrid formulation attractive, and the lack of continuity in surface tractions is a small sacrifice for this.

Some care must be exercised in choosing the order of the displacement and the stress field. Suppose there are n_N stress modes of amplitude P_N, which of course are all independent. Then the natural flexibility $\mathbf{f}_N$ will not be singular, or rank deficient, and can be inverted. If we look at $\mathbf{k}_g$ in (6.49) then whatever the rank of $\mathbf{B}_g$ the rank of $\mathbf{k}_g$ also cannot exceed n_N. But $\mathbf{k}_g$ arises as a measure of the element's internal virtual work $\bar{\boldsymbol{\rho}}_g^t \mathbf{k}_g \boldsymbol{\rho}_g$ and it is rank deficient because $\boldsymbol{\rho}_g$ contains the rigid-body motions, say n_0 of them. So if there are n_ρ chosen nodal displacements, the rank of $\mathbf{k}_g$ should not be less than $n_\rho - n_0$. We must therefore make sure that

$$n_N \geqslant n_\rho - n_0$$

If this inequality is violated then there will be zero element-work modes which give rise to displacement fields other than rigid-body motions. Apart from the convenience of being able to select the order of the stress field n_N separately from the displacements n_ρ, it turns out that there are no significant advantages in letting n_N greatly exceed $n_\rho - n_0$. As the order of the assumed stress field n_N is increased, it is found that the answers tend to the equivalent displacement model if there is one.

6.7 Further generalizations

One of the advantages claimed at the outset for the virtual work approach, compared with the energy extremum school, was that it did not require finding a functional merely to minimize it. Both in the original form and the weighted residual form, the PVD and the PVF are direct formulations. We might rightly expect therefore that the weighted residual or Galerkin technique could be applied to any branch of engineering, physics, etc., where there is a field problem consisting of a differential equation and boundary conditions. If this is so, then perhaps we can discretize the problem and apply the finite-element method as opposed to, say, the finite-difference method. It turns out that this is indeed the case. Moreover, some of the problems are then so akin to structural mechanics in form that virtually identical finite-element programmes can be used to solve them. It would be incomlete not to discuss this feature (even in a structural analysis text) and highlight the similarities and the differences. Moreover, some of the problem areas do interact with structures anyway – potential flow and heat conduction, for instance. Actually to apply finite-element methods to other continuous field problems must surprise the early pioneers of structural analysis who used physical and geometrical arguments for structures composed of real separate elements like bars and beams. It is not the first time in science that a simplifying unification comes about which had escaped the many separate workers in their different disciplines. We shall look at a further finite-element formulation in this concluding section, but that is not to say that finite differences do not also have their advocates. The principles of virtual work can readily be discretized in the finite-difference form but the complications of structural geometry usually make this unattractive; nevertheless, there are one or two systems for geometrically simple structures which use a variational finite-difference form.[70]

The Galerkin form can of course be tried with any differential equation, even the nonlinear variety, although success and convergence are not automatically guaranteed. Even in linear differential equations, particularly those involving directional transport, the Galerkin approach may have to be tailored to suit the flow. To clarify the type of differential equation where success is guaranteed, we recall the governing structural equations:

$$\mathbf{D}\boldsymbol{\sigma} + \mathbf{p}_v = \mathbf{0}; \qquad \boldsymbol{\varepsilon} = \mathbf{D}^t\mathbf{u}; \qquad \boldsymbol{\sigma} = \boldsymbol{\kappa}\boldsymbol{\varepsilon}$$

Therefore

$$\mathbf{D}\boldsymbol{\kappa}\mathbf{D}^t\mathbf{u} + \mathbf{p}_v = \mathbf{0}$$

The symmetric form of the combination of matrix operators in this last equation is characteristic of linear *self-adjoint* systems, and is the deciding factor which which produces a familiar finite-element solution. The field problems like potential flow, heat conduction, groundwater movement, and so on, all involve forms of the above with various types of boundary conditions involving prescribed values of the unknown or its derivatives. The steady-state heat-conduction problem is typical and involves all the possible combinations of interior and surface complexities, so we consider this. Figure 6.13 shows an arbitrary body with a temperature $T(x, y, z)$ varying throughout the body but not with time. The internal heat source, if any, is q per unit volume, and the surface S may lose heat to a surrounding fluid due to the temperature difference $(T - T_f)$ or gain it from a prescribed heat flux source Q. The governing differential equation relating heat generation q to the consequent thermal gradients can be shown to be

$$\mathbf{D}\boldsymbol{\kappa}\mathbf{D}^t T + q = 0 \tag{6.50}$$

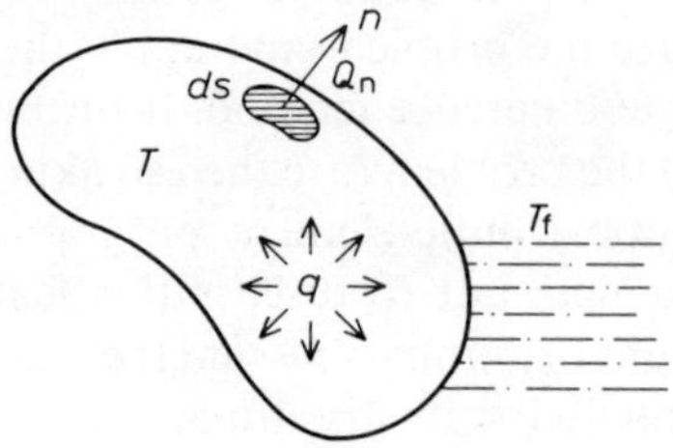

Fig. 6.13 The heat conduction problem

where $\mathbf{D}$ this time is the gradient operator $[\partial/\partial x \quad \partial/\partial y \quad \partial/\partial z]$ and the thermal conductivity $\boldsymbol{\kappa}$ is the diagonal matrix of the three directional conductivities

$$\boldsymbol{\kappa} = [\kappa_x \quad \kappa_y \quad \kappa_z]$$

The boundary conditions for normal heat flow can be written as

$$(\mathbf{D}n)\boldsymbol{\kappa}\mathbf{D}^{\mathrm{t}}T - Q + h(T - T_f) = 0 \tag{6.51}$$

where T_f is the fluid temperature and h is an appropriate heat transfer coefficient. Simpler potential problems like inviscid flow are isotropic versions of (6.50) and (6.51). The Galerkin form of this problem, applied to these last two equations, can be written as

$$\int_V (\mathbf{D}\boldsymbol{\kappa}\mathbf{D}^{\mathrm{t}}T + q)\,\overline{T}\mathrm{d}v - \int_S [(\mathbf{D}n)\boldsymbol{\kappa}\mathbf{D}^{\mathrm{t}}T - Q + h(T - T_f)]\,\overline{T}\mathrm{d}s = 0 \tag{6.52}$$

where $\overline{T}$ is a 'virtual temperature' weighting. Because the operator is of our symmetrical form we may apply the Gauss identity, which can be written as

$$\int_V [(\mathbf{D}T) + (\mathbf{D}\boldsymbol{\kappa}\mathbf{D}^{\mathrm{t}}T)\,\overline{T}]\,\mathrm{d}v = \int_S (\mathbf{D}n)\boldsymbol{\kappa}(\mathbf{D}^{\mathrm{t}}T)\,\overline{T}\mathrm{d}s$$

So (6.52) becomes

$$\int_V [-(\mathbf{D}T)\boldsymbol{\kappa}\mathbf{D}^{\mathrm{t}}\,\overline{T} + q\,\overline{T}]\,\mathrm{d}v - \int_S [-Q + h(T - T_f)]\,\overline{T}\mathrm{d}s = 0 \tag{6.53}$$

Now construct a finite-element model in terms of nodal temperatures

$$T = \boldsymbol{\omega}\mathbf{T}_g \tag{6.54}$$

where $T_g = \{T_1 \quad T_2 \quad \ldots \quad T_N\}$ at N nodes. The element integrals in (6.53) can now be performed and they become

$$\mathbf{T}_g^{\mathrm{t}}\,[-\mathbf{k}_g\mathbf{T}_g + P_g] \tag{6.55}$$

where

$$\mathbf{k}_g = \mathbf{k}_v + \mathbf{k}_s$$

The interior stiffness has a familiar ring,

$$\mathbf{k}_v = \int_{V_g} \boldsymbol{\alpha}^t \boldsymbol{\kappa} \boldsymbol{\alpha}\, \mathrm{d}v; \qquad \text{where} \quad \boldsymbol{\alpha} = \mathbf{D}^t \boldsymbol{\omega}$$

The surface 'stiffness', if any, for the element

$$\mathbf{k}_s = \int_{S_g} \boldsymbol{\omega}^t h \boldsymbol{\omega}\, \mathrm{d}s$$

The equivalent 'load'

$$\mathbf{P}_g = \int_{V_g} \boldsymbol{\omega}^t q\, \mathrm{d}v + \int_{S_g} \boldsymbol{\omega}^t (Q + hT_f)\, \mathrm{d}s$$

The summation of (6.55) for all elements takes the usual form once we have related nodal values T_g to a global list $\mathbf{T}$ by $\mathbf{T}_g = \mathbf{a}_g \mathbf{T}$, whence

$$\mathbf{KT} = \mathbf{R}$$

where

$$\mathbf{K} = \sum \mathbf{a}_g^t \mathbf{k}_g \mathbf{a}_g \qquad \text{and} \qquad \mathbf{R} = \sum \mathbf{a}_g^t \mathbf{P}_g$$

The assembly of the global stiffness and loading is therefore identical to the structural finite-element model and the major part of the computer coding will be the same except for the special surface stiffness term $\mathbf{k}_s$ which has to be incorporated into $\mathbf{k}_g$ whenever elements have a surface on S. It is in fact easier to organize the assembly by dropping $\mathbf{k}_s$ from $\mathbf{k}_g$ and simply adding these stiffnesses to $\mathbf{K}$ as if there were separate elements of zero thickness.

6.8 Summary

Equilibrium elements

Assume $\boldsymbol{\sigma} = \boldsymbol{\beta} \mathbf{P}_N$, satisfying $\mathbf{D}\boldsymbol{\beta} = \mathbf{0}$

$$\mathbf{f}_N = \int_{V_g} \boldsymbol{\beta}^t \boldsymbol{\phi} \boldsymbol{\beta}\, \mathrm{d}v$$

Choose nodal forces and find $\mathbf{P}_g = [\mathbf{B}_N \quad \mathbf{B}_0]\{\mathbf{P}_N \quad \mathbf{P}_0\}$

Force method	*Simulated stiffness method*
Invert, or use $\boldsymbol{\rho}_g = \mathbf{A}_N \boldsymbol{\rho}_N$	$\mathbf{k}_g = \mathbf{B}_N^t \mathbf{f}_N^{-1} \mathbf{B}_N$
to find $\mathbf{P}_N = \mathbf{A}_N^t \mathbf{P}_g$	$\boldsymbol{\rho}_g = \mathbf{a}_g \mathbf{r}$
then $\mathbf{f}_g = \mathbf{A}_N^t \mathbf{f}_N \mathbf{A}_N$	$\mathbf{K} = \sum \mathbf{a}_g^t \mathbf{k}_g \mathbf{a}_g$
Then use force method (3.43) with $\mathbf{b}_{xg}$ and $\mathbf{b}_{0g}$.	Solve $\mathbf{Kr} = \mathbf{R}$, using (6.24) for $\mathbf{R}$.

Mixed elements

Assume separately $\mathbf{u} = \boldsymbol{\omega}\boldsymbol{\rho}_g$, $\boldsymbol{\sigma} = \boldsymbol{\beta}\mathbf{P}_g$, then,

$$\boldsymbol{\rho}_g = \mathbf{a}_g\mathbf{r}, \qquad \mathbf{P}_g = \mathbf{b}_g\mathbf{P}$$

and

$$\begin{bmatrix} \mathbf{A}^t & \mathbf{0} \\ -\mathbf{F} & \mathbf{A} \end{bmatrix} \begin{bmatrix} \mathbf{P} \\ \mathbf{r} \end{bmatrix} = \begin{bmatrix} \mathbf{R} \\ \mathbf{H} \end{bmatrix}$$

where these matrices are defined in (6.35) and (6.31)–(6.34);

or: relax continuity of $\mathbf{P}_g$ and solve as a displacement model with $\mathbf{k}_g$ from (6.38);

or: use an equilibrium field, $\mathbf{D}\boldsymbol{\beta} = \mathbf{0}$, and the hybrid stiffness $\mathbf{k}_g$ from (6.49).

Problems

P 6.1

A right-angled isosceles triangle is to be modelled as an equilibrium plane-stress element using a uniform stress field for the three components σ_{xx}, σ_{yy}, σ_{xy}. Find the natural flexibility matrix $\mathbf{f}_N$ and invert it. Choosing two displacements at the centre of each side, generate the $\mathbf{B}_N$ matrix and form $\mathbf{k}_g$. Repeat this with nodes at the three corners and examine the consequence.

P 6.2

A rectangular plane-stress element is to be modelled using four corner nodes. Choose as an equilibrium stress field a truncated form of equation (6.9), namely

$$\boldsymbol{\beta} = \begin{bmatrix} 1 & 0 & 0 & \zeta_2 & 0 \\ 0 & 1 & 0 & 0 & \zeta_1 \\ 0 & 0 & 1 & 0 & 0 \end{bmatrix}$$

Construct $\mathbf{f}_N$ and compare its elements with those of the larger flexibility in equation (6.12). Construct $\mathbf{B}_N$ and then repeat this stage by evaluating $\mathbf{B}_g$ (6.48) in the hybrid formulation assuming a linear displacement field between nodes.

P 6.3

It is required to develop the stiffness for an equilibrium model of a triangular plate-bending element, assuming a constant-moment field. That is, M_{xx}, M_{yy}, and M_{xy} (per unit length) are all constant, and hence satisfy the equilibrium equation (2.43) with (2.40). The general solution [68] is algebraically complicated, so assume the triangle is an isosceles right-angle configuration, as in the figure. The natural stress field,

$$\mathbf{M} = \{M_{xx} \quad M_{yy} \quad M_{xy}\} = \boldsymbol{\beta}\mathbf{P}_{N}$$

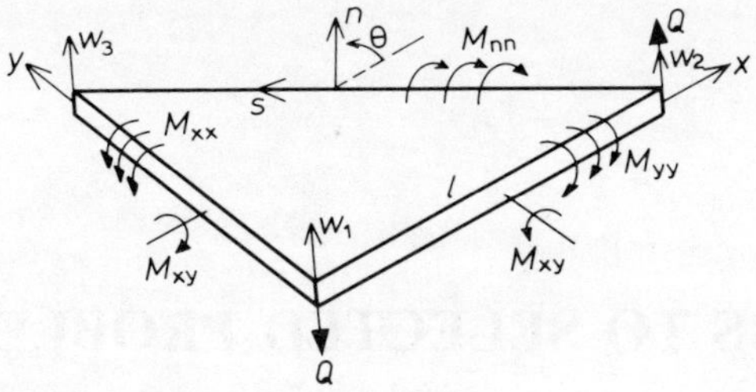

Fig. P. 6.3

is of order 3, so together with three rigid-body movements (w, θ_x, and θ_y) the displacements $\boldsymbol{\rho}_g$, and $\mathbf{k}_g$, are of order 6. We choose as degrees of freedom, the three corner displacements as shown in the figure, together with three rotations $\partial w/\partial x$, $\partial w/\partial y$, and $\partial w/\partial n$ at the mid-side nodes.

Equations (2.39) and (2.40) can be inverted to give the virtual work in the form

$$\int_A \bar{\mathbf{M}}^t \boldsymbol{\phi} \mathbf{M} \, dA, \qquad \text{where}$$

$$\boldsymbol{\phi} = \frac{12}{Et^3} \begin{bmatrix} 1 & -\nu & 0 \\ -\nu & 1 & 0 \\ 0 & 0 & 2(1+\nu) \end{bmatrix}$$

Find the flexibility matrix $\mathbf{f}_N$ and then construct the $\mathbf{B}_N$ necessary to form $\mathbf{k}_g$. You should note (equation (2.42)) that the moments on an inclined face are:

$$M_{nn} = M_{xx} \cos^2 \theta + M_{yy} \sin^2 \theta + 2M_{xy} \sin \theta \cos \theta$$
$$M_{sn} = (M_{yy} - M_{xx}) \sin \theta \cos \theta + M_{xy} (\cos^2 \theta - \sin^2 \theta)$$

Also you should notice that the twisting moment on (say) $y = 0$ has to be equilibrated as two equal and opposite nodal shears of magnitude Q given by $Ql = M_{xy} l$.

P 6.4

A beam of length $2l$ is clamped at both ends and loaded by a uniformly distributed load. One half of the beam has a rigidity $2EI$ and, the other half EI. Using the mixed finite-element of section 6.5 find the bending-moment distribution and compare with the discontinuous distribution predicted by the displacement finite-element of equation (5.5).

SOLUTIONS TO SELECTED PROBLEMS

S 1.4

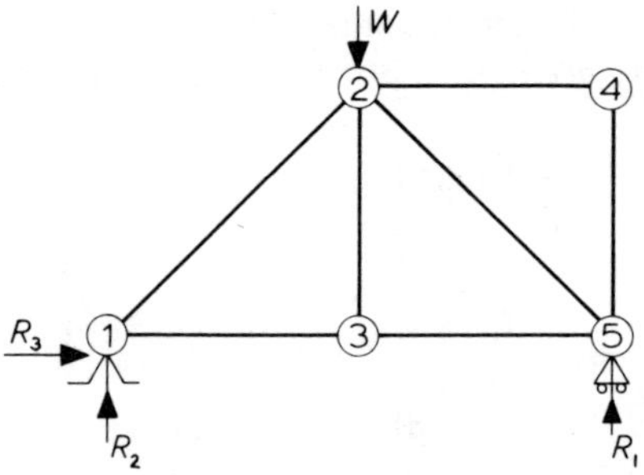

Fig. S. 1.4

$j = 5$, $r = 3$, $m = 7$, $m + r - 2j = 0$, $\therefore$ possibly statically determinate. Moment equilibrium about 1 and 5 gives $R_1 = W/2$, $R_2 = W/2$. Horizontal equilibrium gives $R_3 = 0$. Equilibrium of joints 4, 5, 3, 2 successively gives all the bar forces shown in the figure. Note that the stress system is symmetrical – unlike the displacement system. Now use compatibility (1.26) and stress–strain law for all seven bars:

$$(1\text{–}2):\ (u_2 - 0)\frac{1}{\sqrt{2}} + (v_2 - 0)\frac{1}{\sqrt{2}} = \left(-\frac{W}{\sqrt{2}}\right)\frac{1}{AE}(\sqrt{2}l)$$

$$u_2 + v_2 = -\sqrt{2}\,Wl/AE$$

$$(1\text{–}3):\ u_3 - 0 = \left(\frac{W}{2}\right)\frac{1}{AE}(l) \qquad u_3 = Wl/2AE$$

$$(3\text{–}5):\ u_5 - u_3 = \left(\frac{W}{2}\right)\frac{1}{AE}(l) \qquad u_5 = Wl/AE$$

$$(2\text{–}5):\ (u_5 - u_2)\frac{1}{\sqrt{2}} + (0 - v_2)\left(-\frac{1}{\sqrt{2}}\right) = \left(-\frac{W}{\sqrt{2}}\right)\frac{1}{AE}(\sqrt{2}l)$$

$$-u_2 + v_2 = -(\sqrt{2} + 1)\,Wl/AE$$

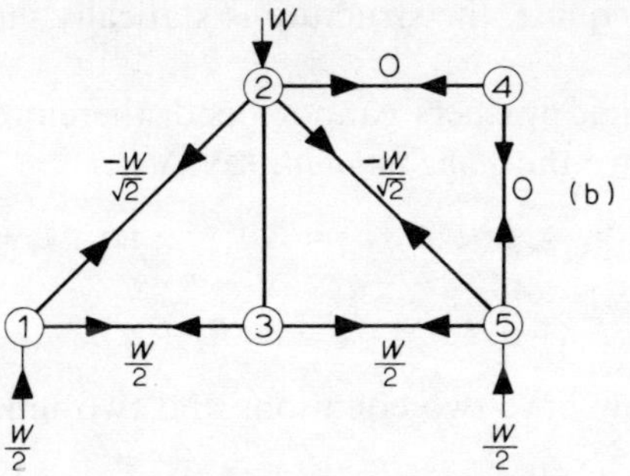

Using (2–5) and (1–2), $u_2 = Wl/2AE$; $\qquad v_2 = -(\sqrt{2} + 1/2)\,Wl/AE$

(2–3): $v_2 - v_3 = 0 \qquad v_3 = v_2 = -1.914\,Wl/AE$

(4–2): $u_4 - u_2 = 0 \qquad u_4 = u_2 = Wl/2AE$

(4–5): $v_4 - 0 = 0 \qquad v_4 = 0$

The exaggerated sketch of displacements is shown.

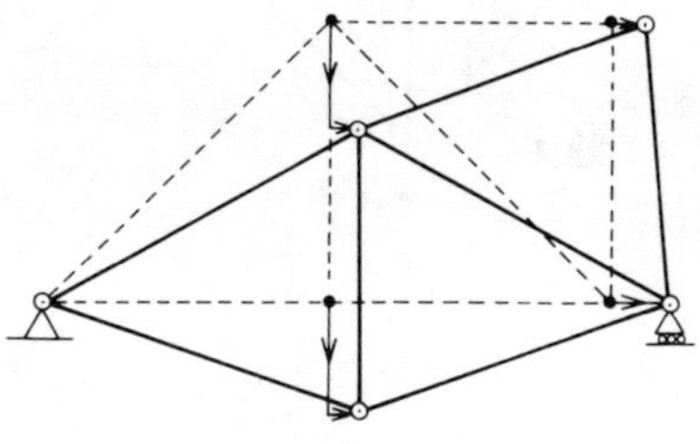

Fig. S. 1.4(c)

S 1.5

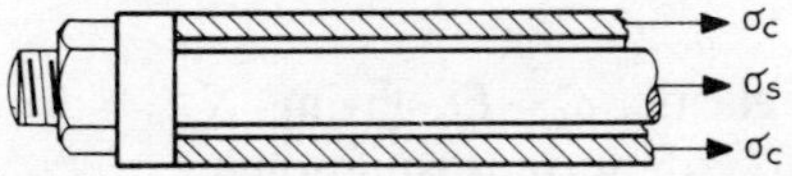

Fig. S. 1.5

Denote the cross-sectional areas and stresses, etc., in copper and steel as A_c, A_s, σ_c, σ_s, etc.

Equilibrium There happens to be no external force in this problem, but we must still enforce equilibrium even though there is no resultant:

$$\sigma_c A_c + \sigma_s A_s = 0$$

This equation is inadequate, the structure is statically indeterminate, so use:

Compatibility The rigid washers cannot bend, therefore the ends of the two components must move the *same amount*, say Δ.

$$\therefore \quad \varepsilon_c = \Delta/l; \qquad \varepsilon_s = \Delta/l; \qquad \text{i.e.} \quad \varepsilon_c = \varepsilon_s$$

Stress–strain $\varepsilon_c = \sigma_c/E_c + \alpha_c T; \qquad \varepsilon_s = \sigma_s/E_s + \alpha_s T$

Putting $\varepsilon_c = \varepsilon_s$, we now have two equations and two unknowns. Solving,

$$\sigma_c = -\sigma_s \frac{A_s}{A_c}; \qquad \sigma_s\left[1 + \frac{E_s A_s}{E_c A_c}\right] = E_s\,[\alpha_c - \alpha_s]\,T$$

giving $\sigma_s = 85.5\,\text{MN/m}^2$, $\sigma_c = -30.8\,\text{MN/m}^2$.

S 1.8

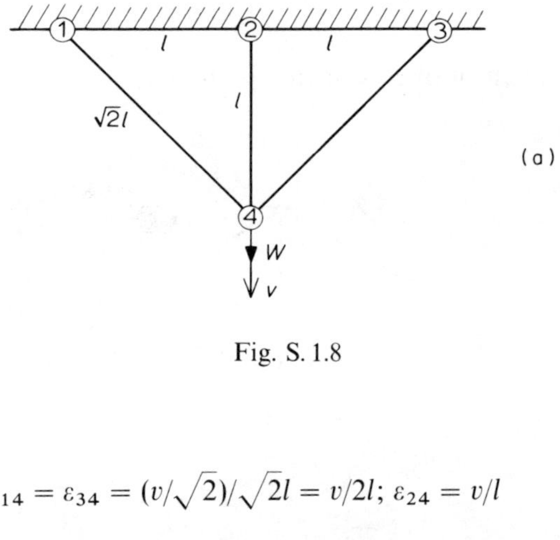

Fig. S. 1.8

Compatibility $\varepsilon_{14} = \varepsilon_{34} = (v/\sqrt{2})/\sqrt{2}l = v/2l; \varepsilon_{24} = v/l$

Equilibrium $N_{14} = N_{34}; 2N_{14}/\sqrt{2} + N_{24} = W$

Stress–strain

(1) If all bars are elastic, $\sigma = E\varepsilon$, giving $N_{14} = AEv/2l$, $N_{24} = AEv/l$, and substituting into equilibrium equations:

$$2AE\frac{v}{2l}\frac{1}{\sqrt{2}} + AE\frac{v}{l} = W; \qquad \frac{v}{l}\left(\frac{1}{\sqrt{2}} + 1\right) = \frac{W}{AE}$$

The maximum strain is clearly $\varepsilon_{24} = v/l$, therefore this solution is valid up to $v/l = 0.001$, at which point

$$\frac{W}{AE} = 0.001\left(1 + \frac{1}{\sqrt{2}}\right) = 0.001\,707 \qquad \text{(i)}$$

(2) For $v/l > 0.001$, stress in bar (2–4) remains at $E/1000$ and so equilibrium equation becomes

$$2AE\frac{v}{2l}\frac{1}{\sqrt{2}} + \frac{AE}{1000} = W; \qquad \text{or} \qquad \frac{v}{l}\left(\frac{1}{\sqrt{2}}\right) = \frac{W}{AE} - 0.001 \qquad \text{(ii)}$$

This is valid until strain in (1–4) (3–4) also becomes $v/2l = 0.001$, and equation (ii) delivers:

$$\frac{W}{AE} = 0.001 + 0.002/\sqrt{2} = 0.002\,414$$

Beyond this strain no further increase in stress is possible and the maximum structural strength has been reached (the *collapse* load).

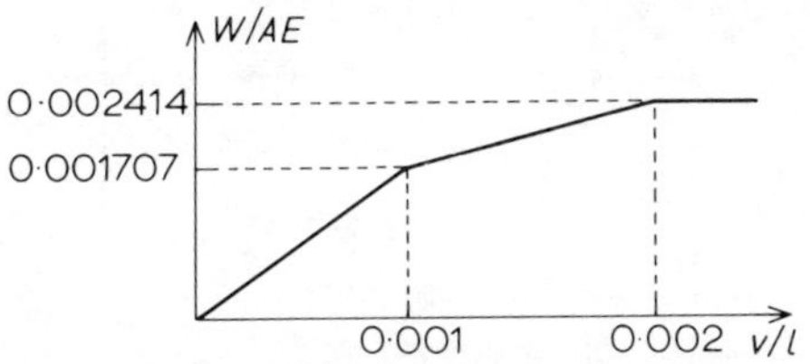

(b)

S 1.10

As shown previously, $\Delta_{14} = \Delta_{34} = v/2$; $\Delta_{24} = v$. The PVD is

$$W\bar{v} = \sum N\bar{\Delta}$$

and using the quoted stress–strain law:

$$W\bar{v} = 2 \cdot 100 \cdot 72 \cdot \frac{v}{600}\left[1 - \left(\frac{v}{6}\right)^2\right]\frac{\bar{v}}{\sqrt{2}} + 100 \cdot 72 \cdot \frac{v}{300}\left[1 - \left(\frac{v}{3}\right)^2\right]\bar{v}$$

$$\therefore \quad v\left[\frac{1}{\sqrt{2}} - \frac{v^2}{36\sqrt{2}} + 1 - \frac{v^2}{9}\right] = \frac{W \cdot 600}{100 \cdot 72 \cdot 2}$$

Collecting terms:

$$v\,[1 - 0.0766v^2] = 0.0244W$$

If $W = 30$ kN, the right-hand side is 0.732. The linear-elastic solution, discarding v^2, is clearly $v = 0.732$. The true deflection must be slightly greater, so trying $v = 0.8$, the left-hand side is 0.761; then $v = 0.75$ gives 0.718; so finally $v = 0.77$ mm gives a left-hand side of 0.735, which is close enough. On Substituting $v = 0.77$ into

$$N_{24} = 100 \cdot 72 \cdot \frac{v}{300}\left[1 - \left(\frac{v}{3}\right)^2\right]$$

we find $N_{24} = 17.26\,\text{kN}$, or 57 per cent of W.
If $W = 50\,\text{kN}$, we similarly find

$$v = 1.46\,\text{mm} \qquad \text{and} \qquad N_{24} = 26.74\,\text{kN}.$$

S 2.2

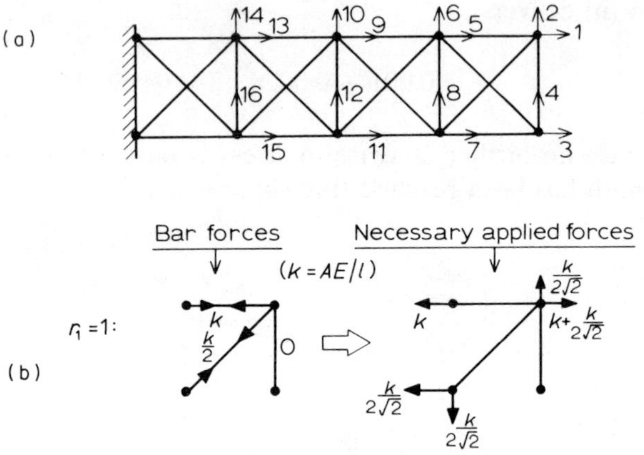

$$\{R_1 \;\; R_2 \;\; R_5 \;\; R_7 \;\; R_8\} = \{k_{11} \;\; k_{21} \;\; k_{51} \;\; k_{71} \;\; k_{81}\}$$
$$= k\left\{1+\frac{1}{2\sqrt{2}} \;\; \frac{1}{2\sqrt{2}} \;\; -1 \;\; \frac{-1}{2\sqrt{2}} \;\; \frac{-1}{2\sqrt{2}}\right\}$$

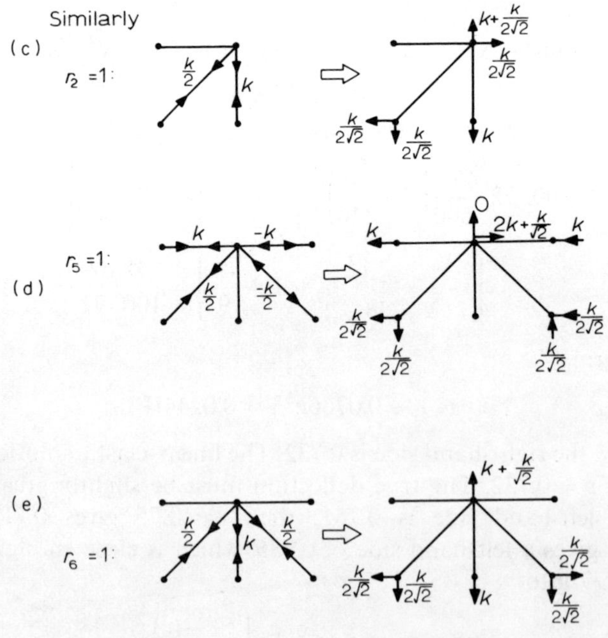

Fig. S. 2.2

On inserting the values into each row of **K** we generate the complete stiffness matrix $\mathbf{K}/k$:

row \ column	1	2	3	4	5	6	7	8	9	10	11	12
1	$1+\frac{1}{4}\sqrt{2}$	$\frac{1}{4}\sqrt{2}$			-1		$-\frac{1}{4}\sqrt{2}$	$-\frac{1}{4}\sqrt{2}$				
2	$\frac{1}{4}\sqrt{2}$	$1+\frac{1}{4}\sqrt{2}$		-1			$-\frac{1}{4}\sqrt{2}$	$-\frac{1}{4}\sqrt{2}$				
3			$1+\frac{1}{4}\sqrt{2}$	$-\frac{1}{4}\sqrt{2}$	$-\frac{1}{4}\sqrt{2}$	$\frac{1}{4}\sqrt{2}$	-1					
4		-1	$-\frac{1}{4}\sqrt{2}$	$1+\frac{1}{4}\sqrt{2}$	$\frac{1}{4}\sqrt{2}$	$-\frac{1}{4}\sqrt{2}$						
5	-1		$-\frac{1}{4}\sqrt{2}$	$\frac{1}{4}\sqrt{2}$	$2+\frac{1}{\sqrt{2}}$				-1		$-\frac{1}{4}\sqrt{2}$	$-\frac{1}{4}\sqrt{2}$
6			$\frac{1}{4}\sqrt{2}$	$-\frac{1}{4}\sqrt{2}$		$1+\frac{1}{\sqrt{2}}$		-1			$-\frac{1}{4}\sqrt{2}$	$-\frac{1}{4}\sqrt{2}$
7	$-\frac{1}{4}\sqrt{2}$	$-\frac{1}{4}\sqrt{2}$	-1				$2+\frac{1}{\sqrt{2}}$		$-\frac{1}{4}\sqrt{2}$	$\frac{1}{4}\sqrt{2}$	-1	
8	$-\frac{1}{4}\sqrt{2}$	$-\frac{1}{4}\sqrt{2}$				-1		$1+\frac{1}{\sqrt{2}}$	$\frac{1}{4}\sqrt{2}$	$-\frac{1}{4}\sqrt{2}$		
9					-1		$-\frac{1}{4}\sqrt{2}$	$\frac{1}{4}\sqrt{2}$	$2+\frac{1}{\sqrt{2}}$			
10							$\frac{1}{4}\sqrt{2}$	$-\frac{1}{4}\sqrt{2}$		$1+\frac{1}{\sqrt{2}}$		-1
11					$-\frac{1}{4}\sqrt{2}$	$-\frac{1}{4}\sqrt{2}$	-1				$2+\frac{1}{\sqrt{2}}$	
12					$-\frac{1}{4}\sqrt{2}$	$-\frac{1}{4}\sqrt{2}$				-1		$1+\frac{1}{\sqrt{2}}$

Notice there are two quick checks. Firstly, the stiffness matrix should be symmetrical. Secondly, for each unit displacement pattern, the applied forces **R** must be self-equilibrating, so all entries in each column must add up to zero unless a bar is connected to a support (since reactions are not included in the above). Only six columns of course can be checked thus far.

S 2.5

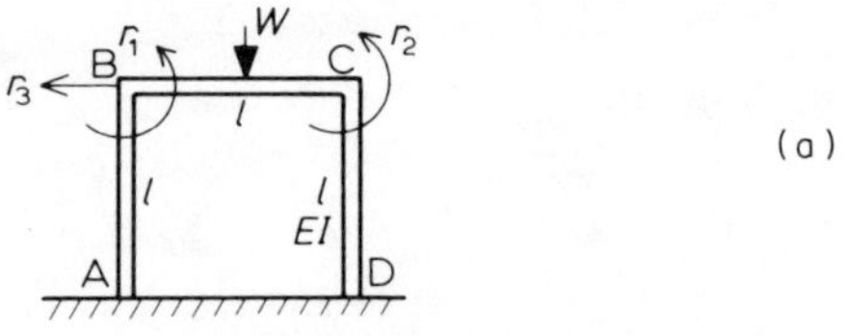

Fig. S. 2.5

The kinematically equivalent loads at B and C from equation (2.29a) are two moments $R_1 = -Wl/8$ and $R_2 = Wl/8$. The stiffness matrix is given in (2.28) so the equations $\mathbf{Kr} = \mathbf{R}$ become

$$\frac{2EI}{l^3}\begin{bmatrix} 4l^2 & l^2 & -3l \\ & 4l^2 & -3l \\ \text{symm} & & 12 \end{bmatrix}\begin{bmatrix} r_1 \\ r_2 \\ r_3 \end{bmatrix} = \frac{Wl}{8}\begin{bmatrix} -1 \\ 1 \\ 0 \end{bmatrix}$$

Solving, $r_1 = -r_2 = -Wl^2/48EI$; $r_3 = 0$. The bending moments at the joints are found by substituting back into (2.24) with $\rho_3 = \rho_1 = 0$ in every case. We find:

$$\text{In AB:}\quad M_A = \frac{2EI}{l}[0 + r_1] = -\frac{Wl}{24}$$

$$M_B = \frac{2EI}{l}[2r_1 + 0] = -\frac{Wl}{12}$$

$$\text{In BC:}\quad M_B = \frac{2EI}{l}[2r_1 + r_2] = -\frac{Wl}{24}$$

$$M_C = \frac{2EI}{l}[2r_2 + r_1] = \frac{Wl}{24}$$

$$\text{In CD:}\quad M_C = \frac{2EI}{l}[2r_2 + 0] = \frac{Wl}{12}$$

$$M_D = \frac{2EI}{l}[0 + r_2] = \frac{Wl}{24}$$

The moment notation corresponds to the rotation convention so that a positive moment in, say, BC, causes compression in the upper surface at C, but at B causes tension. If we plot the moment distribution with positive moment causing tension in the outside of the portal frame, the result is as shown in Fig. S 2.5 (b). The discontinuity at the joints of $Wl/12 + Wl/24 = Wl/8$ matches the applied moment. The full solution can be recovered by adding the self-equilibrating fixed-end solution – in this case only BC has such a correction (c). The total of (b) + (c) is shown together with the deflected shape whose curvature must match the bending moment diagram, of course.

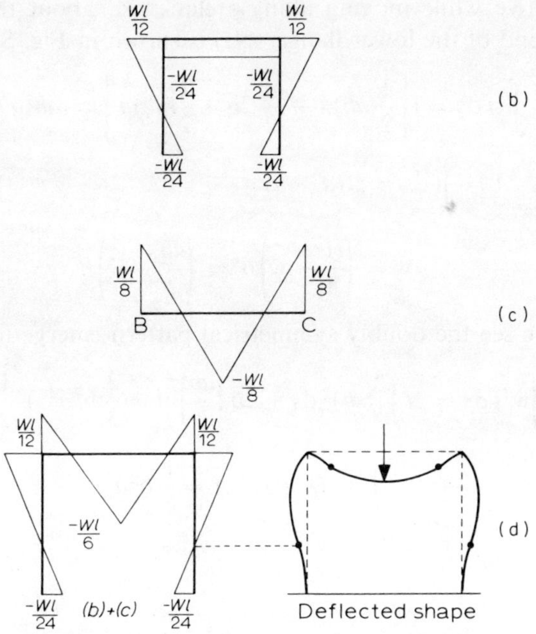

S 2.8

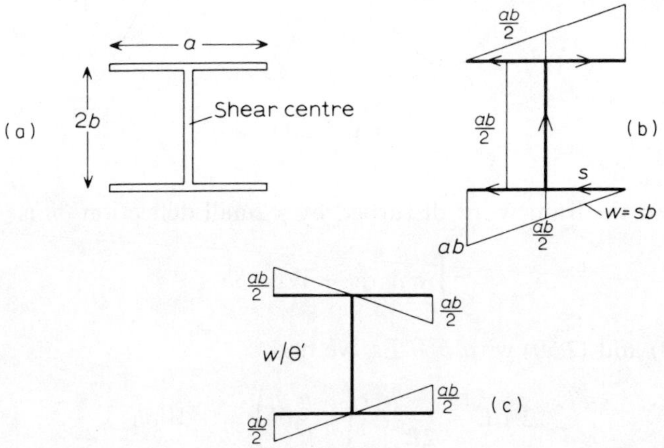

Fig. S. 2.8

The shear centre, or centre of twist, is clearly at the centre of the web.

$$\hat{w}(s) = \int_0^s d\,\mathrm{d}s$$

where d is positive while moving along s clockwise about the shear centre. Starting at the end of the lower flange. $\hat{w}(s)$ is shown in Fig. S 2.8 (b).

$$H = \int \hat{w} t \, ds = t\left[\frac{1}{2}(ab)a + \frac{ab}{2}2b + \frac{1}{2}(ab)a\right] = tab(a+b)$$

$$A = \int t \, ds = 2t(a+b)$$

$$\therefore \quad w = \left(\frac{H}{A} - \hat{w}\right)\theta' = \left(\frac{ab}{2} - \hat{w}\right)\theta'$$

Plotting w/θ', we see the doubly symmetrical pattern emerge (c).

$$Q = \int \hat{w}^2 t \, ds = 2t\int_0^a (sb)^2 \, ds + 2b\left(\frac{ab}{2}\right)^2 t = \frac{2}{3}tb^2a^3 + \frac{1}{2}ta^2b^3$$

$$\therefore \quad \Gamma = Q - H^2/A = \frac{1}{6}tb^2a^3$$

S 2.11

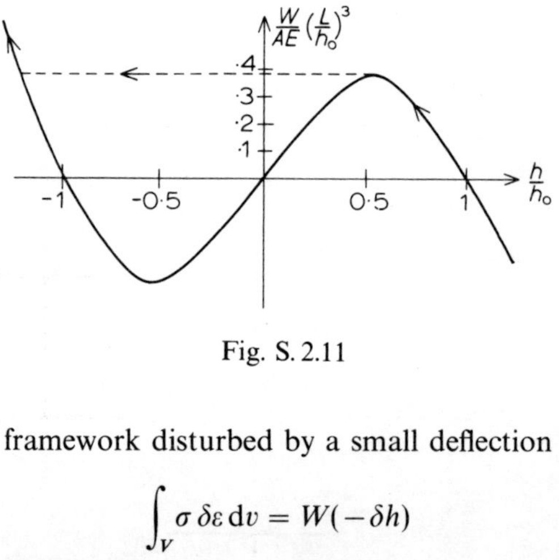

Fig. S. 2.11

The PVD for the framework disturbed by a small deflection δh is

$$\int_V \sigma \, \delta\varepsilon \, dv = W(-\delta h)$$

Using (2.60) and (2.59) with $\sigma = E\varepsilon$, we find

$$2AlE\frac{(h_0^2 - h^2)}{2L^2}\left(-\frac{h}{L^2}\delta h\right) = -W\delta h$$

or

$$\left(1 - \frac{h^2}{h_0^2}\right)\left(\frac{h}{h_0}\right) = \frac{W}{AE}\left(\frac{L}{h_0}\right)^3\left(\frac{L}{l}\right)$$

If $h \ll L$, $L/l = 1/\sqrt{(1 + h^2/L^2)} \simeq 1$, so the last term is unity. The cubic relationship between W and h is shown in the figure. The unloaded position is at

(1,0) and, as W increases, a maximum load is reached, at $h/h_0 = 1/\sqrt{3}$, of $(W/AE)(L/h_0)^3 = 0.38$. For values of W in excess of these there is clearly a 'snap-through' to a negative value of h/h_0 beyond -1. If $(W/AE)(L/h_0)^3$ is a large number, either positive or negative, the curve remains monotonic but does not become linear (which it should) in this solution which has assumed $h/L \ll 1$.

S 3.5

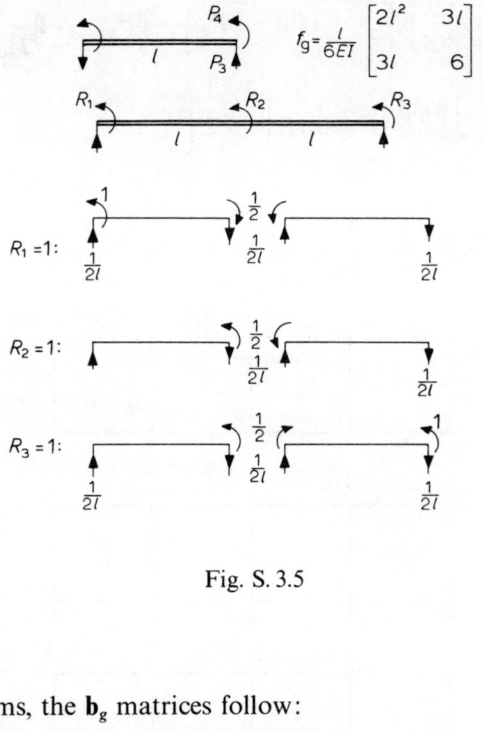

Fig. S. 3.5

From the diagrams, the $\mathbf{b}_g$ matrices follow:

$$g=1:\ \mathbf{b}_1 = \begin{bmatrix} -\frac{1}{2l} & -\frac{1}{2l} & -\frac{1}{2l} \\ -\frac{1}{2} & \frac{1}{2} & \frac{1}{2} \end{bmatrix} \qquad g=2:\ \mathbf{b}_2 = \begin{bmatrix} -\frac{1}{2l} & -\frac{1}{2l} & \frac{1}{2l} \\ 0 & 0 & 1 \end{bmatrix}$$

$$\mathbf{b}_1^t \mathbf{f} \mathbf{b}_1 = \frac{l}{12EI}\begin{bmatrix} 7 & -2 & -2 \\ -2 & 1 & 1 \\ -2 & 1 & 1 \end{bmatrix} \qquad \mathbf{b}_2^t \mathbf{f} \mathbf{b}_2 = \frac{l}{12EI}\begin{bmatrix} 1 & 1 & -2 \\ 1 & 1 & -2 \\ -2 & -2 & 7 \end{bmatrix}$$

$$F = \sum_{g=1,2} \mathbf{b}_g^t \mathbf{f}_g \mathbf{b}_g = \frac{l}{12EI}\begin{bmatrix} 8 & -1 & -4 \\ -1 & 2 & -1 \\ -4 & -1 & 8 \end{bmatrix}$$

S 3.7

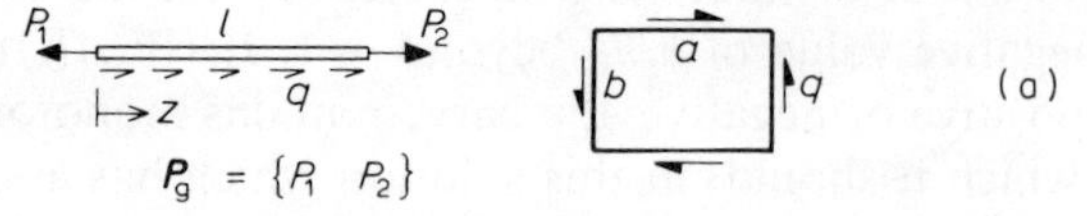

Fig. S. 3.7

Boom:

$$\sigma = \frac{1}{A}(P_1 - qz) = \frac{1}{A}\left[\left(1 - \frac{z}{l}\right)P_1 + \frac{z}{l}P_2\right] = \boldsymbol{\beta}\mathbf{P}_g; \qquad \boldsymbol{\beta} = \frac{1}{A}\begin{bmatrix} 1 - \frac{z}{l} & \frac{z}{l} \end{bmatrix}$$

$$\phi = \frac{1}{E}; \qquad \mathbf{f}_g = \int \boldsymbol{\beta}^t \phi \boldsymbol{\beta}\, dv = \frac{l}{6AE}\begin{bmatrix} 2 & 1 \\ 1 & 2 \end{bmatrix}$$

Shear panel:

$$\sigma = \frac{q}{t}, \qquad \phi = \frac{1}{G}, \qquad \beta = \frac{1}{t}, \qquad f_g = \frac{1}{Gt^2}tab = \frac{ab}{Gt}$$

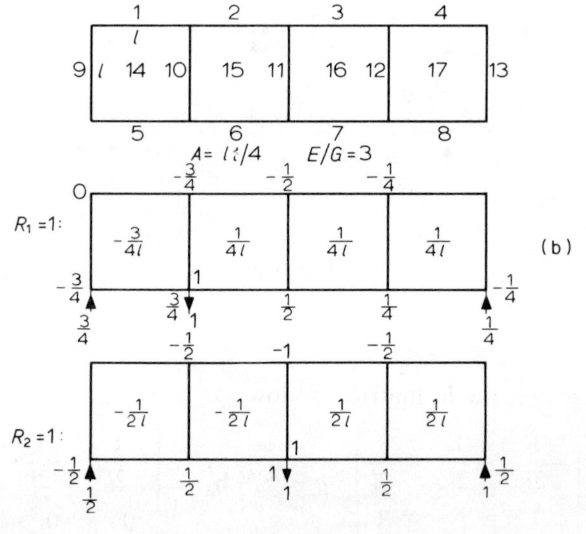

$$\mathbf{b}_1 = \begin{bmatrix} 0 & 0 \\ -\frac{3}{4} & -\frac{1}{2} \end{bmatrix} \quad \mathbf{b}_2 = \begin{bmatrix} -\frac{3}{4} & -\frac{1}{2} \\ -\frac{1}{2} & -1 \end{bmatrix} \quad \mathbf{b}_3 = \begin{bmatrix} -\frac{1}{2} & -1 \\ -\frac{1}{4} & -\frac{1}{2} \end{bmatrix} \quad \mathbf{b}_4 = \begin{bmatrix} -\frac{1}{4} & -\frac{1}{2} \\ 0 & 0 \end{bmatrix}$$

$$\mathbf{b}_5 = \begin{bmatrix} 0 & 0 \\ \frac{3}{4} & \frac{1}{2} \end{bmatrix} \quad \mathbf{b}_6 = \begin{bmatrix} \frac{3}{4} & \frac{1}{2} \\ \frac{1}{2} & 1 \end{bmatrix} \quad \mathbf{b}_7 = \begin{bmatrix} \frac{1}{2} & 1 \\ \frac{1}{4} & \frac{1}{2} \end{bmatrix} \quad \mathbf{b}_8 = \begin{bmatrix} \frac{1}{4} & \frac{1}{2} \\ 0 & 0 \end{bmatrix}$$

$$\mathbf{b}_9 = \begin{bmatrix} 0 & 0 \\ -\frac{3}{4} & -\frac{1}{2} \end{bmatrix} \quad \mathbf{b}_{10} = \begin{bmatrix} 0 & 0 \\ 1 & 0 \end{bmatrix} \quad \mathbf{b}_{11} = \begin{bmatrix} 0 & 0 \\ 0 & 1 \end{bmatrix} \quad \mathbf{b}_{12} = \begin{bmatrix} 0 & 0 \\ 0 & 0 \end{bmatrix}$$

$$\mathbf{b}_{13} = \begin{bmatrix} 0 & 0 \\ -\frac{1}{4} & -\frac{1}{2} \end{bmatrix} \mathbf{b}_{14} = \begin{bmatrix} -\frac{3}{4l} & -\frac{1}{2l} \end{bmatrix} \mathbf{b}_{15} = \begin{bmatrix} \frac{1}{4l} & -\frac{1}{2l} \end{bmatrix} \mathbf{b}_{16} = \begin{bmatrix} \frac{1}{4l} & \frac{1}{2l} \end{bmatrix}$$

$$\mathbf{b}_{17} = \begin{bmatrix} \frac{1}{4l} & \frac{1}{2l} \end{bmatrix}$$

$$\mathbf{F} = \sum \mathbf{b}^t \mathbf{f} \mathbf{b}$$

Top and lower booms both produce:

$$\frac{l}{24AE}\begin{bmatrix} 18 & 22 \\ 22 & 32 \end{bmatrix}$$

Vertical booms:

$$\frac{l}{24AE}\begin{bmatrix} 13 & 4 \\ 4 & 12 \end{bmatrix}$$

Shear panels:

$$\frac{1}{4Gt}\begin{bmatrix} 3 & 2 \\ 2 & 4 \end{bmatrix} = \frac{3l}{16AE}\begin{bmatrix} 3 & 2 \\ 2 & 4 \end{bmatrix}$$

$$\text{Total} \quad \mathbf{F} = \frac{l}{24AE}\begin{bmatrix} 62.5 & 57 \\ 57 & 94 \end{bmatrix}$$

The contributions from the vertical boom deformations and from the shear panel deformations are not very large. In more slender beams they become even less significant and are often ignored.

S 3.11

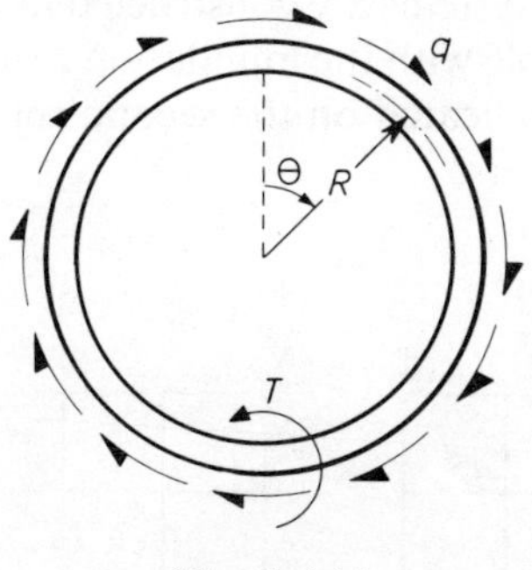

Fig. S. 3.11

Normally such a problem has three redundancies, but in this case because the loading is antisymmetrical about the diameter $\theta = 0$, there will be no axial force or moment at the section $\theta = 0$. Thus the shear X_1 is the only redundancy. Taking moments about θ, we have:

$$M_0 = \int_{\phi=0}^{\phi=\theta} (qR\,\mathrm{d}\phi)\,R\,[1 - \cos(\theta - \phi)] = \frac{T}{2\pi}[\theta - \sin\theta]$$

$$M_1 = R\sin\theta$$

from (3.31),

$$\delta_{10} = \frac{2}{EI}\frac{TR}{2\pi}\int_0^{\pi}(\theta - \sin\theta)\sin\theta R d\theta = \frac{TR^2}{2EI}$$

$$\delta_{11} = \frac{2}{EI}\int_0^{\pi} R^2 \sin^2\theta R\, d\theta = \frac{R^3\pi}{EI}$$

from (3.30) $$\therefore \quad X_1 = -\delta_{10}/\delta_{11} = -T/2\pi R$$

$$M(\theta) = M_0 + X_1 M_1 = \frac{T}{\pi}[\theta/2 - \sin\theta]$$

S 3.12

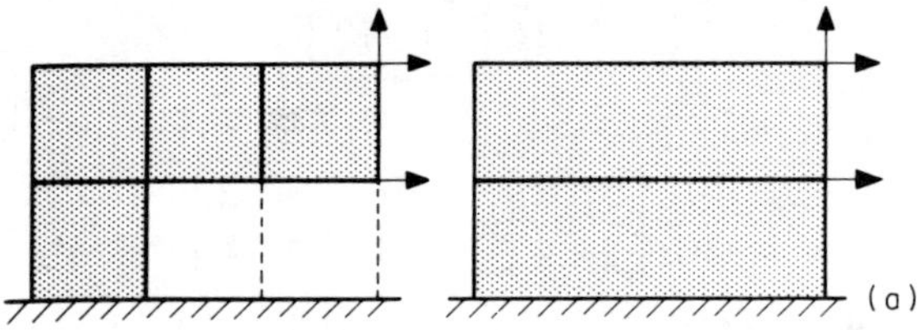

Fig. S. 3.12

A simple statically determinate system could be achieved by cutting two booms and shear panels as shown (i.e. four redundancies). However, a more regular and 'stiffer' $\mathbf{b}_{0g}$ system, suitable for the loads shown, would be simply as shown in the second figure. The four $\mathbf{b}_{xg}$ systems are constructed like Fig. 3.31, noting that half of one such system is possible with the foundation reacting both boom loads and panel shears. The values indicated on the second set of figures are the nonzero boom loads and panel shears.

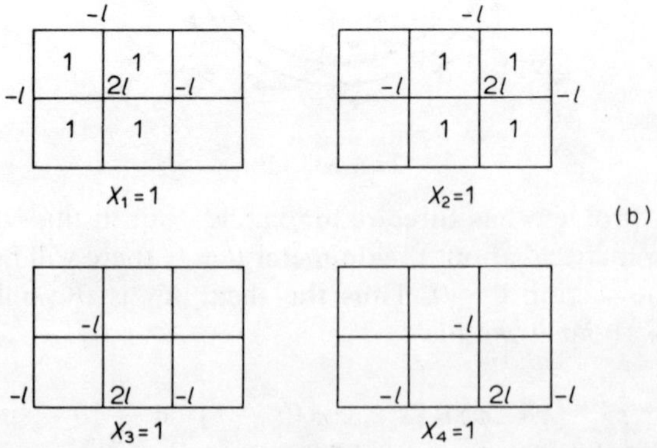

S 4.2

$$v = \sum v_n \sin\frac{n\pi z}{2l}; \qquad \bar{v} = \sin\frac{m\pi z}{2l}$$

$$\int_0^l \left[2EI\sum\frac{n^2\pi^2}{4l^2}v_n \sin\frac{n\pi z}{2l}\right]\frac{m^2\pi^2}{4l^2}\sin\frac{m\pi z}{l}\,dz$$

$$+\int_l^{2l}\left[EI\sum\frac{n^2\pi^2}{4l^2}v_n\sin\frac{n\pi z}{2l}\right]\frac{m^2\pi^2}{4l^2}\sin\frac{m\pi z}{l}\,dz = -W\sin\left(\frac{m\pi}{2l}\frac{3l}{2}\right)$$

Now

$$\int_0^l \sin\frac{n\pi z}{2l}\sin\frac{m\pi z}{2l}\,dz = \frac{l}{\pi}\left[\frac{\sin(n-m)\pi/2}{(n-m)} - \frac{\sin(n+m)\pi/2}{(n+m)}\right]$$

$$\text{or} \qquad = \frac{l}{2} \quad \text{if } n = m$$

similarly,

$$\int_l^{2l} = -\frac{l}{\pi}\left[\frac{\sin(n-m)\pi/2}{(n-m)} - \frac{\sin(n+m)\pi/2}{(n+m)}\right] \qquad \text{or} \qquad = \frac{l}{2} \quad \text{if } n = m$$

With $n = 1$, 2, and 3, we generate three equations:

$m = 1$:

$$\frac{\pi^4}{16l^4}2EI\left[v_1\frac{l}{2} + v_2\frac{4l}{\pi}\left(1+\frac{1}{3}\right) + v_3\frac{9l}{\pi}(0)\right]$$

$$+\frac{\pi^4}{16l^4}EI\left[v_1\frac{l}{2} - v_2\frac{4l}{\pi}\left(1+\frac{1}{3}\right) + 0\right] = -W\frac{1}{\sqrt{2}}$$

or

$$v_1\left(\frac{3}{2}\right) + v_2\left(\frac{16}{3\pi}\right) = -W\frac{16l^3}{\pi^4\sqrt{2}EI} = -\frac{1}{\sqrt{2}}\beta$$

where

$$\beta = 16Wl^3/\pi^4 EI$$

Similarly, for $m = 2$:

$$v_1\left(\frac{16}{3\pi}\right) + v_2(24) + v_3\left(\frac{144}{5\pi}\right) = \beta$$

For $m = 3$:

$$v_2\left(\frac{144}{5\pi}\right) + v_3\left(\frac{243}{2}\right) = -\frac{1}{\sqrt{2}}\beta$$

Solving these three equations:

$$v_1 = -0.570\,\beta; \qquad v_2 = 0.087\,\beta; \qquad v_3 = -0.012\,\beta \text{ (a fair convergence)}$$

Back-substituting to find $M(z)$,

$$M = -EIv''(z) = \frac{EI(z)\pi^2}{4l^2}\left[v_1 \sin\frac{\pi z}{2l} + 4v_2 \sin\frac{2\pi z}{2l} + 9v_3 \sin\frac{3\pi z}{2l}\right]$$

or

$$\frac{M}{Wl} = \frac{EI(z)4}{I\pi^2}\left[-0.57 \sin\frac{\pi z}{2l} + 0.348 \sin\frac{2\pi z}{2l} - 0.108 \sin\frac{3\pi z}{2l}\right]$$

This convergence is not as impressive, although the answer shown in the figure makes a brave attempt to cope with the discontinuity.

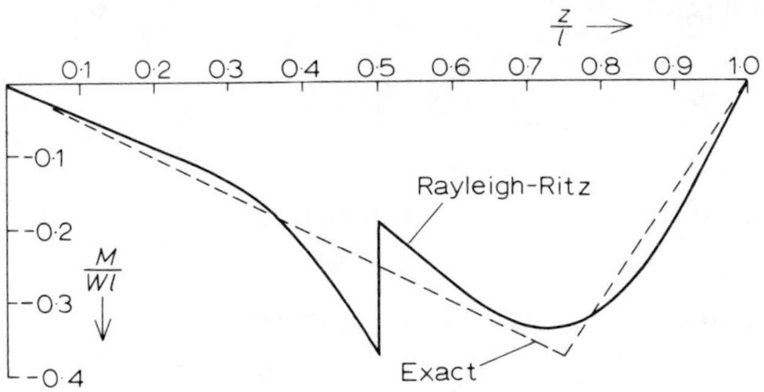

Fig. S. 4.2

S 4.7

$$v = \sum v_n \sin\frac{n\pi z}{l} \qquad \text{satisfies} \quad v(0) = 0 \quad \text{and} \quad v(l) = 0$$

and as an equilibrium bonus satisfies $v''(l) = 0$ also. We must also satisfy the kinematic boundary condition

$$v'(0) = 0 = \sum v_n \frac{n\pi}{l} \qquad \text{or} \qquad \sum n v_n = 0$$

The incremental PVD for initial buckling is

$$\int_0^l EIv''\,\delta v''\,\mathrm{d}z = P\int_0^l v'\,\delta v'\,\mathrm{d}z + M_0\,\delta v'(0)$$

where M_0 is the support moment at the clamped end. Putting $\delta v = \sin m\pi z/l$,

$$\int_0^l\left[\left(EI\sum -\frac{n^2\pi^2}{l^2} v_n \sin\frac{n\pi z}{l}\right)\left(-\frac{m^2\pi^2}{l^2}\sin\frac{m\pi z}{l}\right)\right.$$
$$\left. - P\left(\sum\frac{n\pi}{l} v_n \cos\frac{n\pi z}{l}\right)\left(\frac{m\pi}{l}\cos\frac{m\pi z}{l}\right)\right]\mathrm{d}z = M_0\frac{m\pi}{l}$$

Both trigonometric products are orthogonal;

$$\therefore \quad v_n \left(EI \frac{n^4\pi^4}{l^4}\frac{l}{2} - P\frac{n^2\pi^2}{l^2}\frac{l}{2} \right) = M_0 \frac{n\pi}{l}$$

i.e.

$$nv_n \left[\frac{\pi^2 EI}{l^2} n^2 - P \right] = M_0 \frac{2}{\pi}$$

But

$$\sum nv_n = 0, \qquad \therefore \quad \sum \frac{M_0 2/\pi}{\pi^2 EI/l^2 n^2 - P} = 0$$

or

$$\sum \frac{1}{n^2 - \lambda} = 0, \qquad \text{where} \qquad \lambda = P \Bigg/ \frac{\pi^2 EI}{l^2}.$$

This summation cannot terminate at $n = 1$ since $nv_n = 0$ gives $v_1 = 0$; but at $n = 2, 3, 4$ we find the lowest root λ of the above series to be 2.5; 2.333; and 2.2.

S 4.11

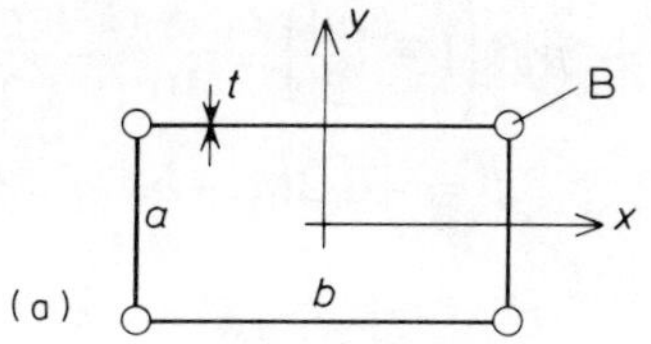

Fig. S. 4.11

$$\sigma_{zz} = \frac{2x}{b} f(z) \text{ in } y = a/2; \qquad \sigma_{zz} = \frac{2y}{a} f(z) \text{ in } x = b/2$$

In $y = a/2$, satisfying equilibrium:

$$\frac{\partial \sigma_{zz}}{\partial z} + \frac{\partial \sigma_{zx}}{\partial x} = 0; \qquad \sigma_{zx} = -\frac{x^2}{b} f'(z) + f_1(z).$$

Similarly in $x = b/2$:

$$\sigma_{zy} = -\frac{y^2}{a} f'(z) + f_2(z)$$

Equilibrium at the booms $(b/2, a/2)$:

$$\sigma_{zx} t + \sigma_{zy} t = \frac{\partial}{\partial z} (B\sigma_{zz})$$

$$\therefore \quad f_1(z) + f_2(z) = f'(z) \left(\frac{B}{t} + \frac{b}{4} + \frac{a}{4} \right) \qquad \text{(i)}$$

Torque equilibrium:

$$T = \int_{-b/2}^{b/2} \sigma_{zx} ta \,dx - \int_{-a/2}^{a/2} \sigma_{zy} tb \,dy$$

$$\therefore \quad f_1(z) - f_2(z) = \frac{T}{tab} + \frac{f'(z)}{12}(b - a) \qquad \text{(ii)}$$

Hence we have $f_1(z)$ and $f_2(z)$ from (i) and (ii), and now we know:

in $y = a/2$,

$$\sigma_{zx} = \frac{T}{2tab} + \frac{f'(z)}{12}\left(2b + a + \frac{6B}{t} - \frac{12x^2}{b}\right)$$

in $x = b/2$,

$$-\sigma_{zy} = \frac{T}{2tab} - \frac{f'(z)}{12}\left(2a + b + \frac{6B}{t} - \frac{12y^2}{a}\right)$$

We can generate $\bar{\boldsymbol{\sigma}}$ by putting $T = 0$ and replacing $f(z)$ by $\bar{f}(z)$. Substituting into the PVF and integrating over the cross-section, we have:

$$\int_0^\infty \left(3f\bar{f} + \frac{T}{bt}(\mu - 1)(1 + 3\beta)\bar{f}' + C_1 f'\bar{f}'\right) dz = 0$$

where

$$\frac{C_1}{a^2} = \frac{\mu}{2} + \beta(1 + \mu)^2\left(1 + \frac{3\beta}{2}\right) + \frac{3}{10}\frac{(1 + \mu^3)}{(1 + \mu)}, \qquad \mu = b/a$$

$$\beta = 4B/2(a + b)t$$

Integrating by parts:

$$\int_0^\infty (3f - C_1 f'')\bar{f}\,dz + \left\{\left[\frac{T}{bt}(\mu - 1)(1 + 3\beta) + C_1 f'\right]\bar{f}\right\}_0^\infty = 0$$

From the integral we have the differential equation whose solution is

$$f(z) = C_2 \exp(-\alpha z/a), \qquad \frac{\alpha^2}{a^2} = \frac{3}{C_1}$$

From the second term, at $z = 0$, $\bar{f}$ is arbitrary, hence

$$\frac{T}{bt}(\mu - 1)(1 + 3\beta) - C_1 C_2 \alpha/a = 0$$

or

$$C_2 = \frac{(\mu - 1)(1 + 3\beta)\alpha T}{3abt}$$

Putting $b = 2a$ ($\mu = 2$), we find at $z = 0$,

$$\frac{\sigma_{zx}}{(T/2tab)} = 1 - \frac{(1 + 3\beta)}{(1.9 + 9\beta + 27\beta^2/2)}\left(\frac{5}{6} + \frac{3\beta}{2} - \frac{x^2}{a^2}\right) \qquad \text{(iii)}$$

and

$$-\frac{\sigma_{zy}}{(T/2tab)} = 1 + \frac{(1+3\beta)}{(1.9+9\beta+27\beta^2/2)}\left(\frac{2}{3}+\frac{3\beta}{2}-\frac{2y^2}{a^2}\right) \quad \text{(iv)}$$

The first term is the simple Bredt–Batho prediction and the rest is the effect of the constraint. For heavy booms, as $\beta \to \infty$, (iii) tends to $\frac{2}{3}$ and (iv) tends to $\frac{4}{3}$. These discontinuous shears are also predicted by classical theory[44] even when there are no booms, because it does not take σ_{zz} into account as here. As $\beta \to 0$ we do not have the classical discontinuity in shear: the behaviour is shown in the figure.

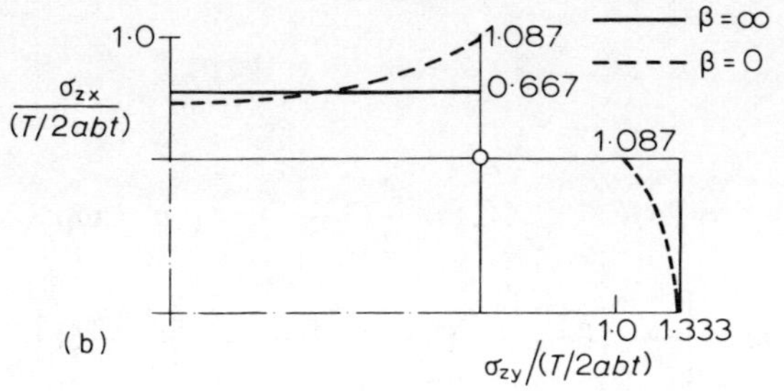

S 4.13

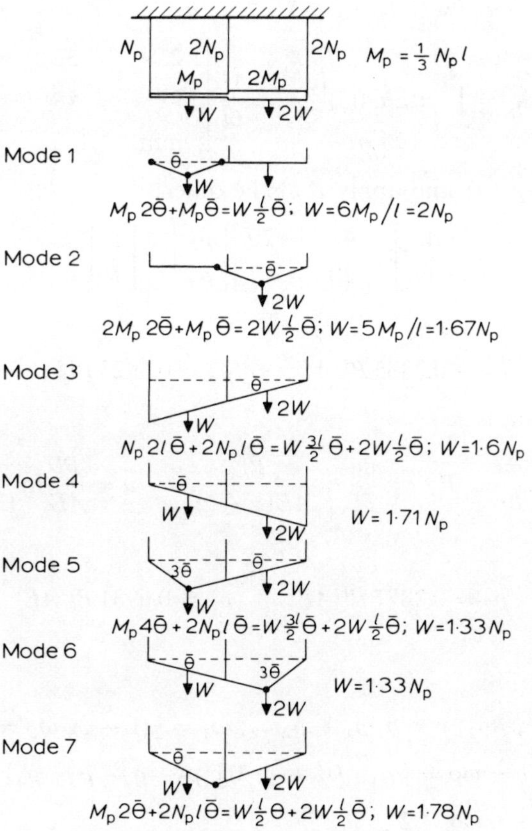

Fig. S. 4.13

S 5.1

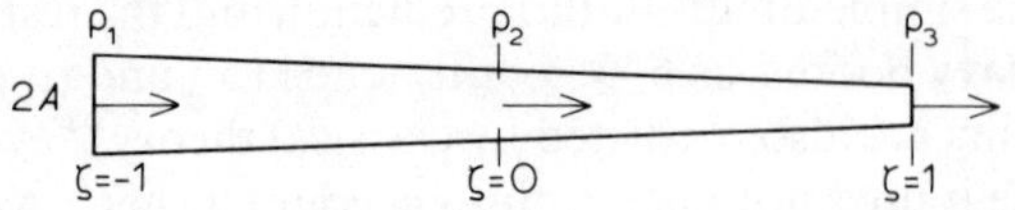

Fig. S. 5.1

$$\zeta = 2x/l; \qquad A(\zeta) = \frac{A}{2}(3-\zeta); \qquad u = \boldsymbol{\omega\rho}$$

$$\boldsymbol{\omega} = \begin{bmatrix} -\frac{\zeta}{2}(1-\zeta) & (1-\zeta)(1+\zeta) & \frac{\zeta}{2}(1+\zeta) \end{bmatrix}$$

$$\alpha = \frac{\partial \boldsymbol{\omega}}{\partial x} = \frac{1}{l}[-1+2\zeta \quad -4\zeta \quad 1+2\zeta]$$

$$\mathbf{k}_g = \int_{-1}^{1} \boldsymbol{\alpha}^t E \boldsymbol{\alpha} A(\zeta) \frac{l\mathrm{d}\zeta}{2} = \frac{AE}{6l}\begin{bmatrix} 25 & -32 & 3 \\ & 48 & -20 \\ \text{symm.} & & 17 \end{bmatrix}$$

Hold one end, $\rho_1 = 0$, and apply P at the other:

$$\frac{AE}{6l}\begin{bmatrix} 48 & -20 \\ -20 & 17 \end{bmatrix}\begin{bmatrix} \rho_2 \\ \rho_3 \end{bmatrix} = \begin{bmatrix} 0 \\ P \end{bmatrix}$$

Thus

$$\rho_2 = 0.2885\, Pl/AE; \qquad \rho_3 = 0.6923\, Pl/AE$$

The exact solution is

$$A(\zeta)E\frac{\partial u}{\partial x} = P; \qquad \frac{\partial u}{\partial \zeta} = \frac{Pl}{AE(3-\zeta)}, \qquad u = \frac{Pl}{AE}\ln\frac{4}{(3-\zeta)}$$

i.e.

$$\rho_2 = 0.2877\, Pl/AE; \qquad \rho_3 = 0.6931\, Pl/AE$$

S 5.9

Put $u = \rho_2\omega_1 + \rho_4\omega_2$; $v = \rho_1\omega_1 + \rho_3\omega_2$; $\omega_1 = \frac{1}{2}(1-\zeta)$, $\omega_2 = \frac{1}{2}(1+\zeta)$

$$v = \boldsymbol{\omega\rho} = [\omega_1 \quad 0 \quad \omega_2 \quad 0]\{\rho_1 \quad \rho_2 \quad \rho_3 \quad \rho_4\}$$

$$\frac{\partial \boldsymbol{\omega}}{\partial x} = \frac{2}{l}[\omega_1' \quad 0 \quad \omega_2' \quad 0]$$

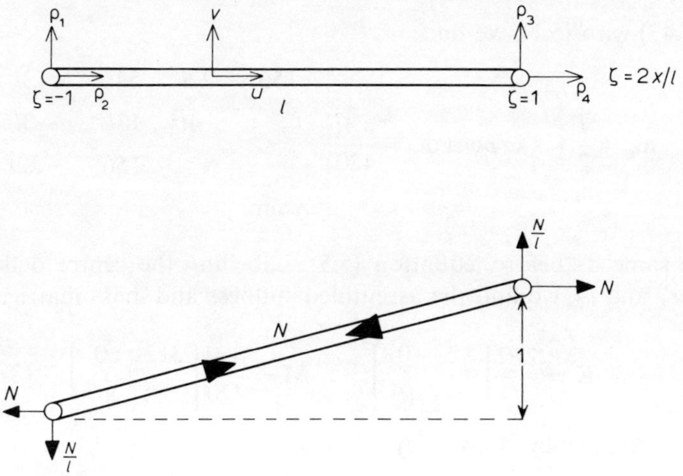

Fig. S. 5.9

Using (5.42),

$$\mathbf{k}_G = \frac{4}{l^2}\int_{-1}^{1} \begin{bmatrix} \omega_1' \\ 0 \\ \omega_2' \\ 0 \end{bmatrix} \left(\frac{N}{A}\right) [\omega_1' \quad 0 \quad \omega_2' \quad 0]\, A\frac{l}{2}\mathrm{d}\zeta$$

$$= \frac{N}{l}\begin{bmatrix} 1 & 0 & -1 & 0 \\ 0 & 0 & 0 & 0 \\ -1 & 0 & 1 & 0 \\ 0 & 0 & 0 & 0 \end{bmatrix}$$

Check by applying unit displacement $\rho_3 = 1$, and maintain in equilibrium as shown in the figure. The extra $\mathbf{P}_g$ needed are clearly

$$\mathbf{P}_g = \left\{-\frac{N}{l} \quad 0 \quad \frac{N}{l} \quad 0\right\}$$

Similarly for $\rho_1 = 1$,

$$\mathbf{P}_g = \left\{\frac{N}{l} \quad 0 \quad -\frac{N}{l} \quad 0\right\}$$

These check the above. Unit displacements $\rho_2 = 1$ and $\rho_4 = 1$ do not realign stresses.

S 5.12

Using (5.43) with (5.1), we find

$$\mathbf{m}_g = \frac{l}{2}\int_{-1}^{1} \boldsymbol{\omega}^t \rho \boldsymbol{\omega} A \, d\zeta = \frac{\rho A l}{420}\begin{bmatrix} 156 & 22l & 54 & -13l \\ & 4l^2 & 13l & -3l^2 \\ & & 156 & -22l \\ \text{symm.} & & & 4l^2 \end{bmatrix}$$

$\mathbf{k}_g$ is the same as before, equation (5.5). Labelling the centre deflection and rotation r_1 and r_2, we find the assembled stiffness and mass matrices are:

$$\mathbf{K} = \frac{2EI}{l^2}\begin{bmatrix} 12 & 0 \\ 0 & 4l^2 \end{bmatrix}; \qquad \mathbf{M} = \frac{\rho A l}{420}\begin{bmatrix} 312 & 0 \\ 0 & 8l^2 \end{bmatrix}$$

The natural frequency is given by

$$\begin{bmatrix} 12\lambda^2 - 312\Omega^2 & 0 \\ 0 & 4\lambda^2 l^2 - 8l^2\Omega^2 \end{bmatrix}\begin{bmatrix} r_1 \\ r_2 \end{bmatrix} = 0$$

where $\lambda^2 = 840\, EI/\rho A l^4$.

$$\therefore \quad 12\lambda^2 - 312\Omega^2 = 0 \qquad \text{with} \qquad r_1 \text{ nonzero}$$

and

$$4\lambda^2 - 8\Omega^2 = 0 \qquad \text{with} \qquad r_2 \text{ nonzero}$$

The smallest frequency corresponds to the symmetric mode with $r_2 = 0$, and is given by:

$$312\Omega^2 = 12\lambda^2 = 12.840\, EI/\rho A l^6$$

or

$$\Omega^2 \rho A l^4/EI = 32.3; \qquad \text{compared with the exact solution, 31.28}$$

If we lump one half of the mass of each element to the two end displacements, the new assembled mass matrix is simply:

$$\mathbf{M} = \rho A l \begin{bmatrix} 1 & 0 \\ 0 & 0 \end{bmatrix}$$

Thus

$$\left| \frac{2EI}{l^3}\begin{bmatrix} 12 & 0 \\ 0 & 4l^2 \end{bmatrix} - \Omega^2 \rho A l \begin{bmatrix} 1 & 0 \\ 0 & 0 \end{bmatrix} \right| = 0$$

The symmetric-mode frequency is therefore given by

$$\left(12 - \frac{\Omega^2 \rho A l^4}{2EI}\right) r_1 = 0, \qquad \text{or} \qquad \frac{\Omega^2 \rho A l^4}{EI} = 24$$

This is lower than the exact value when we normally expect a finite-element solution to be too stiff. The reason is that the effective mass has been increased by this lumping. Normally the effective inertia at, say, node 1 due to unit accelerations at 1 and 2 is seen to be the first row of the mass matrix:

$$\left(\frac{\rho Al}{420}\right)[156 + 54] = \tfrac{1}{2}\rho Al$$

But by lumping *all* the mass $\rho Al/2$ at node 1, instead of $(156/420)\rho Al$, we are overestimating the mass and underestimating the frequency.

APPENDIX A
MATRICES AND LINEAR ALGEBRA

A.1 Matrix manipulation

It is possible to write a fairly thick text on linear algebra – the manipulation of matrices and vectors and the like – and this could be both expensive and counter-productive. We require only a few algebraic rules. Firstly, the reader who has somehow avoided matrix notation only has to realize that it is a convenient form of shorthand to enable us to label a set of discrete numbers. Moreover, provided a computer has been told what to expect, we may write instructions in matrix form and let the computer respond to 'read $K[i,j]$' with as many i and j as we care to feed it.

A *list* of, say, forces or displacements may be written as a column matrix

$$\mathbf{r} = \begin{bmatrix} r_1 \\ r_2 \\ \vdots \\ r_n \end{bmatrix}$$

although in this text we denote such a column by the special brackets

$$\mathbf{r} = \{r_1 \quad r_2 \quad r_3 \quad \cdots \quad r_n\}$$

Such a column matrix is often referred to as a *vector* having n components. In the case of displacement and force components

$$\mathbf{u} = \{u \quad v \quad w\} \qquad \text{and} \qquad \mathbf{p}_v = \{p_{vx} \quad p_{vy} \quad p_{vz}\}$$

they are actual three-dimensional components. If we label the forces $\mathbf{R}$ similarly, we now have to invent a multiplication rule for the work product $R_1r_1 + R_2r_2 + \cdots R_nr_n$. The so-called 'row × column' rule says that in a matrix multiplication $\mathbf{AB}$, each column of a *row* matrix $\mathbf{A}$ is to be multiplied by the equivalently placed row of a *column* matrix and then summed thus:

$$[A_1 \quad A_2 \quad \cdots \quad A_n]\begin{bmatrix} B_1 \\ B_2 \\ \vdots \\ B_n \end{bmatrix} = A_1B_1 + A_2B_2 + \cdots + A_nB_n$$

The *transpose* of a matrix simply interchanges columns and rows, thus:

$$\mathbf{r}^t = [r_1 \quad r_2 \quad \cdots \quad r_n]$$

Therefore we can now form the work product as $\mathbf{R}^t\mathbf{r}$. But it is also clearly equal to $\mathbf{r}^t\mathbf{R}$ and this is shortly explained. Column and row matrices are not enough. We may have a set of equations for displacements and forces like:

$$k_{11}r_1 + k_{12}r_2 + \cdots k_{1n}r_n = R_1$$

followed by:

$$k_{21}r_1 + k_{22}r_2 + \cdots k_{2n}r_n = R_2 \quad \text{etc.}$$

We therefore form the array of $n \times n$ rows and columns

$$\mathbf{K} = \begin{bmatrix} k_{11} & k_{12} & \cdots & k_{1n} \\ \vdots & & & \vdots \\ k_{n1} & & \cdots & k_{nn} \end{bmatrix}$$

and summarize our equations as

$$\mathbf{Kr} = \mathbf{R}$$

where the product $\mathbf{Kr}$ is formed by multiplying *each* row by the column in turn. This rule can be extended to two general matrices **A** and **B**, both arrays, to form $\mathbf{AB} = \mathbf{C}$. The ith row of **A** is multiplied by the jth row of **B** and the summed value placed in the matrix **C** as the element in row i, column j. This rule is only possible if the number of columns in **A** is the same as the number of rows in **B**. If **A** is $l \times m$ then **B** must be of the form $m \times n$ to produce a final product $l \times n$. These products may be continued:

$$\underset{(l\times m)}{\mathbf{A}} \times \underset{(m\times n)}{\mathbf{B}} \times \underset{(n\times a)}{\mathbf{C}} \times \underset{(a\times q)}{\mathbf{D}} = \underset{(l\times q)}{\mathbf{E}}$$

It is obvious that **AB** is not the same as **BA** even if the latter could be formed. So matrices in general are not commutative. But they are associative and additive, so

$$\mathbf{A(BC)} = \mathbf{(AB)C} = \mathbf{ABC}; \qquad \text{and} \qquad \mathbf{A(B + C)} = \mathbf{AB} + \mathbf{AC}$$

By trying a simple (2×2) (2×3) product, the reader should verify that

$$(\mathbf{AB})^t = \mathbf{B}^t\mathbf{A}^t$$

The work product $\mathbf{R}^t\mathbf{r}$ produces a single quantity or *scalar*, so if we take its transpose it remains unchanged.

Thus

$$[\mathbf{R}^t\mathbf{r}]^t = [\mathbf{R}^t\mathbf{r}]$$

therefore

$$\mathbf{r}^t\mathbf{R} = \mathbf{R}^t\mathbf{r}, \qquad \text{as forecast}$$

A matrix is *symmetrical* if any element $K_{ij} = K_{ji}$, and it clearly equals its own transpose. In linear structural mechanics the forces $\mathbf{R} = \mathbf{Kr}$, so the strain energy absorbed by the structure as forces $\mathbf{R}$ increase from $\mathbf{0}$ to $\mathbf{R}$ is simply

$$\tfrac{1}{2}\mathbf{R}^t\mathbf{r} = \tfrac{1}{2}[\mathbf{Kr}]^t\mathbf{r} = \tfrac{1}{2}\mathbf{r}^t\mathbf{K}^t\mathbf{r} = \tfrac{1}{2}\mathbf{r}^t\mathbf{Kr}$$

This is still a scalar and equals its own transpose. The product $\mathbf{r}^t\mathbf{Kr}$ is known as a *congruent transformation.* If any matrix $\mathbf{K}$ produces a positive number from such a transformation for *any* column $\mathbf{r}$, then it is said to be *positive definite.* Because energy is positive then $\mathbf{K}$ must be positive definite.

The *determinant* of a matrix $\mathbf{A}$ with elements a_{ij} is given by

$$|\mathbf{A}| = a_{11}M_{11} - a_{12}M_{12} + a_{13}M_{13} - \cdots - (-1)^n a_{1n}M_{1n}$$

where the *minor* M_{13}, etc., is the determinant of the matrix with row 1 and column 3 missing, and so on. The *cofactor* A_{ij} is defined as $(-1)^{i+j}M_{ij}$. The transpose of the array of cofactors is called the *adjoint* or $\hat{\mathbf{A}}$. Of the many properties of determinants those that are most important are:

(a) Interchanging two rows or two columns changes the sign of $|\mathbf{A}|$.
(b) If all elements in a row or column of $\mathbf{A}$ are zero, then so is $|\mathbf{A}|$.
(c) Multiplication by a constant c of all elements in a row or column of $\mathbf{A}$ multiplies $|\mathbf{A}|$ by c.
(d) Adding a constant multiple of row (or column) i to row (column) j does not change the determinant.
(e) If one row (i) is a linear combination of rows j and k then the determinant must be zero. This follows from (d) since we could subtract this linear combination from row i without changing $|\mathbf{A}|$, but a row of zeros would ensue which makes $|\mathbf{A}| = 0$.

It may be shown that

$$\mathbf{A}\hat{\mathbf{A}} = |\mathbf{A}|\mathbf{I}$$

where $\mathbf{I}$ is a *unit matrix* consisting only of unit elements on the leading diagonal.

$$\text{Writing} \qquad \frac{\mathbf{A}\hat{\mathbf{A}}}{|\mathbf{A}|} = \mathbf{I}$$

this is symbolically written as $\mathbf{AA}^{-1} = \mathbf{I}$,

$$\text{where} \qquad \mathbf{A}^{-1} = \frac{\hat{\mathbf{A}}}{|\mathbf{A}|}$$

This is called the *inverse* of $\mathbf{A}$ in analogy to reciprocals of ordinary numbers, since it has the same effect. Consider for example the set of equations ($\mathbf{A}$ *square*)

$$\mathbf{Ax} = \mathbf{B}$$

where $\mathbf{x}$ is a column vector. Premultiplying by $\mathbf{A}^{-1}$: $\mathbf{A}^{-1}\mathbf{Ax} = \mathbf{A}^{-1}\mathbf{B}$. But

$$\mathbf{A}^{-1}\mathbf{A} = \mathbf{I} \quad \text{and} \quad \mathbf{Ix} = \mathbf{x}, \quad \text{therefore}$$

$$\mathbf{x} = \mathbf{A}^{-1}\mathbf{B}$$

The solution of $\mathbf{Ax} = \mathbf{B}$ is not usually found by forming $\hat{\mathbf{A}} \div |\mathbf{A}|$ as this is expensive. However, this expression does show us that the problem $\mathbf{Ax} = \mathbf{B}$ has no solution if $|\mathbf{A}| = 0$; the matrix $\mathbf{A}$ is said to be *singular* and it happens because of linear dependence. Thus we cannot obtain a solution until the number of equations matches the number of unknowns.

If a square matrix, $n \times n$, has one row (or column) which is a linear combination of others, it is said to be degenerate. There may be l such linear dependencies and the matrix is then said to be degenerate to order l. The *rank* of a matrix is

$$r = n - l$$

and is of course equal to the size of the matrix if it is not singular. If *rank deficiency* is caused by a linear dependence of certain rows, then we can also define rank as the size of the largest nonzero determinant taken from the rows and columns of the matrix. A matrix product must have a rank equal to the smallest rank of its constituent matrices.

A.2 Matrix inversion

The solution of large sets of equations like $\mathbf{Kr} = \mathbf{R}$ may be symbolically written as $\mathbf{r} = \mathbf{K}^{-1}\mathbf{R}$ but fortunately the inverse is never needed for large systems, so the title is nominal. For simply solving the set of simultaneous equations, with perhaps a few load vectors $\mathbf{R}$, there are two commonly used techniques. The *Gauss elimination*, or more often the *Jordan elimination* version is popular and relies on the following idea. Suppose we wish to find $\mathbf{A}^{-1}$, that is solve $\mathbf{Ax} = \mathbf{I}$ for $\mathbf{x}$. The aim is then to operate on $\mathbf{A}$ and to eventually convert it to $\mathbf{I}$. Call this transforming matrix $\mathbf{T}$, so $\mathbf{TA} = \mathbf{I}$ and $\mathbf{T}$ is therefore the desired inverse $\mathbf{T} = \mathbf{A}^{-1}$. However, it is not necessary to construct $\mathbf{T}$ explicitly, but to operate on $\mathbf{Ax} = \mathbf{I}$ simultaneously until we are left with

$$\mathbf{TAx} = \mathbf{TI}$$

so $\mathbf{x} = \mathbf{T}$ emerges as the right-hand side. If the inverse is not required, but merely a solution of $\mathbf{Ax} = \mathbf{B}$, then the operation $\mathbf{TAx} = \mathbf{TB}$: $\mathbf{x} = \mathbf{TB}$ also emerges, and $\mathbf{B}$ of course usually contains far fewer columns than $\mathbf{I}$. The process is conducted numerically on the *augmented* matrix $[\mathbf{A}:\mathbf{B}]$ simply to avoid burdening the computer with an 'equals' sign. The final operator $\mathbf{T}$ is found successively as follows.

Firstly, divide the first row of $\mathbf{A}$ $(a_{11} \quad a_{12} \quad \cdots \quad a_{1n})$ by a_{11} so that the first element is unity, then use this row scaled by a_{i1} to subtract from each of the i rows and produce a first column of unity followed by zeros $\{1 \quad 0 \quad 0 \quad 0 \quad \cdots \quad 0\}$. The process is then repeated using the scaled second row to produce a second column $\{0 \quad 1 \quad 0 \quad 0 \quad 0 \quad \cdots \quad 0\}$. The second row operating on all others will leave the first column untouched since its first element is now zero. A simple example illustrates the process. Solve:

$$\begin{bmatrix} 3 & 2 & 1 \\ 2 & 3 & 2 \\ 1 & 2 & 3 \end{bmatrix} \begin{bmatrix} x_1 \\ x_2 \\ x_3 \end{bmatrix} = \begin{bmatrix} 10 \\ 14 \\ 14 \end{bmatrix}$$

The augmented matrix is

$$\begin{bmatrix} 3 & 2 & 1 & 10 \\ 2 & 3 & 2 & 14 \\ 1 & 2 & 3 & 14 \end{bmatrix}$$

Dividing the first row by 3 and then multiplying by 2 and 1 before subtracting from rows 2 and 3 produces

$$\begin{bmatrix} 1 & \frac{2}{3} & \frac{1}{3} & \frac{10}{3} \\ 0 & \frac{5}{3} & \frac{4}{3} & \frac{22}{3} \\ 0 & \frac{4}{3} & \frac{8}{3} & \frac{32}{3} \end{bmatrix}$$

Repeating this by scaling the *pivot* a_{22} by 3/5 in the second row, and operating on rows 1 and 3 produces

$$\begin{bmatrix} 1 & 0 & -\frac{1}{5} & \frac{2}{5} \\ 0 & 1 & \frac{4}{5} & \frac{22}{5} \\ 0 & 0 & \frac{8}{5} & \frac{24}{5} \end{bmatrix}$$

Finally, scaling the pivot a_{33} by $\frac{5}{8}$, and operating on rows 1 and 2 produces the required final form

$$\begin{bmatrix} 1 & 0 & 0 & 1 \\ 0 & 1 & 0 & 2 \\ 0 & 0 & 1 & 3 \end{bmatrix}$$

and $\{x_1 \quad x_2 \quad x_3\} = \{1 \quad 2 \quad 3\}$.

If at any stage one of the diagonal 'pivotal' elements is zero, then columns to the right must be exchanged which means that the final order of $\mathbf{X}$ becomes rearranged. However, we have seen the largest terms in $\mathbf{K}$ arrange themselves on the leading diagonal. A different method which avoids the need of manipulating pivots, which exploits the symmetric nature of $\mathbf{K}$, and is rather easier to manipulate if the entire matrix will not fit in core, is the *Cholesky decomposition* technique.

Decomposition assumes that the symmetric matrix **A** can be written as the product of a lower and an upper triangular form; a lower triangular matrix **L** being such that element $l_{ij} > 0$ for $j \leqslant i$, but $l_{ij} = 0$ for $j > i$. In particular, the Cholesky decomposition converts $\mathbf{Ax} = \mathbf{B}$ to the form[74]

$$\mathbf{LL}^{t}\mathbf{x} = \mathbf{B}$$

where

$$l_{ii} = \left(a_{ii} - \sum_{k=1}^{i-1} l_{ik}^2\right)^{1/2}; \qquad l_{ij} = \left(a_{ij} - \sum_{k=1}^{j-1} l_{ik}l_{jk}\right)\Big/ l_{jj} \quad (j < i)$$

When these elements are evaluated in turn row by row, it is possible to overwrite the previously used a_{ij}. Once the above form has been constructed, we may put $\mathbf{L}^t\mathbf{x} = \mathbf{y}$ and solve $\mathbf{Ly} = \mathbf{B}$ for **y** immediately, since **L** is a lower diagonal form and we progressively solve y_1, y_2, y_3, etc. Having solved **y** then $\mathbf{L}^t\mathbf{x} = \mathbf{y}$ is solved for **x** since $\mathbf{L}^t$ is upper triangular, and $x_n, x_{n-1}, \ldots$ emerge directly. (This triangularization ruse can even be applied to the *element* matrices **k** before assembling **K**.[75])

Provided **A** is positive definite then l_{ii} will be real, so the inverse of a stiffness matrix presents no problems until the size is large and round-off errors occur. Substructuring can be used to overcome this problem. Ill-conditioning, or a near-singular matrix, can be produced in **K** if the modelling is poor, the structure badly held, or just the labelling of displacements can produce too many off-diagonal terms. The size of the determinant is not a good absolute measure of conditioning; a better one is the ratio of extreme eigenvalues, but expensive to evaluate.

A.3 The eigenvalue problem

The standard eigenvalue problem is of the form $\mathbf{Ax} = \lambda\mathbf{x}$, or

$$(\mathbf{A} - \lambda\mathbf{I})\mathbf{x} = \mathbf{0}$$

The buckling and vibration problems of Chapter 5 are of this form, $(\mathbf{K} - \Omega^2\mathbf{M})\mathbf{r} = \mathbf{0}$, which can be premultiplied by $\mathbf{M}^{-1}$ to produce $(\mathbf{M}^{-1}\mathbf{K} - \Omega^2\mathbf{I})\mathbf{r} = \mathbf{0}$ as the standard form. The trivial solution of the above equation is of course $\mathbf{x} = \mathbf{0}$, but there can be a nonzero solution of $\mathbf{x} = \{x_1 \quad x_2 \quad \ldots \quad x_n\}$ if the equations are not independent and at least one (say the last) is a linear combination of the others. Then we may ignore it, divide all the remaining equations by x_n, and solve the $(n-1)$ equations for $\{x_1 \quad x_2 \quad \ldots \quad x_n\}/x_n$ in terms of the 'right-hand side' $-\{a_{1n} \quad a_{2n} \quad \ldots \quad a_{n-1,n}\}$. The condition for linear dependence is of course

$$|\mathbf{A} - \lambda\mathbf{I}| = 0$$

Considering this in full,

$$\begin{bmatrix} a_{11}-\lambda & a_{12} & a_{13} & \cdots & a_{1n} \\ a_{21} & a_{22}-\lambda & a_{23} & \cdots & a_{2n} \\ \vdots & & & & \vdots \\ a_{n1} & & \cdots & & a_{nn}-\lambda \end{bmatrix}$$

Clearly this determinant produces a *characteristic* polynomial with powers of λ up to λ^n and therefore having n roots; that is, in principle it could be factorized in the form

$$(\lambda-\lambda_1)(\lambda-\lambda_2)(\lambda-\lambda_3)\ldots(\lambda-\lambda_n)=0$$

These roots are called the *eigenvalues.* On injecting any root λ_i back into the equations and solving $\{x_1 \quad x_2 \quad \ldots x_{n-1}\}/x_n$ we get the ratio $x_1{:}x_2{:}x_3{:}\ldots{:}x_n$ which is called the ith *eigenvector*. The absolute values of $\mathbf{x}$ are not of course known. Solving for the eigenvalues from this characteristic polynomial is not feasible for large systems, the number of multiplications is roughly proportional to n^4 just to produce the polynomial. However, one useful property arises from the above: if we put $\lambda = 0$ in the characteristic equation, we get the product of the eigenvalues and doing the same in the determinant produces $|\mathbf{A}|$. Thus the product of the eigenvalues is $|\mathbf{A}|$. A structure which is unsupported (like an aircraft in flight) has a singular $\mathbf{K}$ with perhaps a rank deficiency of 6, corresponding to the six sets of rigid-body movements. The product $\mathbf{M}^{-1}\mathbf{K}$ is therefore rank deficient also and $|\mathbf{M}^{-1}\mathbf{K}| = 0$. There will therefore be possibly six zero eigenvalues corresponding to steady non-oscillatory motion. Another fact from the above is that if $\mathbf{A}$ is diagonal ($a_{ij} = 0$; $i \neq j$), the eigenvalues are simply the diagonal terms. The so-called *transformation methods*, for finding eigenvalues, attempt to convert $\mathbf{A}$ to diagonal or near-diagonal form. Another group of techniques attempts to iterate with initially guessed eigenvectors. The reader is recommended to consult a specialist text[74] for such methods.

APPENDIX B
ENERGY THEOREMS

B.1 Theorem of minimum potential energy

This is the equivalent of the PVD. Detailed proofs and extension of traditional energy theorems may be found in many texts[13,14,15,72] so the following summary is necessarily brief, but includes the vital ingredients.

The *principle of minimum potential energy* is an established universal law of physics which is used extensively in all branches of science and mechanics. It is the only energy theorem in structural mechanics which has a real, physically independent meaning in which the energy is real, and we may consider small changes in real quantities – displacement and strain. Firstly, we define strain energy as the *internal* potential energy due to stress and strain and denote it as U_i. We must not specify the stress–strain law, so it is necessary to take incremental values of strain and then integrate.

Thus

$$\delta U_i = \int_V \boldsymbol{\sigma}^{\mathrm{t}}\, \delta\boldsymbol{\varepsilon}\, \mathrm{d}v \tag{B.1}$$

$$U_i = \int_V \int_0^{\varepsilon} \boldsymbol{\sigma}^{\mathrm{t}}\, \mathrm{d}\boldsymbol{\varepsilon}\, \mathrm{d}v \tag{B.2}$$

The real *external* work W done by the applied forces may also involve nonlinear force–displacement relationships, so we write this in incremental form:

$$\delta W = \int_V \mathbf{p}_v^{\mathrm{t}}\, \delta\mathbf{u}\, \mathrm{d}v + \int_{S_p} \mathbf{p}_s^{\mathrm{t}}\, \delta\mathbf{u}\, \mathrm{d}s \tag{B.3}$$

We should be clear that this is the real change in W due to small changes in displacement $\delta\mathbf{u}$. We are not allowed to have changes $\delta\mathbf{u}$ over the surface S_u where displacements are prescribed. Now by the PVD, putting $\bar{\boldsymbol{\varepsilon}} = \delta\boldsymbol{\varepsilon}$, $\bar{\mathbf{u}} = \delta\mathbf{u}$, we see that (B.1) and (B.3) are equal, thus

$$\delta U_i = \delta W \tag{B.4}$$

We now define the usual *external* potential energy of applied forces as

$$U_e = -\int_V \mathbf{p}_v^t \mathbf{u}\,\mathrm{d}v - \int_{S_p} \mathbf{p}_s^t \mathbf{u}\,\mathrm{d}s$$

So due to an incremental change in displacements everywhere, this changes by

$$\delta U_e = -\int_V \mathbf{p}_v^t\,\delta\mathbf{u}\,\mathrm{d}v - \int \mathbf{p}_s^t\,\delta\mathbf{u}\,\mathrm{d}s = -\delta W \qquad \text{from (B.3)}$$

On substituting into (B.4),

$$\delta U_i = \delta U_e$$

or

$$\delta U = 0 \tag{B.5}$$

where $U = U_i + U_e$ is the *total potential energy.*

If its first variation δU is zero, when the displacements $\delta\mathbf{u}$ and strains $\delta\boldsymbol{\varepsilon}$ undergo small perturbations, then U must be an extremum or turning point. By considering second-order changes $\delta^2 U$, for stable equilibrium, it can be concluded that U is actually a *minimum.* A more elegant proof can be found in the quoted texts.[13,71,72]

Many of the uses we found for the PVD can be derived also from (B.5), but it should be noted that U_i has to be formed before differentiating unless we resort to the virtual-displacement trick. Suppose for example we allow a small displacement δu at a single force P, then $\delta W = P\delta u$ and (B.4) becomes

$$\delta U_i = P\,\delta u$$

or

$$P = \frac{\partial U_i}{\partial u} \tag{B.6}$$

This is the equivalent of the unit displacement theorem. For simple structures like frameworks and beams, and if Hooke's law is valid, the evaluation of strain energy is simple. When stress is proportional to strain, the strain energy in a beam for example is simply

$$U_i = \tfrac{1}{2}\int_V E\varepsilon^2\,\mathrm{d}v = \tfrac{1}{2}\int_0^l \int_A E(yv'')^2\,\mathrm{d}A\,\mathrm{d}z = \tfrac{1}{2}\int_0^l EI(v'')^2\,\mathrm{d}z$$

The theorem can be used in the approximate Rayleigh–Ritz process; for example, we may discretize the beam deflection as

$$v(z) = \sum_n v_n f_n(z), \qquad \text{where } f_n(z) \text{ is assumed}$$

Then U_i is a quadratic function of v_n, and on differentiating with respect to v_n to minimize, there will be the expected set of linear equations in v_n. But we must *not* let $f_n(z)$ violate kinematic boundary conditions since δv must be zero where v is prescribed. If $v(0) = 0$, say, and we cannot choose $f_n(0) = 0$, then equation (B.5) has to be extended by adding a *Langrange multiplier* $\lambda v(0)$ and minimizing with respect to v_n and λ.[13]

B.2 Minimum complementary potential energy

This is the equivalent of the PVF. When we try to formulate the dual of the previous theorem we run into trouble when we switch force and displacement, stress and strain. Firstly, we have to invent a *complementary strain energy* defined as

$$U_i^* = \int_V \int_0^\sigma \boldsymbol{\varepsilon}^t \mathrm{d}\boldsymbol{\sigma}\, \mathrm{d}v \tag{B.7}$$

which has no simple physical meaning. There is immediately a complication because at the limit $\boldsymbol{\sigma} = 0$, there may be an initial strain, that is $\boldsymbol{\varepsilon} = f(\boldsymbol{\sigma}) + \boldsymbol{\eta}$. It is then necessary to incorporate an additional term to supplement the complementary energy of stress by

$$\int_V \int_0^\sigma \boldsymbol{\eta}^t \,\mathrm{d}\boldsymbol{\sigma}\, \mathrm{d}v = \int_V \boldsymbol{\eta}^t \boldsymbol{\sigma}\, \mathrm{d}v$$

Now define complementary external work by

$$\delta W^* = \int_V \delta \mathbf{p}_v^t \mathbf{u}\, \mathrm{d}v + \int_{S_u} \delta \mathbf{p}_s^t \mathbf{u}\, \mathrm{d}s \tag{B.8}$$

where variations in $\mathbf{p}_s$ are not allowed over S_a where $\mathbf{p}_s$ is prescribed. By the PVF, reading $\delta\boldsymbol{\sigma}$ for $\bar{\boldsymbol{\sigma}}$ and $\delta\mathbf{p}$ for $\bar{\mathbf{p}}$, we have, using (B.7) and (B.8),

$$\delta(U_i^* - W^*) = 0 \tag{B.9}$$

provided initial strains are absent. This is usually called the principle of minimum complementary potential energy. If we take a *single* force P, increase it by δP, then $\delta W^* = u\delta P$, where u is the corresponding displacement. Thus

$$\delta U_i^* = u\delta P$$

or

$$u = \frac{\partial U_i^*}{\partial P} \tag{B.10}$$

This is the equivalent of the unit-load theorem. In the special case of linear elasticity when stresses are proportional to strains,

$$U_i^* = \tfrac{1}{2} \int_V \boldsymbol{\sigma}^t \boldsymbol{\varepsilon}\, \mathrm{d}v = \tfrac{1}{2} \int_V \boldsymbol{\varepsilon}^t \boldsymbol{\sigma}\, \mathrm{d}v = U_i$$

Equation (B.10) then becomes:

$$u = \frac{\partial U_i}{\partial P} \tag{B.11}$$

This equation is known as Castigliano's theorem. For structures like simple beams, the strain energy is simply then

$$U_i = U_i^* = \tfrac{1}{2}\int \frac{M^2}{EI}\,\mathrm{d}z, \qquad \text{and} \qquad u = \int \frac{M}{EI}\frac{\partial M}{\partial P}\,\mathrm{d}z$$

The theorem can be used to set up the equations of Chapter 3 for solving redundant frameworks, once we have set

$$\boldsymbol{\sigma} = \boldsymbol{\beta}\mathbf{P}_g = \boldsymbol{\beta}\mathbf{b}_{0g}\mathbf{R} + \boldsymbol{\beta}\mathbf{b}_{xg}\mathbf{X}$$

We evaluate

$$U_i^* = \int_V \frac{\sigma^2}{2E}\,\mathrm{d}v$$

and minimize by setting

$$\frac{\partial U_i^*}{\partial X} = 0$$

This is permissible since $\partial W^*/\partial X = 0$ if the X systems are self-equilibrating.

B.3 Reissner's variational theorem[11]

The Hellinger–Reissner theorem[10] is the usual theorem quoted for setting up mixed elements. It is motivated by the same philosophy of sections B.1 and B.2, namely to find an energy and minimize it. Unfortunately there is no simple energy and the 'functional' – usually labelled Π_R – is simply a mathematical device which delivers equations of equilibrium and compatibility when it is subjected to perturbations in displacement and stress and then put to zero. It is also not a minimum principle, but then we already knew that mixed methods are neither upper nor lower bounds. Reissner's functional is written as

$$\Pi_R = \int_V (\boldsymbol{\sigma}^t\mathbf{D}^t\mathbf{u} - \mathbf{p}_v^t\mathbf{u})\,\mathrm{d}v - U_i^* - \int_{S_p} \mathbf{p}_s^t\mathbf{u}\,\mathrm{d}s - \int_{S_u} [(\mathbf{D}n)\boldsymbol{\sigma}]^t(\mathbf{u} - \tilde{\mathbf{u}})\,\mathrm{d}s \tag{B.12}$$

If this is subjected to a variation in stress $\delta\boldsymbol{\sigma}$ *and* displacement $\delta\mathbf{u}$, we find

$$\delta\Pi_R = \int_V (\delta\boldsymbol{\sigma}^t\mathbf{D}^t\mathbf{u} + \boldsymbol{\sigma}^t\mathbf{D}^t\,\delta\mathbf{u} - \mathbf{p}_v^t\,\delta\mathbf{u} - \boldsymbol{\varepsilon}^t\,\delta\boldsymbol{\sigma})\,\mathrm{d}v$$
$$- \int_{S_p} \mathbf{p}_s^t\delta\mathbf{u}\,\mathrm{d}s - \int_{S_u} [(\mathbf{D}n)\,\delta\boldsymbol{\sigma}]^t(\mathbf{u} - \tilde{\mathbf{u}})\,\mathrm{d}s$$

noting that $\delta\mathbf{u} = 0$ on S_u. Now use Gauss:

$$\int_V \boldsymbol{\sigma}^t\mathbf{D}^t\,\delta\mathbf{u}\,\mathrm{d}v = \int_{S_p} \delta\mathbf{u}^t(\mathbf{D}n)\,\boldsymbol{\sigma}\,\mathrm{d}s - \int_V \delta\mathbf{u}^t\mathbf{D}\boldsymbol{\sigma}\,\mathrm{d}v$$

whence

$$\delta\Pi_{\mathrm{R}} = \int_V \{\delta\boldsymbol{\sigma}^{\mathrm{t}}[\mathbf{D}^{\mathrm{t}}\mathbf{u} - \boldsymbol{\varepsilon}] - \delta\mathbf{u}^{\mathrm{t}}[\mathbf{D}\boldsymbol{\sigma} + \mathbf{p}_v]\}\,\mathrm{d}v + \int_{S_p} \delta\mathbf{u}^{\mathrm{t}}[(\mathbf{D}n)\boldsymbol{\sigma} - \mathbf{p}_s]\,\mathrm{d}s$$

$$- \int_{S_u} [(\mathbf{D}n)\,\delta\boldsymbol{\sigma}]^{\mathrm{t}}(\mathbf{u} - \tilde{\mathbf{u}})\,\mathrm{d}s \qquad \text{(B.13)}$$

We see that this equation is completely equivalent to our two equations (4.31) and (4.32) when we put $\delta\Pi_{\mathrm{R}} = 0$ for arbitrary increments $\delta\boldsymbol{\sigma}$ and $\delta\mathbf{u}$.

The principle of minimum potential energy and Reissner's variational theorem are valid for finite elements even with discontinuous tractions, as was first pointed out by Prager.[73] It is only necessary to realize that the Gauss identity can only be applied *across an element* since $\mathbf{D}\boldsymbol{\sigma}$ is infinite across S_g, so if Π_{R} is summed over all elements we can take the term in $(\mathbf{D}n)\boldsymbol{\sigma}$ in (B.13) to be different for contiguous elements, and so we have the usual weighted residual form

$$\int_{S_g} \delta\mathbf{u}^{\mathrm{t}}[(\mathbf{D}n)\boldsymbol{\sigma}_1 - (\mathbf{D}n)\boldsymbol{\sigma}_2]\,\mathrm{d}s = 0$$

However, to apply Reissner's principle to discontinuous displacements, it is necessary to use a modified functional containing an additional term

$$\Pi_{\mathrm{MR}} = -\int_{S_g} [(\mathbf{D}n)\boldsymbol{\sigma}]^{\mathrm{t}}\mathbf{u}\,\mathrm{d}s \qquad \text{(B.14)}$$

The variation in this extra term produces

$$\delta\Pi_{\mathrm{MR}} = \int_{S_g} \{[(\mathbf{D}n)\,\delta\boldsymbol{\sigma}]^{\mathrm{t}}\mathbf{u} + [(\mathbf{D}n)\boldsymbol{\sigma}]^{\mathrm{t}}\,\delta\mathbf{u}\}\,\mathrm{d}s$$

This combines with the other surface integral to leave just

$$-\int_{S_g} [(\mathbf{D}n)\,\delta\boldsymbol{\sigma}]^{\mathrm{t}}\mathbf{u}\,\mathrm{d}s$$

and this term will satisfy continuity in the mean of the displacement $\mathbf{u}$ provided $(\mathbf{D}n)\,\delta\boldsymbol{\sigma}$ is continuous.

APPENDIX C
THE PRINCIPLE OF VIRTUAL WORK IN POLARS

We have used both the PVD and the PVF knowing that they would deliver equilibrium and compatibility; and have proved this quality, using cartesian stresses and spatial coordinates, for an arbitrary continuum. It is not necessary to justify the principle of virtual work in other coordinates, but for completeness we show the extension. To do both the PVD and PVF together, we adopt a sleight-of-hand and recall the principle of virtual work without specifying which version, thus:

$$\int_V \boldsymbol{\sigma}^t \boldsymbol{\varepsilon} \, dv = \int_V \mathbf{u}^t \mathbf{p}_v \, dv + \int_S \mathbf{u}^t \mathbf{p}_s \, ds \tag{C.1}$$

The Gauss identity is

$$\int_V \boldsymbol{\sigma}^t \mathbf{D}^t \mathbf{u} \, dv = -\int_V \mathbf{u}^t \mathbf{D} \boldsymbol{\sigma} \, dv + \int_S \mathbf{u}^t (\mathbf{D}n) \boldsymbol{\sigma} \, ds \tag{C.2}$$

Comparing (C.1) and (C.2), we deduce that

In V:

$$\boldsymbol{\varepsilon} = \mathbf{D}^t \mathbf{u}, \qquad \mathbf{D}\boldsymbol{\sigma} + \mathbf{p}_v = \mathbf{0}; \qquad \text{and} \qquad (\mathbf{D}n)\boldsymbol{\sigma} = \mathbf{p}_s \quad \text{on the surface } S.$$

Now consider a cylindrical polar system (r, θ, z) with displacements u, v, and w in these directions, and body and surface forces $\mathbf{p}_v$ and $\mathbf{p}_s$ whose components are similarly orientated. The stress and strain components are

$$\boldsymbol{\sigma} = \{\sigma_{rr} \quad \sigma_{\theta\theta} \quad \sigma_{zz} \quad \sigma_{r\theta} \quad \sigma_{\theta z} \quad \sigma_{zr}\}$$

and

$$\boldsymbol{\varepsilon} = \{\varepsilon_{rr} \quad \varepsilon_{\theta\theta} \quad \varepsilon_{zz} \quad \varepsilon_{r\theta} \quad \varepsilon_{\theta z} \quad \varepsilon_{zr}\}$$

Equation (C.1) will still look the same in these new components and is still usable in the PVD or PVF form. It remains therefore to generalize the Gauss identity in the new coordinate system. If we use two new matrices,

$$\mathbf{D} = \begin{bmatrix} \frac{\partial}{\partial r} & 0 & 0 & \frac{1}{r}\frac{\partial}{\partial \theta} & 0 & \frac{\partial}{\partial z} \\ 0 & \frac{1}{r}\frac{\partial}{\partial \theta} & 0 & \frac{\partial}{\partial r} & \frac{\partial}{\partial z} & 0 \\ 0 & 0 & \frac{\partial}{\partial z} & 0 & \frac{1}{r}\frac{\partial}{\partial \theta} & \frac{\partial}{\partial r} \end{bmatrix}$$

and the scalar operators,

$$\mathbf{A} = \begin{bmatrix} 0 & -\frac{1}{r} & 0 & 0 & 0 & 0 \\ 0 & 0 & 0 & \frac{1}{r} & 0 & 0 \\ 0 & 0 & 0 & 0 & 0 & 0 \end{bmatrix}$$

it may be shown that Gauss's identity may be written as

$$\int_V \boldsymbol{\sigma}^t(\mathbf{D}^t - \mathbf{A}^t)\mathbf{u}\,dv = -\int_V \mathbf{u}^t \frac{1}{r}(\mathbf{D} + \mathbf{A})(r\boldsymbol{\sigma})\,dv + \int_S \mathbf{u}^t[(\mathbf{D}n)\boldsymbol{\sigma}]\,ds \quad \text{(C.3)}$$

Comparing (C.3) with (C.2), we deduce the equations of compatibility and equilibrium:

$$\text{In } V: \quad \boldsymbol{\varepsilon} = (\mathbf{D}^t - \mathbf{A}^t)\mathbf{u}$$

$$\frac{1}{r}(\mathbf{D} + \mathbf{A})(r\boldsymbol{\sigma}) = \mathbf{p}_v$$

$$\text{On } S: \quad (\mathbf{D}n)\boldsymbol{\sigma} = \mathbf{p}_s$$

APPENDIX D
NUMERICAL INTEGRATION

Only the simplest element stiffnesses can be cheaply integrated in closed form. Even for those two-dimensional elements which have an integrable stiffness, it is no more expensive to integrate numerically than to store the closed-form expressions and inject the elements' properties. For isoparametric elements and others where mapping is necessary, it is essential to use numerical integration because of the presence of the Jacobian in the integrand. It is assumed here that any element can be so treated, so no generality is lost in restricting the discussion to integrals taken between $\zeta = \pm 1$ in one dimension, and further unit limits in more dimensions.

The integral of any finite function $f(\zeta)$ rests on the simple concept that it can be approximated by a polynomial in the interval $-1 \leqslant \zeta \leqslant 1$ to any degree that turns out to be appropriate. If the chosen polynomial is of order n then it can be fitted to the integrand function $f(\zeta)$ by using its value $f(\zeta_i)$ at $(n+1)$ points $\zeta_0, \zeta_1, \zeta_2, \ldots \zeta_i, \ldots, \zeta_n$. The *Newton–Cotes* series of formulae takes these points equally spaced at intervals $h = 2/n$ from the first value $f(-1) = f_0$ to the last $f(+1) = f_n$. The integral is then written as

$$\int_{-1}^{1} f(\zeta)\,\mathrm{d}\zeta = Ah \sum_{i=0}^{n} w_i f_i \tag{D.1}$$

where the $(n+1)$ weights w_i are to be determined, and A is merely a scaling factor. By putting

$$f(\zeta) = \sum_{0}^{n} a_n \zeta^n$$

then *any* polynomial up to order n can be integrated exactly, so inserting single terms $f(\zeta) = \zeta^n$ into (D.1) we can generate $(n+1)$ equations for the weights w_i. The results for curves up to order three are as follows:

n	A	w_0	w_1	w_2	w_3
1	$\frac{1}{2}$	1	1		
2	$\frac{1}{3}$	1	4	1	
3	$\frac{3}{8}$	1	3	3	1

The first algorithm is the obvious trapezoidal rule, and the second is the equally familiar Simpson's rule which is exact for polynomial integrands up to order 2. It can be shown that the error in (D.1), due to truncating a polynomial, and therefore approximating a higher-order variation with a lower-order one, is of order

$$h^{k+1}\frac{\partial^k f(\zeta)}{\partial \zeta^k}$$

Thus subdividing the interval into small portions h can rapidly reduce the error provided higher derivatives of the approximated function do not exist. It is not common to use Newton–Cotes formulae of higher order than in the above table.

A much more popular routine in the finite-element world is the *Gauss* or *Gauss–Legendre* weighting scheme, often called *Gaussian quadrature.* Here instead of taking n equal intervals we choose to evaluate both the weighting terms w_i *and the sampling points* $\zeta_0, \zeta_1, \zeta_2, \ldots, \zeta_n$ as variables such that the integral up to a certain degree polynomial is exact. So if we have n points and weights, the approximation

$$\int_{-1}^{1} f(\zeta)\,\mathrm{d}\zeta = \sum_{0}^{n} w_i f(\zeta_i)$$

has $2n$ variables w_i and ζ_i and will integrate exactly a $(2n-1)$ order polynomial with $2n$ terms ζ^m for $m = 0$ to $2n-1$. The Gauss points ζ_i are symmetrically placed about $\zeta = 0$, and together with the weights w_i can be shown to have the following values:

Number of Gauss points	ζ_i	w_i
2	$\pm 0.577\,35$	1.0
3	0	$\frac{8}{9}$
	$\pm 0.774\,59$	$\frac{5}{9}$
4	$\pm 0.339\,98$	0.652 15
	$\pm 0.861\,14$	0.347 85

The Gauss integration therefore only involves about a half of the number of ordinates as Simpson's rule for comparable accuracy, and the only penalty paid is the awkward position of the Gauss points. However, it is much cheaper to evaluate $f(\zeta_i)$ at awkward points than twice the number at more convenient points $f(ih)$ in Simpson's rule. For further details on accuracy and the use of higher-order integrals, the reader is referred to standard texts.[64]

REFERENCES

1. Mach, E. (1893) *The Science of Mechanics*, trans. and pub. Court, Illinois (1960).
2. Timoshenko, S. P. *The History of the Strength of Materials*, McGraw-Hill (1953).
3. Maxwell, J. C. 'On the calculation of the equilibrium and the stiffness of frames', *Phil. Mag. Ser. 4*, **27**, 294 (1864).
4. Engesser, Fr. 'Über statisch unbestimmte Träger bei beliebigen Formänderungsgesetze und über den Satz von der Kleinsten Ergänzungsarbeit', *Zeitschrift des Architekten und Ingenieur–Vereins zu Hanover*, **35**, (1889).
5. Rayleigh, J. W. S. *The Theory of Sound*, Macmillan (1877).
6. Ritz, W. 'Über eine neue Methode zur Lösung gewisser Randwertaufgeben', *Göttingener Nachsichten Math-Phys. Klasse*, 236 (1908).
7. Galerkin, B. G. 'Series solutions in rods and plates', *Vestnik Inzhenerov i Tekhnikov*, **19**, (1915).
8. Westergaard, H. M. 'On the method of complementary energy', *Trans. Am. Soc. Eng.*, **107**, 765–93 (1942).
9. Libove, C. 'Complementary energy method for finite deformations', *Jnl. Am. Soc. Civ. Eng.*, **90**(EM6), 49–71 (1964).
10. Hellinger, E. 'Die allgemeinen Ansätze der Mechanik der Kontinua', *Enzyclopädie der Mathematischen Wissenschaften IV*, **4**, 654–5 (1914).
11. Reissner, E. 'On a variational theorem in elasticity', *Jnl. Math. Phys.*, **29**, 90–5 (1950).
12. Reissner, E. 'On a variational theorem for finite elastic deformations', *Jnl. Math. Phys.*, **32**, (1953).
13. Washizu, K. *Variational Methods in Elasticity and Plasticity*, Pergamon (1960).
14. Argyris, J. H., and Kelsey, S. *Energy Theorems and Structural Analysis*, Butterworths (1954).
15. Neal, B. G. *Structural Theorems and their Applications*, Pergamon (1964).
16. Timoshenko, S. P. *Theory of Elasticity*, McGraw-Hill (1934).
17. Courant, R. *Differential and Integral Calculus*, Vol. 2, Blackie and Son (1962).
18. Wagner, H. 'Verdrehung und Knickung von offenen Profilen', *Zeitschrift 25 Jahre Technische Hochschule Danzig*, Verlag A. W. Kafferman, Danzig (1929).
19. Argyris, J. H. 'The open tube', *Aircraft Engineering*, Vol. XXVI, April (1954).
20. Schreyer, H. L., and Masur, E. F. 'Buckling of shallow arches', *Jnl. Eng. Mech. Div. Am. Soc. Civ. Eng.*, **92**(EM4), August (1966).
21. Timoshenko, S. P., and Woinowsky-Krieger, S. *Theory of Plates and Shells*, McGraw-Hill (1959).
22. Argyris, J. H., and Dunne, P. C. 'Structural analysis', Part II of *Structural Principles and Data*, Pitmans (1952).
23. Wittrick, W. H. 'A generalisation of Macaulay's method with applications in structural mechanics', *A.I.A.A. Jnl.*, **3**, 2 (1965).
24. Müller-Breslau, H. *Die neueren Methoden der Festigkeitslehre und der Statik der Baukonstruktionen*, Körner, Leipzig (1886).
25. Denke, P. H. 'A general digital computer analysis of statically indeterminate structures', NASA TN D-1666 (1962).
26. Denke, P. H. 'A computerised static and dynamic structural analysis system, Pt. 3', *Soc. Auto. Eng., International Cong. and Exposition*, Detroit (1965).
27. Robinson, J. *Integrated Theory of Finite Element Methods*, Wiley (1973).

28. McGuire, W., and Gallagher, R. H. *Matrix Structural Analysis*, Wiley (1979).
29. Allman, D. J. 'A theory for elastic stresses in Adhesive bonded lap joints', RAE (Farnborough) TR 76024 (1976).
30. Kuhn, P. *Stresses in Aircraft and Shell structures*, McGraw-Hill (1956).
31. Gallagher, R. H. *Correlation Study of Methods of Matrix Structural Analysis*, Pergamon (1964).
32. Rayleigh, J. W. S. *The Theory of Sound* (1877), reprinted by Dover (1945).
33. Ritz, W. 'Theorie der Transversalschwingungen einer Quadratischer Platte', *Ann. d. Physik*, **28** (1909).
34. Tsui, T. Y. 'Wave propagation in finite length bars with varying cross-section', *Jnl. Appl. Mech.*, **35** (1968).
35. Macaulay, W. H. 'Note on the deflection of beams', *Messenger of Maths*, **48**, (1919).
36. Crandall, S. H. *Engineering Analysis*, McGraw-Hill (1956).
37. Finlayson, B. A., and Scriven, L. E. 'The method of weighted residuals – a review', *App.. Mech. Reviews*, **19**, 735–48 (1966).
38. Reissner, E. 'On some variational theorems in elasticity', *Problems in Continuum Mechanics – Muskhelishvili Anniversary Volume*, Philadelphia, McGraw-Hill (1958).
39. Timoshenko, S. P. *Theory of Elastic Stability*, McGraw-Hill (1936).
40. Kantorovitch, L. V. 'A direct method of solving the problem of the minimum of a double integral', *Izv. Akad. Nauk. SSSR*, **5** (1933).
41. Kanotorovitch, L. V., and Krylov, V. L. *Approximate Methods of Higher Analysis*, Interscience (New York) and P. Noordhoff (Holland) (1958).
42. Timoshenko, S. P., and Goodier, J. N. *Theory of Elasticity*, McGraw-Hill (1951).
43. Kerr, A. D., and Alexander, H. 'An application of the extended Kantorovitch method to the stress analysis of a clamped rectangular plate', *Acta. Mech.*, **6** (1968).
44. Argyris, J. H., and Dunne, P. C. 'The general theory of conical and cylindrical tubes', *Jnl. Roy. Aero. Soc.* (1947).
45. Neal, B. G. *The Plastic Methods of Structural Analysis*, Chapman & Hall (1956).
46. Horne, M. R. *Plastic Theory of Structures*, Pergamon (1979).
47. Courant, R. 'Variational methods for the solution of problems of equilibrium and vibration', *Bull. Am. Math. Soc.*, **49** (1943).
48. Zienkiewicz, O. C. *The Finite Element Method* (3rd edn), McGraw-Hill (1977).
49. Turner, M., Clough, R., Martin, H., and Topp, L. 'Stiffness and deflection analysis of complex structures', *Jnl. Aero. Sci.*, **23**, 9 (1956).
50. Pestel, E. C. 'Dynamic stiffness matrix formulation by means of Hermitian polynomials', *First Conf. Matrix Meth. Structural Mechanics*, Wright-Patterson Air Force Base, Dayton (1965).
51. Gallagher, R. H. *Finite Element Analysis – Fundamentals*, Prentice-Hall (1975).
52. Argyris, J. H. Technical Notes 1–21 on The Finite Element Method (Monthly), *Jnl. Roy. Aero. Soc.* (September 1965 to May 1969).
53. Henshell, R. D., and Shaw, K. G. 'Crack elements are unnecessary', *Int. Jnl. Num. Meth. Eng.*, **9** (1975).
54. Walz, J. E., Fulton, R. E., and Cyrus, N. T. 'Accuracy and convergence of finite element approximations', *Second Conf. on Matrix Meth. AFFDR TR* (1968).
55. Irons, B. M. 'Engineering applications of numerical integration in stiffness methods', *A.I.A.A. Jnl.*, **4**, 11 (1966).
56. Zienkiewicz, O. C., Too, J., and Taylor, R. L. 'Reduced integration techniques in general analysis of plates and shells', *Int. Jnl. Num. Meth. Eng.*, **3** (1971).
57. Clough, R. W., and Penzien, J. *Dynamics of Structures*, McGraw-Hill (1975).
58. Guyan, R. J. 'Reduction of stiffness and mass matrices', *A.I.A.A. Jnl.*, **3**, (1965).
59. Hughes, T. J. R., Taylor, R. L., and Kanoknukulachai, W. 'A simple and efficient finite element for plate bending', *Int. Jnl. Num. Meth. Eng.*, **11** (1977).
60. Fraijs de Veubeke, B. 'Upper and lower bounds in matrix structural analysis', *AGARDograph 72*, Pergamon (1964).

61. Ashwell, D. G. 'Strain elements with applications to arches, rings, and cylindrical shells', *in Finite-Element Thin Shell Analysis*, Wiley (1976).
62. Argyris, J. H. *Recent Advances in Matrix Methods of Structural Analysis*, Pergamon (1964).
63. Fraijs de Veubeke, B. 'Displacement and equilibrium models in the finite element method', *Stress Analysis* (ed. Zienkiewicz and Holister), Wiley (1965).
64. Bajpai, A. C., Mustoe, L. R., and Walker, D. *Advanced Engineering Mathematics*, Wiley (1977).
65. Morley, S. V. 'A comparative study of stiffness and force methods in plane stress elements', MSc Thesis, Aeronautics Department, Imperial College, London (1974).
66. Allman, D. J. 'A simple cubic displacement element for plate bending', RAE (Farnborough) TR 74176 (1974).
67. Pian, T. H. H. 'Derivation of element stiffness matrices by assumed stress distributions', *A.I.A.A. Jnl.*, **2**, 1333–6 (1964).
68. Morley, L. S. D. 'The constant moment plate bending element', *Jnl. Strain Analysis*, **6**, 1, (1971).
69. Malkus, D. S., and Hughes, T. J. R. 'Mixed finite element techniques – reduced and selective integration techniques, a unification of concepts', *Comp. Meth. in Appl. Mech. Eng.*, **15** (1978).
70. Bushnell, D. 'Finite difference energy models versus finite element models', *Numerical and Computer Methods in Structural Mechanics*, (Ed. Fenves *et al.*), Academic Press (1973).
71. Biezeno, C. B., and Grammel, R. *Engineering Dynamics*, **1**, Van Nostrand (1954).
72. Richards, T. H. *Energy Methods in Stress Analysis*, Ellis Horwood (1977).
73. Prager, W. 'Variational principles of linear elastostatics for discontinuous displacements, strains, and stresses', *Recent Progress in Applied Mechanics* (*the Folke Odquist Volume*), Wiley (1967).
74. Jennings, A. *Matrix Computation for Engineers and Scientists*, Wiley (1977).
75. Argyris, J. H., and Bronlund, O. E. 'The natural factor formulation of the stiffness for the matrix displacement method', *Comp. Meth. Appl. Mech. Eng.*, **5** (1975).

INDEX